RHODIUM CATALYZED HYDROFORMYLATION

Catalysis by Metal Complexes

Volume 22

The titles published in this series are listed at the end of this volume.

RHODIUM CATALYZED HYDROFORMYLATION

Edited by

PIET W.N.M. VAN LEEUWEN

Institute of Molecular Chemistry,
University of Amsterdam,
Amsterdam, The Netherlands

and

CARMEN CLAVER

Department de Quimica Física i Inorgánica,
Universitat Rovira i Virgili,
Tarragona, Spain

KLUWER ACADEMIC PUBLISHERS

DORDRECHT / BOSTON / LONDON

A C.I.P. Catalogue record for this book is available from the Library of Congress.

ISBN 1-4020-0421-4
Transferred to Digital Print 2001

Published by Kluwer Academic Publishers,
P.O. Box 17, 3300 AA Dordrecht, The Netherlands.

Sold and distributed in North, Central and South America
by Kluwer Academic Publishers,
101 Philip Drive, Norwell, MA 02061, U.S.A.

In all other countries, sold and distributed
by Kluwer Academic Publishers,
P.O. Box 322, 3300 AH Dordrecht, The Netherlands.

Printed on acid-free paper

TABLE OF CONTENTS

Preface xi

1 Introduction to hydroformylation 1
Piet W. N. M. van Leeuwen
1.1 History of phosphorus ligand effects 1
1.2 Hydroformylation 6
1.3 Ligand parameters 8

2 Hydroformylation with unmodified rhodium catalysts 15
Raffaello Lazzaroni, Roberta Settambolo and Aldo Caiazzo
2.1 Introduction 15
2.2 Regioselectivity in the rhodium-catalyzed hydroformylation of vinyl and vinylidenic substrates 16
2.2.1 Catalyst precursors 17
2.2.2 Influence of the alkene structure on the regioselectivity 17
2.2.3 Influence of temperature 21
2.2.4 Influence of CO and H_2 partial pressures 22
2.3 Mechanism of the hydroformylation of vinyl and vinylidenic alkenes 22
2.3.1 Activation of the catalyst precursor 24
2.3.2 Behavior of the isomeric alkyl-metal intermediates via deuterioformylation 24
2.3.3 In situ IR investigation of the formation and reactivity of acylrhodium intermediates 28
2.4 Origin of the regioselectivity 29
2.4.1 Influence of the nature of the substrate 29
2.4.2 Influence of the reaction parameters 31

3 Rhodium phosphite catalysts 35
Paul C. J. Kamer, Joost N. H. Reek, and Piet W. N. M. van Leeuwen
3.1 Introduction 35

3.2 Monophosphites 37
3.2.1 Catalysis 37
3.2.2 Mechanistic and kinetic studies 40
3.3 Diphosphites 44
3.3.1 Catalysis 44
3.3.2 Mechanistic and kinetic studies 48
3.4 Hydroformylation of internal alkenes 55
3.4.1 Hydroformylation of less reactive internal and functionalized alkenes 55
3.4.2 Formation of linear aldehydes starting from internal alkenes 57
3.5 Calixarene based phosphites 59

4 Phosphines as ligands 63
Piet W. N. M. van Leeuwen, Charles P. Casey, and Gregory T. Whiteker
4.1 Monophosphines as ligands 63
4.1.1 Introduction 63
4.1.2 The mechanism 64
4.1.3 Ligand effects 66
4.1.4 In situ studies 68
4.1.5 Kinetics 69
4.1.6 Regioselectivity 72
4.1.7 Conclusion 75
4.2 Diphosphines as ligands 76
4.2.1 Introduction 76
4.2.2 Ferrocene based diphosphine ligands 78
4.2.3 BISBI ligands and the natural bite angle 82
4.2.4 Xantphos ligands: tunable bite angles 87
4.2.5 The mechanism, regioselectivity, and the bite angle. Concluding remarks 96

5 Asymmetric hydroformylation 107
Carmen Claver and Piet W.N.M. van Leeuwen
5.1 Introduction 107
5.2 Rhodium systems with chiral diphosphite ligands 109
5.2.1 C_2 Symmetric chiral diphosphite ligands 109
5.2.2 Catalyst preparation and hydroformylation 111
5.2.3 Characterisation of $RhH(L)(CO)_2$ intermediates. Solution structures of hydroformylation catalysts 113
5.2.4 Structure versus stability and enantioselectivity 115

5.2.5 Chiral cooperativity and effect of substituents in diastereomeric diphosphite ligands 116
5.2.6 C_1 Sugar backbone derivatives. Diphosphinite and diphosphite ligands 121
5.3 Phosphine-phosphite rhodium catalysts 124
5.3.1 Introduction 124
5.3.2 Rhodium complexes with BINAPHOS and related ligands 124
5.3.3 $RhH(CO)_2$(BINAPHOS) complexes; models for enantioselectivity 127
5.3.4 Separation studies for the BINAPHOS system 129
5.3.5 Chiral phosphine-phosphite ligands containing a stereocenter in the backbone 129
5.4 Diphosphine rhodium catalysts 131
5.4.1 Introduction 131
5.4.2 C_1 Diphosphines as chiral ligands 131
5.4.3 C_2 Diphosphines as chiral ligands 132
5.4.4 The Rh/BDPP system. HPNMR and HPIR studies under hydroformylation conditions 136
5.5 Mechanistic considerations 138
5.5.1 Regioselectivity 138
5.5.2 Enantioselectivity and conclusions 140

6 Hydroformylation in organic synthesis 145
Sergio Castillón and Elena Fernández
6.1 Introduction 145
6.2 Hydroformylation of unfuctionalized alkenes 146
6.3 Hydroformylation of functionalized alkenes 149
6.4 Substrate directed stereoselectivity 155
6.5 Control of the regio- and stereoselectivity by heteroatom directed hydroformylation 160
6.6 Consecutive processes under hydroformylation conditions 164
6.6.1 Hydroformylation-acetalization (intramolecular) 165
6.6.2 Hydroformylation-acetalization (intermolecular) 166
6.6.3 Hydroformylation-amination (intramolecular) 168
6.6.4 Hydroformylation-amination-reduction. Hydroaminomethylation 172
6.6.5 Consecutive hydroformylation-aldol reaction 175
6.6.6 Consecutive hydroformylation-Wittig reaction 177
6.7 Alkyne hydroformylation 178
6.8 Conclusion 182

7 Aqueous biphasic hydroformylation 189
Jürgen Herwig and Richard Fischer
7.1 Principles of biphasic reactions in water 189
7.1.1 Why two-phase catalysis? Scope and Limitations 189
7.1.2 Concepts for two-phase hydroformylation 190
7.2 Hydroformylation of propene and butene 191
7.2.1 Historic overview of two-phase hydroformylation technology 191
7.2.2 Ligand developments 191
7.2.3 Kinetics and catalyst pre-formation 193
7.2.4 Process description 196
7.2.5 Status of the operated plants 197
7.2.6 Economics 198
7.3 Reaction of various alkenes 199
7.3.1 Ethylene to propanal: why not applied? 199
7.3.2 Long-chain alkenes 200

8 Process aspects of rhodium-catalyzed hydroformylation 203
Peter Arnoldy
8.1 Introduction 203
8.2 Economics 204
8.3 Catalyst selectivity and activity 206
8.3.1 Catalyst selectivity 206
8.3.2 Catalyst activity 207
8.4 Catalyst stability; degradation routes, losses and recovery 208
8.4.1 Rhodium loss routes 208
8.4.2 Ligand loss routes 209
8.4.3 Catalyst recovery processes 210
8.5 Process concepts 211
8.5.1 Type I: Stripping reactor process/Rh containment in reactor 212
8.5.2 Type II: Liquid recycle process/use of distillative separation 213
8.5.3 Type III: Two-phase reaction/extraction process 215
8.5.4 Type IV: Extraction after one-phase reaction 216
8.6 Survey of commercialized processes and new developments 220
8.6.1 Hydroformylation of butenes 220
8.6.2 Branched higher alkenes to mainly plasticizer alcohols 223

8.6.3 Linear higher alkenes to mainly detergent alcohols 223
8.6.4 1,4-Butanediol 225
8.6.5 Nylon monomers 226

9 Catalyst preparation and decomposition 233
Piet W. N. M. van Leeuwen
9.1 Introduction 233
9.2 Catalyst preparation 233
9.3 Catalyst decomposition 235
9.3.1 Metal plating or cluster formation 235
9.3.2 Oxidation of phosphorus ligands 235
9.3.3 Phosphorus-carbon bond breaking in phosphines 237
9.3.4 Decomposition of phosphites 243
9.3.5 Formation of dormant sites 247
9.4 Concluding remarks 249

10 Novel developments in hydroformylation 253
Joost N. H. Reek, Paul C. J. Kamer, and Piet W. N. M. van Leeuwen
10.1 Introduction 253
10.2 New bimetallic catalysts 253
10.3 Novel developments in catalyst separation 256
10.3.1 Micellar catalysis 256
10.3.2 Supported aqueous phase catalysis (SAPC) 260
10.3.3 Hydroformylation in supercritical fluids 262
10.3.4 Fluorous Biphase catalysis 265
10.3.5 Dendrimer supported catalysts 267
10.3.6 Novel developments in polymer supported catalyst 269
10.4 Supramolecular catalysis 274
10.5 Conclusions 277

Subject index 281

Preface

This book covers the developments in rhodium catalyzed hydroformylation of the last decade, one of the most important reactions in industry catalyzed by homogeneous catalysts. The work includes many of the advances that have been made by academic and industrial researchers. The field has undergone drastic changes, both in its industrial applications and in our understanding. Clearly, the new advances pose new problems and set new targets for future research.

In spite of the importance of the field, the last reviews covering a broad area in hydroformylation are outdated (Falbe 1980, Pruett 1977) and it was felt timely to bring together the recent developments. Only in the area of aqueous biphasic hydroformylation there are several exhausting reviews available. This is the first monograph on hydroformylation of this type and for other processes there not many examples.

The aim of the book is to review the mainstream of the activities in the field and not to present a complete coverage of the literature, not even the recent literature. Several thousands of papers and patents deal with rhodium-catalyzed hydroformylation and a complete review would be impossible. We have chosen for a more didactic approach, in which we have tried to avoid one-liners about publications. In the book one will find typical examples about kinetics, applications in organic chemistry, industrial processes, mechanistic understanding, etc. In the mainstream activities we have tried to include industrial developments. We may have missed new catalyst systems that are as yet small but may turn out to be of major importance later, but that can hardly be avoided. New and important developments involving other metals, such as cobalt, platinum, and palladium will also be absent.

While writing we had a broad audience in mind: chemists and engineers in industry and academia with an interest in homogeneous catalysis, whose backgrounds may be as varied as those of the present authors: inorganic, organic, organometallic, catalytic, chemical engineering. It is hoped that specialists in one area will read with interest the chapters on the neighbouring expertise. The book is also meant for PhD-students and advanced students interested in this area.

The combination of topics we have chosen is rather unique, connecting studies on ligand effects, catalyst characterization, industrial requirements regarding stability and separation, catalyst decomposition, and applications in fine and bulk chemistry. The reader will notice the importance of one discipline for the other. In many cases these relationships have already been established, but for other cases the book might assist future developments. The key roles that ligands may play in selectivity may be an eye-opener for organic chemists and it will further enhance the large number of new applications and reactions that are being discovered. The comments in several chapters on catalyst preparation and feed purification may be useful for scientists who are not specialized in homogeneous catalysis using transition metal complexes.

Hydroformylation is also a model reaction system in homogeneous catalysis as it contains so many aspects such as ligand effects (electronic, steric, bite angle), *in situ* studies, complicated kinetics, and effects of conditions and impurities. All this, combined with its practical value, makes it an ideal topic in education.

The editors are very grateful to the authors for the good work they did and the prompt responses. The writing took only a few months, as did the production by the publisher. Writing the book has been rewarding, because we learnt many things. Most of all perhaps, we obtained a clearer view on what we still don't fully understand.

Amsterdam, Tarragona
Piet van Leeuwen, Carmen Claver

Chapter 1

Introduction to hydroformylation

Phosphorus ligands in homogeneous catalysis

Piet W. N. M. van Leeuwen

Institute of Molecular Chemistry, University of Amsterdam, Nieuwe Achtergracht 166, 1018 WV, Amsterdam, The Netherlands

1.1 History of phosphorus ligand effects

In this chapter we will briefly review "phosphorus ligand effects" in homogeneous catalysis and hydroformylation more in particular. First we will have a look at a few historical landmarks in homogeneous catalysis concerned with the use of phosphorus ligands, then focus on the history of rhodium catalyzed hydroformylation, and subsequently summarize a few basic concepts. Since phosphorus ligands are the only ligands used in hydroformylation in addition to carbon monoxide, we will not discuss ligands containing other donor atoms. In later chapters we will see that in hydroformylation, as it is today, bidentate phosphorus ligands are of great importance. In the introduction we show that in the early history the positive effect of bidentates on selectivities and rates of catalytic reactions was not fully recognized [1].

The favorable effects of phosphine ligands in catalysis have been known for more than half a century. One of the first reports involves the use of triphenylphosphine in the "Reppe" chemistry, the reactions of alkynes, alcohols and carbon monoxide to make acrylic esters [2]. An early example of a phosphine-modified catalytic process is the Shell process for alkene hydroformylation using a cobalt catalyst containing an alkylphoshine [3].

Hydrocyanation as applied by Du Pont is another early example of an industrially applied catalytic reaction employing ligands [4]. The nickel catalyzed reaction uses aryl phosphite ligands for the production of adiponitrile from butadiene and hydrogen cyanide. The development of this process has played a key-role in the introduction of the now common study

P.W.N.M. van Leeuwen and C. Claver (eds.), *Rhodium Catalyzed Hydroformylation*, 1–13.

of "ligand effects" in the field of homogeneous catalysis using organometallic complexes [5].

Both academia and industries made important contributions to the new field in the early sixties with the appearance of the first phosphine modified and other hydrogenation catalysts. An early example of a phosphine-free ruthenium catalyst was published by Halpern [6]. In 1963 Cramer (Du Pont) reported a triphenylphosphine-modified platinum-tin catalyst for the hydrogenation of alkenes [7]. In the same year Breslow (Hercules) included a few phosphine complexes of late transition metals in a hydrogenation study employing metal salts reduced by aluminum alkyls, but interestingly the systems containing phosphine were less active [8]!

Rhodium catalyzed hydrogenation was discovered in the mid-sixties by Wilkinson and coworkers [9]. The mechanism of this reaction using $RhCl(PPh_3)_3$ as the catalyst was studied in great detail. These studies by Wilkinson and many others have been a major stimulant for workers in this area. Substitution at the aromatic ring revealed an electronic effect on the reaction rate, electron donors giving higher rates [10]. A few months later Vaska published his first work on the rhodium and iridium catalyzed hydrogenation of alkenes [11].

Rhodium-catalyzed hydroformylation using catalysts modified with alkylphosphines and arylphosphines was reported by Wilkinson's group [12]. Phosphine ligand variation hardly affected the rate and selectivity under the circumstances used (70 °C and 100 bar). Pruett (Union Carbide Corporation) found that phosphites can also be used, and the type of phosphite had a profound effect on rates and selectivities [13].

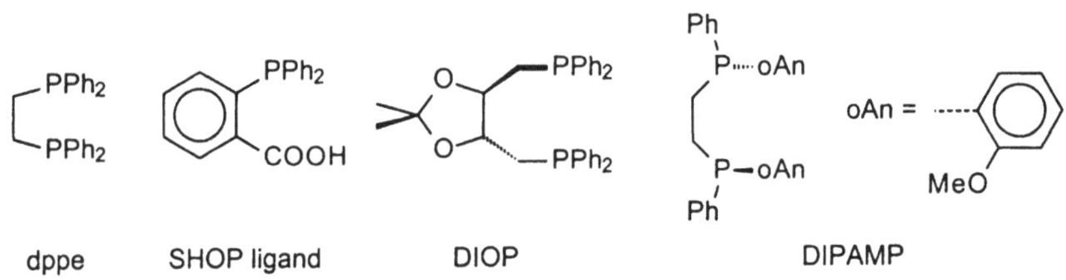

Figure 1. Structures of dppe, SHOP ligand, DIOP, and DIPAMP

Bidentate ligands have played an important role in the development of the chemistry of metal organic complexes. The synthesis of dppe was reported as early as 1959 [14]. Chatt and Hieber [15] explored the coordination chemistry of several diphosphines with an ethane bridge, but it took a while before diphosphines became routinely included in catalysis studies. In the early sixties diphosphines were mentioned in patents, but specific advantages are not apparent. In their exploration of carbonyl chemistry of cobalt related to carbonylation catalysis, Heck and Breslow

[16] reported that $HCo(CO)_4$ gave unidentifiable complexes with dppe. The use of dppe in cobalt catalyzed hydroformylation was reported by Slaugh [17], but compared to PBu_3 it had little effect on the rate and the selectivity of the cobalt carbonyl catalyst. Copolymerization of butadiene and propylene oxide using nickel bromide and dppe was published in 1965 [18].

In the late sixties at Shell Development, Keim and coworkers discovered that certain bidentates containing an oxygen and a phosphorus donor atom formed excellent catalysts with nickel for the oligomerization of ethene [19]. Typical ligands are diphenylphosphinoacetic acid or 2-diphenylphosphinobenzoic acid (SHOP ligand, Figure 1). The ligand required a relatively laborious ligand synthesis for those days. In addition it was the first process utilizing the concept of two-phase catalysis. This discovery led to the Shell Higher Olefins Process that came on stream in 1977.

Hata [20] reported a phosphine-free iron catalyst for the codimerization of butadiene and ethene in 1964. A year later this was followed by phosphine-free rhodium catalysts [21]. The oldest publication describing *advantageous* results for diphosphines we found is by Iwamoto and Yuguchi (1966) who studied the same reaction using iron catalysts containing a range of diphosphines varying in bridge lengths [22]. In many instances the activity of catalysts containing dppe instead of PPh_3 is lower. For example, the hydrogenation of styrene using rhodium(I) chloride and dppe is 70 times slower as compared to the PPh_3 based system [23]. The strong chelating power of the diphosphine was held responsible for this. Thus, initially the use of dppe and other bidentate phosphines in catalysis found little support as they were supposed to lead mostly to more stable complexes, rather than more active or selective catalysts.

Theoretical work of Thorn and Hoffmann [24] explained why migration reactions in complexes containing for instance dppe were slow. The constrained P-M-P angle would slow down the migration reaction, since ideally the phosphine ligand coordinated in the position cis to the migrating group, would have a tendency to widen the P-M-P angle in the process to "pursue" the migrating group.

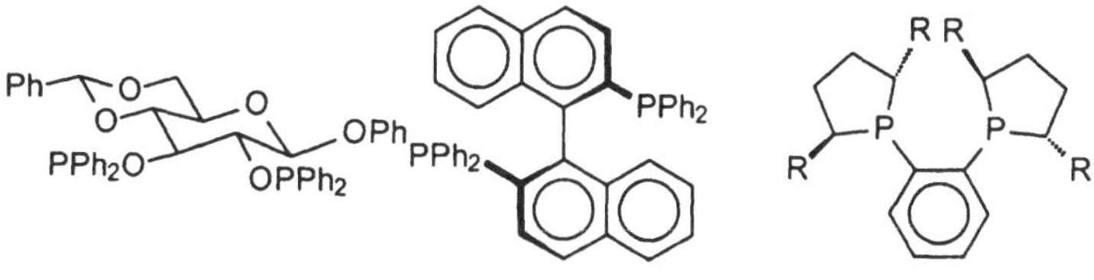

β-GLUP BINAP DuPhos

Figure 2. Structures of Phenyl-β-GLUP, BINAP, and DuPhos

A beneficial use of bidentate diphosphines was discovered in 1971 by Kagan [25] who reported the use of DIOP modified rhodium for the hydrogenation of N-acetylphenylalanine. Monophosphines for asymmetric hydrogenation of similar substrates were reported by Knowles [26]. His discovery of the P-chiral diphosphine DIPAMP, also a bidentate ligand, led to the commercial application of the asymmetric hydrogenation of the Levodopa precursor. For the same process Selke developed another ligand, a sugar based bisphosphonite Phenyl-β–GLUP [27]. The company VEB-Isis applied this ligand for many years in Germany.

In the area of asymmetric hydrogenation chiral diphosphines have played a center role since and many applications have been developed. Important new ligands that have been introduced comprise Noyori's BINAP [28], DuPhos (Burk) [29], Takaya's BINAPHOS [30], and C_1-symmetric ferrocene-based ligands introduced by Togni [31]. Industrial products, of which the synthesis uses enantioselective phosphine-derived metal-catalysts are for instance menthol, metolachlor, biotin, and several alcohols, e.g. (R)-1,2-propanediol, For details about the applications the reader is referred to reviews and references therein [32, 33]. Substituents and backbones have an enormous influence on the performance of the ligands, but usually rationalizations are lacking.

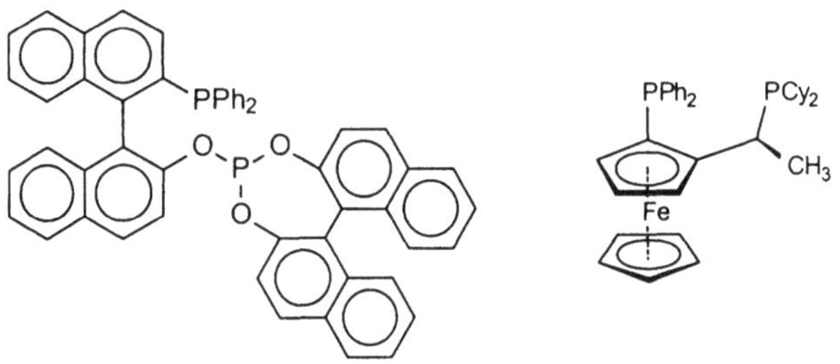

BINAPHOS Josiphos

Figure 3. Asymmetric phosphine ligands BINAPHOS and Josiphos

In carbonylation chemistry using phosphine or phosphite complexes of palladium or rhodium a number of breakthroughs achieved in the seventies and eighties should be mentioned; hydroformylation will be reviewed in section 1.2. Here we will concentrate on those that have found industrial application or may find application in the near future. In the early eighties Sen [34] and Drent [35] discovered that ethene and carbon monoxide can polymerized in an alternating fashion leading to polyketones. The catalyst is a palladium complex containing phosphines and non-coordinating anions in methanol as the solvent. Drent's bidentate phosphine containing catalysts proved by far to be the fastest ones. Especially diphosphines having a propane bridge give a fast reaction to high molecular weight products.

Shell's process-related patents often use the propane bridged 1,3-bis-(di-2-anisylphosphino)propane as the ligand (dapp) [36]. Carilon, Shell's trade name for the terpolymer of ethene, CO and propene – added for lowering the processing temperature of the product – has been in commercial production on a relatively small scale in the late nineties.

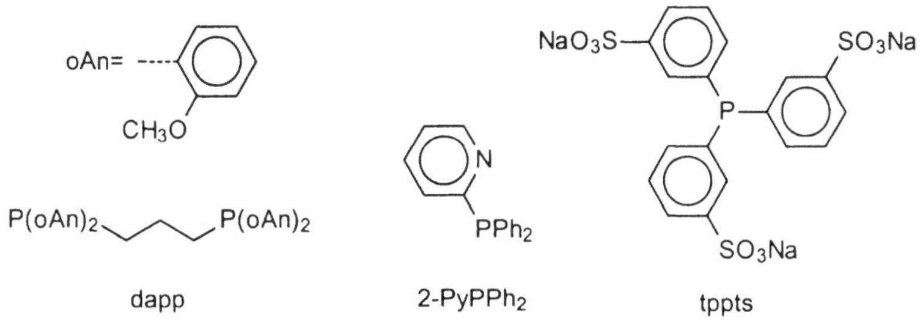

Figure 4. Ligands for bulk chemical processes

Another impressive ligand effect reported by Drent and coworkers [37] concerns the methoxycarbonylation of propyne to form methyl methacrylate. Triphenylphosphine modified palladium catalysts give low rates, but using 2-pyridyldiphenylphosphine instead gives very high rates and selectivities. The mechanism is still a matter of debate [38].

Tris-m-sulfonatophenylphosphine (tppts) plays an important role in the history of homogeneous catalysis [39], mainly due to its use in the Ruhrchemie/Rhône-Poulenc hydroformylation process [40], now operated by Celanese (see 1.2 and Chapter 7). It is also used in a number of fine chemical processes, such as selective hydrogenation with ruthenium [41], carbon-carbon bond formation with rhodium [42], and the Heck reaction [43]. Monosulfonated triphenylphosphine (tppms) is used for the preparation of nonadienol [44] (see Figure 5).

In C-C, C-O, and C-N bond formation reactions catalyzed by palladium and nickel, ligand effects have been explored in an extremely wide area [33]. The data available on ligand effects for these reactions are numerous. In asymmetric allylic alkylation the "embracing" effect of the bidentate ligand explains the efficacy of the ligand in many instances [45]; the longer the backbone, - i.e. the larger the bite angle (vide infra) - and the more effective the ligand interacts with the substrate. For the Heck reaction the ligand size seems to be a dominant factor, as bulky phosphines [46], phosphites [47], and amidites [48] were found to lead to highly effective catalysts. For amidites it was shown that the bulky ligands lead to mono-ligand complexes which are effectively more prone to substrate coordination than bis-ligand complexes. This effect was first observed by us for the same bulky phosphites in rhodium catalyzed hydroformylation [49] (Figure 5).

1.2 Hydroformylation

The first generation of hydroformylation catalysts was based on cobalt carbonyl without phosphine ligand [50]. The conditions were harsh, as the reactivity of cobalt is low. The process was used both for lower as for higher alkenes, and notably also internal alkenes give mainly linear product aldehyde. Initially rhodium catalyzed reaction seemed slow, because the formation of rhodium hydrides requires high pressures of hydrogen [51]. An early commercial application of phosphine-free rhodium was by Mitsubishi for the hydroformylation of higher 1-alkenes in 1970. The kinetics of rhodium carbonyl catalyzed hydroformylation were studied for the first time in the sixties, but in the last decade the studies by Lazzaroni and Garland have revealed interesting aspects that will be dealt with in Chapter 2.

Since Shell's report on the use of phosphines in this process [3], many industries started applying phosphine ligands in the rhodium process as well [52]. While alkylphosphines are the ligands of choice for cobalt, they lead to slow catalysis when applied in rhodium catalysis. In the mid-sixties the work of Wilkinson showed that arylphosphines should be used for rhodium and that even at very mild conditions very active catalysts can be obtained [9].

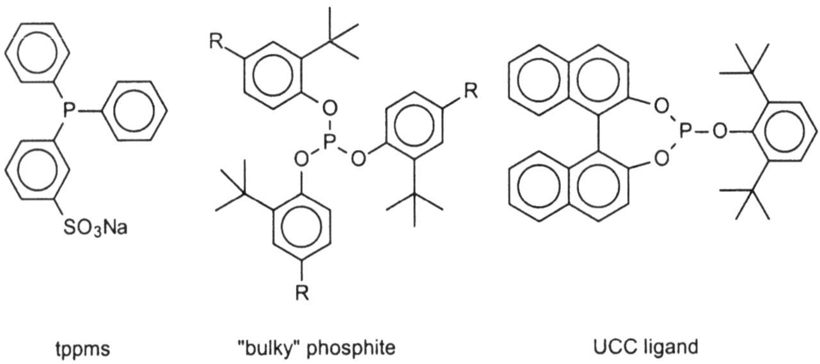

Figure 5. Structures of ttpms, van Leeuwen's "bulky phosphite", and a highly stable, bulky phosphite from UCC

The second generation processes use rhodium as the metal and the first ligand-modified process came on stream in 1974 (Celanese) and more were to follow in 1976 (Union Carbide Corporation) and in 1978 (Mitsubishi Chemical Corporation), all using triphenylphosphine (tpp). The UCC process has been licensed to many other users and it is often referred to as the LPO process. Not only are rhodium catalysts much faster - which is translated into milder reaction conditions -, but also their feedstock utilization is much better than that of cobalt catalysts. For example, the cobalt-alkylphosphine catalyst may give as much as 10% of alkane formation. Since the mid-seventies the rhodium catalysts started to replace the cobalt catalysts in propene and butene hydroformylation. For detergent alcohol production though, even today, the cobalt systems are still in use, because there is no good alternative yet for the hydroformylation of internal higher alkenes.

The third generation process concerns the Ruhrchemie-RhonePoulenc process utilizing a two-phase system containing water-soluble rhodium-tppts in one phase and the product butanal in the organic phase. The process has been in operation in Oberhausen since 1984 by Celanese, as the company is called today. The system will be discussed in Chapter 7. Since 1995 this process is also used for the hydroformylation of 1-butene.

In the late sixties phosphites have also been considered as candidate ligands for rhodium hydroformylation, but tpp turned out to be the ligand of choice. A renewed interest in phosphites started in the eighties after we had discovered the peculiar effect of bulky monophosphites giving very high rates [49]. Bryant and coworkers at Union Carbide expanded this work enormously, first by making more stable bulky monophosphites [53], later by focusing on diphosphites [54]. There is only one relatively small commercial application of "bulky monophosphite" by Kuraray for the hydroformylation of 3-methylbut-3-en-1-ol [55]. A large amount of research has been devoted to diphosphites in the last decade aiming at a variety of applications. The results will be discussed in Chapter 3.

Diphosphines have also become very popular ligands since the late eighties in rhodium hydroformylation, e.g. Eastman's BISBI. Chapter 4 focuses on diphosphines. The two-phase system has undergone considerable improvements involving diphosphines [39].

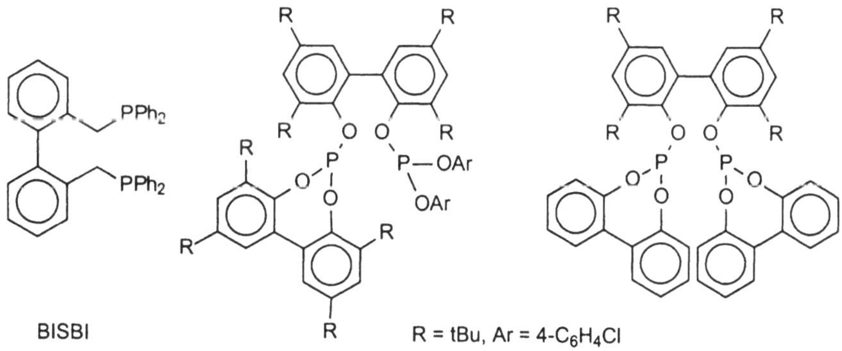

Figure 6. Eastman's BISBI and typical diphosphites from Union Carbide Corporation

In recent years the interest for hydroformylating higher alkenes with catalysts other than cobalt has increased. Platinum and palladium based catalysts have been studied and the results of the latter [56] seem very promising. Platinum has been known for many years to have a high preference for the formation of linear products, but ligand decomposition hampers applications [57]. Palladium and platinum will not be discussed, but recent advances for rhodium have been collected in Chapter 10.

A ligand with great potential for hydroformylation of higher alkenes in a one-phase system that is worked up by adding water to separate catalyst and product afterwards is the monosulfonated triphenylphosphine, tppms, that was studied by Abatjoglou, also at Union Carbide [58] (Chapter 8).

The fourth generation process for large-scale application still has to be selected from the potential processes that have been "nominated". In the chapters to follow several of these candidates will be discussed. The fourth generation will concern higher alkenes only, since for propene hydroformylation there are hardly wishes, if any, left [59] (a cheaper catalyst would be on my shopping list!). Many new phosphite-based catalysts have been reported that will convert internal alkenes to terminal products and recently also a new diphosphines have been reported that will do this [60].

The most interesting ligand discovered for asymmetric hydroformylation is undoubtedly BINAPHOS, introduced by Takaya [30]. Diphosphites have also been studied to this end by UUC [61] and by us [62]; Babin and Whiteker reported the first successful ones [61]. Asymmetric rhodium catalysts are discussed in Chapter 5.

Ligand design for fine chemical applications has been very limited and usually the ligands designed for large-scale applications are also tested for more complicated organic molecules. Tpp has been the workhorse in fine chemicals hydroformylation since Wilkinson's first examples [63, 64], but also bulky phosphite [49], tppts and tppms [41-43] turned out to be very useful, and recently diphosphites have been studied [65] (see Chapter 6).

1.3 Ligand parameters

Tolman reviewed ligand effects for the first time [5]. Prior to his studies [66] the effects of phosphorus ligands on reactions or properties of metal complexes were rationalized mainly in terms of electronic effects. Systematic studies had shown, however, that steric effects are at least as important as electronic effects, and in terms of stability of complexes can even be dominant. Since then, numerous studies have appeared using both the electronic parameter χ and the steric parameter for the cone angle, θ.

The electronic parameter χ is a measure for the overall effect of electron donating and accepting properties of the phosphorus ligand L. It is measured as the symmetric stretching frequency of the carbonyls in $Ni(CO)_3L$, similar to the methods proposed by Strohmeier and Horrocks [67]. High χ-values stand for strong π-acceptors and low χ-values stand for strong σ-donor ligands. An expansion of Tolman's list of χ-values was published by Bartik [68]. NMR methods for the measurement of substituent effects on phosphorus have been reported [69]. χ_i values are used for one substituent.

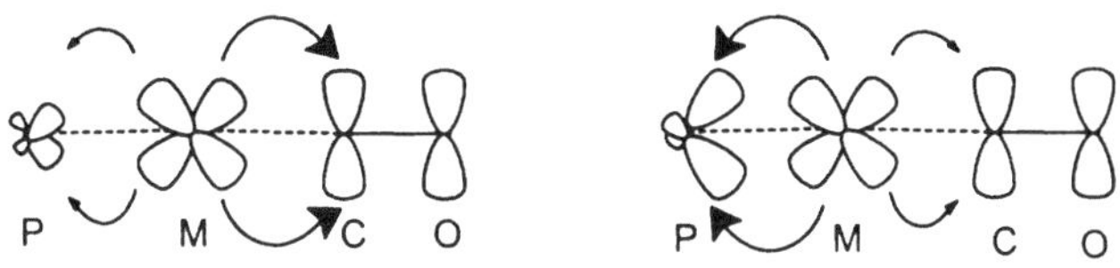

Figure 7. A simple picture of Tolman's χ-value. On the left strong back donation to CO leading to a low stretching frequency. On the right weak back-donation to CO leading to a high stretching frequency [5]

For monodentate phosphorus ligands the cone angle θ is defined as the apex angle of a cylindrical cone, centered at 2.28 Å from the center of the P atom, which touches the outermost atoms of the model (actually, this is done using CPK models). For phosphines containing different substituents an average for the three substituents is taken. Crystal structure determinations have shown that the actual angles in the complexes are smaller than those expressed by the θ-value due to an intermeshing of the substituents. The relative order of cone angles parallels with many properties that have been measured, such as equilibrium constants and rate constants [70].

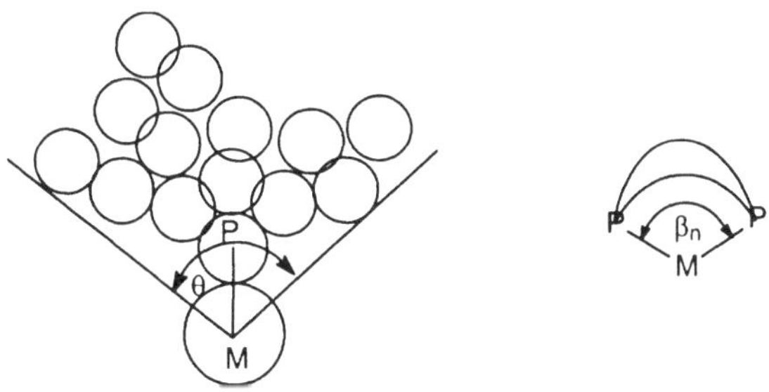

Figure 8. Tolman's cone angle θ [5]. The distance M-P amounts to 2.24 Å. Casey's natural bite angle β_n as calculated by Molecular Mechanics [79, 80].

Several studies have been conducted aiming at the separation of steric and electronic effects [71-73]. For a single step process such as an oxidative addition or one-electron change in electrochemical processes this may be useful, but for multi-step reactions as we are dealing with in catalysis, this technique will encounter many problems. There will be different effects on the distinct steps and linear free-energy relationships will be an exception rather than the rule. When for instance both an oxidative addition step and a reductive elimination step are involved "volcano" curves must be expected for reactivity versus a ligand property, as in a series of metal oxides when plotting oxidation activity versus position in the periodic table. In addition, multiple metal-ligand equilibria will affect the schemes. Enthalpy studies can contribute to the solution of these complex matters [74].

During the seventies the attention for bidentate phosphines as ligands in catalysis had been growing and so did the need for including them into the electronic and steric mapping. Tolman extended the cone angle for monophosphines to diphosphines, which was defined as the average cone angle measured for the two substituents and the angle between the M-P bond and the bisector of the P-M-P angle. Even today, this looks like a good approximation for defining a cone angle of a bidentate. Other parameters for bidentate ligands have been reported, such as the solid angle [75], pocket angle [76], repulsive energy [77], and the accessible molecular surface [78].

In the present study we take a different approach, which is less elaborate than the methods mentioned above. The steric properties of diphosphines are determined by the four substituents at the two phosphorus atoms and the length of the bridge. In general, the most stable complexes are obtained when a five-membered ring can form, i.e. when the bridge between the two phosphorus donor atoms consist of two carbon atoms as in dppe. This is true for octahedral and square-planar complexes in which the "metal-preferred" [79] P-M-P angle is ~ 90°. The vast majority of chelate complexes have been synthesized from bidentate ligands possessing relatively short, bridging backbones. Tetrahedral complexes will prefer P-M-P angles of 109°, and bis-equatorial coordination in a trigonal bipyramid requires an angle of 120°. During catalytic processes transitions between different coordination modes may be needed. The "natural" preference of a ligand for a certain coordination mode can influence a reaction of a catalytic cycle in several ways: stabilization or destabilization of the initial, transition, or final state. In addition the flexibility of a bidentate ligand may be important in order to accelerate certain transitions. In a one-step reaction the effect of the bite angle may be very clear-cut, but a catalytic cycle involves more steps and equilibria and in many instances the effect on catalysis may not be characteristic of the bite angle.

A means to predict the "ligand-preferred" [79] P-M-P angle using molecular mechanics has been developed by Casey and Whiteker [80]. They introduced the concepts of natural bite angle (β_n) and flexibility range for diphosphine ligands. Computer modeled geometries can be used to estimate ligand bite angles. The calculations can even be performed before ligands are synthesized. If computer modeling is employed to design new ligands, it is more important to calculate a correct trend rather than perfect geometries.

The bite angle not only induces steric effects, but also electronic effects, as the P-M-P angles clearly affect the binding in the complex or intermediate states [81].

References

1 van Leeuwen, P. W. N. M.; Kamer, P. C. J.; Reek, J. N. H.; Dierkes, P. *Chem. Rev.* **2000**, *100*, in press.
2 Reppe, W.; Schweckendiek, W. J. *Annalen* **1948**, *104*, 560.
3 Slaugh, L. H.; Mullineaux, R. D. U.S. Pat. 3,239,569 and 3,239,570, 1966 (to Shell); *Chem. Abstr.* **1964**, *64*, 15745 and 19420. *J. Organomet. Chem.* **1968**, *13*, 469.
4 (a) Brown, E. S. *Aspects of Homogeneous Catalysis*; Ugo, R., Ed.; Reidel: **1974;** Vol. 2; pp 57-78. (b) Drinkard, W. C.; Lindsey, R. V. U. S. Pat. 3,655,723, **1970** (to Du Pont); *Chem. Abstr.* **1972**, *77*, 4986.
5 Tolman, C. A. *Chem. Rev.* **1977**, *77*, 313.
6 Halpern, J.; James, B. R. *J. Am. Chem. Soc.* **1961**, *83*, 753.
7 Cramer, R. D.; Jenner, E. L.; Lindsey, R. V.; Stolberg, U. G.; *J. Am. Chem. Soc.* **1963**, *85*, 1691.
8 Sloan, M. F.; Matlack, A. S.; Breslow, D. S. *J. Am. Chem. Soc.* **1963**, *85*, 4015.
9 Young, J. F.; Osborn, J. A.; Jardine, F. A.; Wilkinson, G. *J. Chem. Soc. , Chem. Comm.* **1965**, 131.
10 O'Connor, C.; Wilkinson, G., *Tetrahedron Lett.* **1969**, *18*, 1375.
11 Vaska, L.; Rhodes, R. E. *J. Am. Chem. Soc.* **1965**, *87*, 4970.
12 (a) Jardine, F. H.; Osborn, J. A.; Wilkinson, G.; Young, J. F. *Chem. and Ind. (London)* **1965**, 560. (b) Evans, D.; Osborn, J. A.; Wilkinson, G. *J. Chem. Soc. A.* **1968**, 3133.
13 Pruett, R. L.; Smith, J. A. *J. Org. Chem.* **1969**, *34*, 327.
14 Issleib, K.; Müller, D. W. *Chem. Ber.* **1959**, *92*, 3175.
15 (a) Chatt J.; Hart, W. *J. Chem. Soc.* **1960**, 1378. (b) Hieber, W.; Freyer, W. *Chem.Ber.* **1960**, *93*, 462.
16 Heck, R. F.; Breslow, D. S. *J. Am. Chem. Soc.* **1962**, *84*, 2499.
17 Cannel, L. G.; Slaugh, L. H.; Mullineaux, R. D. Ger. Pat. 1,186,455 (priority date 1960, to Shell); *Chem. Abstr.* **1965**, *62*, 16054.
18 United States Rubber Co, Neth. Appl. 6,507,223; *Chem. Abstr.* **1965**, *62*, 16013.
19 Bauer, R. S.; Glockner, P. W.; Keim, W.; van Zwet, H.; Chung, H. U.S. Pat. 3,644,563, 1972.
20 Hata, G. *J. Am. Chem. Soc.* **1964**, *86*, 3903.
21 Cramer, R. *J. Am. Chem. Soc.* **1965**, *87*, 4717 and **1967**, *89*, 1633.
22 Iwamoto, M.; Yuguchi, S. *J. Org. Chem.* **1966**, *31*, 4290.
23 Dang, T. P.; Kagan, H. B. *Chem. Commun.* **1971**, 481 and *J. Am. Chem. Soc.* **1972**, *92*, 6429.
24 Thorn D. L.; Hoffmann R. *J. Am. Chem. Soc.* **1978**, *100*, 2079.
25 Poulin, J. C.; Dang, T. P.; Kagan, H. B. *J. Organometal. Chem.* **1975**, *84*, 87.
26 (a) Knowles, W. S.; Sabacky, M. J. *J. Chem. Soc., Chem. Commun.* **1971**, 481 (b) Knowles, W. S.; Sabacky, M. J.; Vineyard, B. D., and Weinkauff, J. *J. Am. Chem. Soc.* **1975**, *97*, 2567.
27 Selke, R. and Pracejus, H.; *J. Mol. Catal.*, **1986**, *37*, 213.
28 Miyashita, A.; Yasuda, A.; Tanaka, H.; Toriumi, K.; Ito, T.; Souchi, T.; Noyori, R. *J. Am. Chem. Soc.*, **1980**, *102*, 7932.
29 Burk, M. J. *J. Am. Chem. Soc.* **1991**, *113*, 8518.
30 Sakai, N.; Mano, S.; Nozaki, K.; Takaya, H. *J. Am. Chem. Soc.* **1993**, *115*, 7033.
31 Togni, A. *Angew. Chem.* **1996**, *108*,1581.
32 *Transition Metals for Organic Synthesis; Building Blocks and Fine Chemicals*, ed. M. Beller and C. Bolm, Wiley-VCH, 1998, Vol 1 and 2.

33 *Metal-catalyzed Cross-coupling Reactions*, eds. F. Diederich and P.J. Stang, Wiley-VCH, Weinheim, 1998.

34 Sen, A.; Lai, T. W. *J. Am. Chem. Soc.* **1982**, *104*, 3520.

35 (a) Drent, E. Eur. Pat. Appl. 121,965, 1984 (to Shell); *Chem. Abstr.* **1985**, *102*, 46423. (b) Drent, E.; Budzelaar, P. H. M. *Chem. Rev.* **1996**, *96*, 663.

36 (a) Mastenbroek, B.; Petrus, L.; Rosenbrand, G. G. Can. Pat. Appl., 2,030,896, 1990 (to Shell); *Chem. Abstr.* **1991**, *115*, 184575. (b) Mastenbroek, B.; Tummers, C. H. M.; De Smedt, P. J. M. M. Brit. Pat. Appl., 2,278,366, 1994 (to Shell); *Chem. Abstr.* **1995**, *123*, 170553.

37 Drent, E.; Arnoldy, P.; Budzelaar, P. H. M. *J. Organomet. Chem.* **1994**, *475*, 57.

38 Scrivanti, A.; Beghetto, V.; Campagna, E.; Zanato, M.; Matteoli, U. *Organometallics*, **1998**, *17*, 630.

39 Aqueous Phase Organometallic Catalysis-Concepts and Applications, eds. B. Cornils and W. A. Herrmann, Wiley-VCH, Weinheim, 1998.

40 Cornils, B.; Kuntz, E. G. *J. Organomet. Chem. 502*, **1995**, 177.

41 Grosselin, J. M.; Mercier, C.; Allmang, G.; Grass, F. *Organometallics,* **1991**, *10*, 2126.

42 Mercier, C.; Chabardes, P. *Pure Appl. Chem.*, **1994**, *66*, 1509.

43 Haber, S.; Kleiner, H.-J. WO 9705104, 1997 (to Hoechst); *Chem. Abstr.* **1997**, *126*, 199349.

44 Yoshimura, N.; Tokito, Y. Eur. Pat. Appl. **1987**, 223,103 (to Kuraray); *Chem. Abstr.* **1987**, *107*, 154896.

45 Trost, B. M.; Murphy, D. J. *Organometallics* **1985**, *4*, 1143.

46 Beller, M.; Riermeijer, T. H.; Haber, S.; Kleiner, H. J.; Herrmann, W.A. *Chem. Ber.* **1996**, *129*, 1259.

47 Albisson, D. A.; Bedford, R. B.; Lawrence, S. E.; Scully, P. N. *J. Chem. Soc. Chem. Commun.*, **1998**, 2095.

48 van Strijdonck, G. P. F. ; Boele, M. D. K.; Kamer, P. C. J.; de Vries, J. G.; van Leeuwen, P. W. N. M. *Eur. J. Inorg. Chem.* **1999**, 1073.

49 (a) van Leeuwen, P. W. N. M.; Roobeek, C. F. *J. Organometal. Chem.* **1983**, *258*, 343; Brit. Pat. 2,068,377, US Pat. 4,467,116 (to Shell Oil); *Chem. Abstr.* **1984**, *101*, 191142. (b) Jongsma, T.; Challa, G.; van Leeuwen, P. W. N. M. *J. Organometal. Chem.* **1991**, *421*, 121. (c) van Rooy, A.; Orij, E. N.; Kamer, P. C. J.; van Leeuwen, P. W. N. M. *Organometallics*, **1995**, *14*, 34.

50 Frohning, C. D.; Kohlpaintner, C. W. page 29, in Applied Homogeneous Catalysis with Organometallic Compounds, eds. Cornils B.; Herrmann, W. A., VCH, Weinheim, **1996**, Vol 1 and 2.

51 Vidal, J. L.; Walker, W. E. *Inorg. Chem.***1981**, *20,* 249.

52 Onoda, T. *ChemTech.* **1993**, Sept. 34.

53 Billig, E.; Abatjoglou, A. G.; Bryant, D. R.; Murray, R. E.; Maher, J. M. U.S. Pat. 4,599,206 **1986** (to Union Carbide Corp.); *Chem. Abstr.* **1989**, *109,* 233177.

54 Billig, E.; Abatjoglou, A. G.; Bryant, D. R. U.S. Pat. 4,769,498, U.S. Pat. 4,668,651; U.S. Pat. 4748261, **1987** (to Union Carbide Corp.); *Chem. Abstr.* **1987**, *107*, 7392r.

55 Yoshinura, N.; Tokito, Y. Eur. Pat. 223,103, **1987** (to Kuraray).

56 Drent, E.; Jager, W. W. U.S. Pat. 5,780,684, **1998** (to Shell); *Chem. Abstr.* **199x**, *129*, 137605.

57 Hayashi T.; Kawabata Y.; Isoyama T.; Ogata, I. *Bull. Chem. Soc. Jpn.* **1981**, *54*, 3438.

58 Abatjoglou, A. G.; Bryant, D. R.; Peterson, R. R. U.S. Pat. 5,180,854) **1993** (to Union Carbide Corp.; *Chem. Abstr.* **1991**, *113*, 80944.

59 Wiebus, E.; Cornils, E. *Chem. Ing. Tech.* **1994**, *66*, 916; Cornils, B.; Wiebus, E. *ChemTech,* **1995**, *25*, 33,; Cornils, B.; Wiebus, E. *Recl. Trav. Chim. Pays-Bas,* **1996**, *115*, 211.
60 van der Veen, L. A.; Kamer, P. C. J.; van Leeuwen, P. W. N. M. *Angew. Chem. Int. Ed. Engl.* **1999**, *38*, 336.
61 Babin, J. E.; Whiteker, G. T. U.S. Pat. 5,360,938 **1994** (to Union Carbide Corp.); *Chem. Abstr.* **1995**, *122*, 186609.
62 Buisman, G. J. H.; Kamer, P. C. J.; van Leeuwen, P. W. N. M. *Tetrahedron: Asymmetry*, **1993**, *4*, 1625.
63 Brown, C. K.; Wilkinson, G. *J. Chem. Soc. (A)* **1970**, 2753.
64 Eilbracht, P.; Bärfacker, L.; Buss, C.; Hollmann, C.; Kitsos-Rzychon, B. E.; Kranemann, C. L.; Rische, T.; Roggenbuck, R.; Schmidt A. *Chem. Rev.* **1999**, *99*, 3329.
65 Cuny, G. D.; Buchwald, S. L. *J. Am. Chem. Soc.* **1993**, *115*, 2066.
66 Tolman, C. A. *J. Am. Chem. Soc.* **1970**, *92*, 2953 and 2956.
67 (a) Strohmeier, W.; Müller, F. J. *Chem. Ber.* **1967**, *100*, 2812. (b) Horrocks Jr., W. D.; Taylor, R. C. *Inorg. Chem.* **1963**, *2*, 723.
68 Bartik, T.; Himler, T.; Schulte, H.-G.; Seevogel, K. *J. Organomet. Chem.* **1984**, *272,* 29.
69 Grim, S. O.; Yankowsky, A. W. *J. Org. Chem.* **1977**, *42*, 716.
70 Trogler, W. C.; Marzilli, J. G. *J. Am. Chem. Soc.* **1974**, *96*, 7589.
71 Heimbach, P.; Kluth, J.; Schlenkluhn, H.; Weimann, B. *Angew. Chem.* **1980**, *92*, 567.
72 Fernandez A.; Reyes C.; Wilson M. R.; Woska D. C.; Prock A.; Giering, M. P. *Organometallics,* **1997**, *16*, 342.
73 Joerg S.; Drago R. S.; Sales J. *Organometallics,* **1998**, *17*, 589.
74 Haar, C. M.; Nolan, S. P.; Marshall, W. J.; Moloy, K. G.; Prock, A.; Giering, W. P. *Organometallics*, **1999**, *18*, 474.
75 (a) Hirota M.; Sakakibara K.; Komatsuzaki T.; Akai, I. *Comp. Chem.* **1991**, *15,* 241. (b) White D.; Taverner, B. C.; Coville, N. J.; Wade, P. W. *J. Organomet. Chem.* **1995**, *495*, 41. (c) White D.; Taverner, B. C.; Leach, P. G. L.; Coville N. J. *J. Organomet. Chem.* **1994**, *478*, 205.
76 Koide, Y.; Bott, S. G.; Barron, A. R. *Organometallics,* **1996**, *15*, 2213.
77 a) Brown, T. L. *Inorg. Chem.* **1992**, *31*, 1286. b) Choi, M.-G.; White, D.; Brown, T. L. *Inorg. Chem.* **1993**, *32*, 5591.
78 Angermund, K.; Baumann, W.; Dinjus, E.; Fornika, R.; Gorls, H.; Kessler, M.; Krüger, C.; Leitner, W.; Lutz, F. *Chem. Eur. J.* **1997**, *3*, 755.
79 Dierkes, P.; van Leeuwen, P. W. N. M. *J. Chem. Soc. Dalton Trans.* **1999**, 1519.
80 Casey, C. P.; Whiteker, G. T. *Isr. J. Chem.* **1990**, *30*, 299.
81 Kranenburg, M.; Kamer, P. C. J.; van Leeuwen, P. W. N. M.; Vogt, D.; Keim, W. *J. Chem. Soc. Chem. Commun.* **1995**, 2177; Marcone, J. E.; Moloy, K. G. *J. Am. Chem. Soc.* **1998**, *120*, 8527.

Chapter 2

Hydroformylation with unmodified rhodium catalysts

Reaction mechanism and origin of regioselectivity

Raffaello Lazzaroni[†], Roberta Settambolo[•] and Aldo Caiazzo[†]

[†]*Dipartimento di Chimica e Chimica Industriale, Via Risorgimento 35, 56126 Pisa, Italy;* [•]*Istituto di Chimica Quantistica ed Energetica Molecolare del CNR, Area della Ricerca, Via Alfieri 1, 56010 Ghezzano (PI), Italy.*

2.1 Introduction

The first investigations on rhodium-catalyzed hydroformylation were carried out at the end of 1950's [1], about 20 years after the discovery by Roelen of the cobalt-catalyzed "oxo" reaction [2]. Initially, simple catalyst precursors, such as $RhCl_3$ and Rh/Al_2O_3, were employed. Even at the beginning it was clear that the rhodium-based catalysts were much more active than the cobalt based ones and were much more tolerant of the presence of other functional groups in the unsaturated substrates [3]. The synthesis and the spectroscopic characterization of rhodium-hydride complexes containing triphenylphosphine by Wilkinson's group [4] and their use in the hydrogenation and hydroformylation processes opened the way to the research on phosphine modified rhodium catalysts [5]. There has been an enormous amount of research on the synthesis and use of phosphorus- and sulfur-containing ligands with various steric and electronic characteristics [6], including optically active ones for use in enantioselective processes [7]. So the phosphorus modified catalysts have been used much more extensively than the corresponding unmodified ones [5a, 8].

Nevertheless, unmodified Rh catalytic precursors such as $Rh(CO)_2(acac)$, $[Rh(COD)(OAc)]_2$ and $Rh_4(CO)_{12}$ are still the subject of detailed investigations. As recently reported in the fundamental review of Cornils (1995): "*This is due to their easy availability, their well-known properties and their rather unproblematic handling. Additionally they serve as much simpler models than modified catalysts. But the main reason for their*

P.W.N.M. van Leeuwen and C. Claver (eds.), Rhodium Catalyzed Hydroformylation, 15–33.

persistent topicality is the fact that the mechanism of hydroformylation is far from being well understood" [9].
In this regard we can observe that:

1) Hydroformylations with unmodified catalyst precursors can be carried out under mild reaction conditions, some of the above catalysts being active even at room temperature.
2) With most substrates the reaction is completely chemoselective for the formation of aldehydes. Under mild reaction conditions, the other typical side processes, such as hydrogenation, isomerization, polymerization and aldol condensation, are negligible.
3) The above catalyst precursors allow an extensive investigation of the influence of the reaction parameters as well as of steric and electronic factors connected to the alkene structure on the regioselectivity of the reaction, avoiding any interference from phosphine ligands.

As far as the investigation of reaction mechanism is concerned it is to be noted that:

4) Unmodified catalyst precursors, especially $Rh_4(CO)_{12}$, have been successfully employed by Garland in fundamental IR spectroscopic investigations carried out under typical hydroformylation conditions. Information on the kinetic aspects of the process and on the reactivity of the acyl-rhodium intermediates has been obtained [10].
5) The same catalyst precursor, $Rh_4(CO)_{12}$, has been extensively employed in deuterioformylation experiments at partial substrate conversion. ^{2}H-NMR analyses carried out directly on the crude reaction mixture provide information on the behavior of alkyl-rhodium intermediates under oxo conditions [11].

This chapter will discuss the factors affecting the reaction regioselectivity of several alkenyl substrates and the related mechanistic aspects arising from in situ IR spectroscopic studies and deuterioformylation experiments.

2.2 Regioselectivity in the rhodium-catalyzed hydroformylation of vinyl and vinylidenic substrates

It is well known that the main goal in rhodium-catalyzed hydroformylation of unsaturated, especially vinyl, substrates, concerns the control of reaction regioselectivity, i.e. of the regioisomeric ratio **b**:**l** (branched to linear) between the isomeric aldehydes produced.

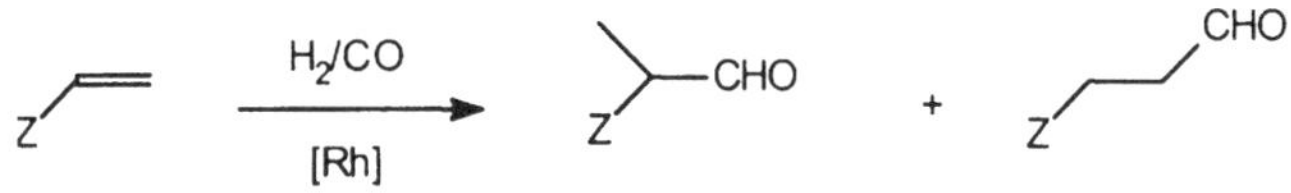

b l

Figure 1. Regioisomeric aldehydes obtained from the rhodium-catalyzed hydroformylation of vinyl alkenes

The influence of substrate structure as well as of reaction parameters on the reaction regioselectivity, in the presence of unmodified rhodium catalysts, has been extensively investigated. The main results will be discussed in the following sections.

2.2.1 Catalyst precursors

The following unmodified catalyst precursors are those most frequently employed in the hydroformylation of vinyl and vinylidenic substrates: $Rh_4(CO)_{12}$ [12], $[Rh(CO)_2(Cl)]_2$ [13], $[Rh(CO)_2(acac)]_2$ [14], $[Rh(cod)(OAc)]_2$ [15], and zwitterionic complexes such as $[BPh_4]^-$ $[Rh(COD)]^+$ (Rh^{zw}) [16]. These last ones, extensively used by Alper and coworkers, can be included in the above group, since they are generally used without any ancillary ligand.

As reported by several authors [8, 10], $Rh_4(CO)_{12}$ generates, under hydroformylation conditions, a rhodium-carbonyl hydride $[HRh(CO)_3]$, which constitutes the catalytic active species of the reaction. A similar process probably occurs with the other catalyst precursors. No information on the course of reaction with Rh^{zw} systems has been reported [16]. Nevertheless it is likely that an analogous hydride species is formed also with this precursor.

2.2.2 Influence of the alkene structure on the regioselectivity

The rhodium-catalyzed hydroformylation of many simple and functionalized substrates in the presence of phosphorus ligands has been extensively investigated. The comparative results are reported in several reviews [7c, 9]. By contrast the results for alkenyl substrates hydroformylated with unmodified rhodium-based precursors are relatively few, the mechanistic aspects of the reaction and not the synthetic goal being the main purpose of these studies.

These substrates can be divided into three different classes, on the basis of their structure:

a) vinyl substrates; b) allyl substrates; c) vinylidenic substrates.

In order to examine the influence of the alkene structure on the regioselectivity, a comparison of the results obtained with different substrates and different catalyst precursors under mild reaction conditions (20-50 °C, 50-160 bar of CO/H_2, 1:1) will be made.

a) Vinyl substrates

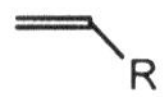

(R = Ar, OR', *t*-Bu, CF_3, OCOR')

These substrates cannot isomerize into internal alkenes, due to the absence of hydrogens in α position with respect to the double bond.

The most investigated alkene belonging to this group is styrene. The hydroformylation of this substrate under mild temperature always provides a large predominance of the branched isomer over the linear one (**b**:**l** = 95/5-98/2), as shown in Table 1. The sole branched isomer is obtained also in the hydroformylation of fluorinated alkenes, namely, fluoroethene, 3,3,3-trifluoropropene and pentafluorostyrene in the presence of $Rh_4(CO)_{12}$ [12b]. Heteroaromatic vinyl substrates, such as vinylfurans [17], vinylthiophenes [18], vinylpyrroles [19] and vinylpyridines [14, 20], behave in a very similar manner to styrene, giving a high prevalence of the branched aldehyde. Isomeric vinylpyridines show a quite singular behavior: whilst 3-vinylpyridine show the typical trend of vinylaromatic alkenes, 4-vinylpyridine gives rise to the hydrogenation product 4-ethylpyridine with more than 96 % selectivity [20]. As far as oxygen containing substrates are concerned, vinyl acetate exclusively gives the branched isomer (**b**:**l** ≈ 99/1) [15], (vinyl)alkyl ethers provide a regioisomeric ratio **b**:**l** = 80/20 [11d], whilst the presence of a phenyl group directly bonded to the oxygen enhances the regioselectivity towards the branched aldehyde (**b**:**l** = 95/5 for $PhO\text{–}CH\text{=}CH_2$) [21]. α,β-Unsaturated aldehydes and ketones give hydrogenation of the double bond only under oxo conditions [5a]. By contrast, alkyl alkenes containing a tertiary carbon atom bonded to the vinyl group, such as 3,3-dimethylbut-1-ene, show the opposite regioselectivity, giving the linear aldehydic isomer almost exclusively [15].

b) Allyl substrates

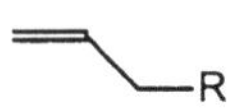

(R = alkyl, OR')

It is well known that alkenes with α hydrogens (linear 1-alkenes, allylethers, etc.) can give more than two regioisomers when migration of the double bond into internal positions occurs and hydroformylation of the resulting internal alkene takes place [22]. However, under mild reaction conditions, particularly at room temperature, there is no isomerization of the starting substrate, hence only two regioisomers are observed in all cases.

Very similar amounts of branched and linear aldehydes are obtained for all the linear 1-alkenes (i.e. **b**:**l** = 48/52 for 1-hexene at 20 °C and 100 bar CO/H_2, 1:1) [22], while an oxygen in β position to the vinyl group favors the formation of the branched isomers as observed for (allyl)ethyl ether (**b**:**l** = 70/30) and similar substrates [11d]. It should be noted that the linear isomer largely predominate over the branched ones in the hydroformylation of 3-alkyl substituted allyl alkenes (i.e. 3-methylbut-1-ene) [5a].

c) Vinylidenic alkenes

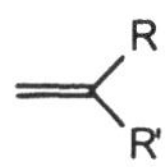

(R, R' = Ar, alkyl)

The hydroformylation rate in the case of vinylidenic alkenes is very low at room temperature, so the reaction is usually carried out at temperatures higher than 80 °C. Whatever kind of Z substituent is present (dialkyl, arylalkyl, diaryl) the linear isomer is almost exclusively produced [11f, 23]. Only when one of the substituents is a 2-pyridyl group does the branched isomer predominates over the linear one; in these cases a high amount of hydrogenation products is obtained [24].

In conclusion, unsaturated vinyl substrates can give opposite regioselectivities depending on the steric and electronic nature of the substituent bonded to the alkenyl moiety. When this substituent is a phenyl or an oxygen the branched aldehydic isomers predominates. By contrast, bulky groups favor the formation of the linear aldehyde, which is observed also in the case of vinylidenic substrates.

Table 1. Isomeric composition of aldehydic products arising from styrene hydroformylation carried out in the presence of unmodified rhodium-based precursors[a].

Catalyst precursor	T (°C)	P (bar)[b]	Reaction times (h)	b:l[c]	Reference
$Rh_4(CO)_{12}$	60	150	5	95/5	12c
$Rh_4(CO)_{12}$	25	150	9.5	98/2	12c
$[Rh(COD)(OAc)]_2$	25	50	16	96/4	15
$[Rh(COD)(Cl)]_2$	80	40	1.5	95/5	13
$[Rh^{zw}]$[d]	47	14	22	98/2	16
$Rh(CO)_2(acac)$	30	90	16	97/3	14

[a] At complete substrate conversion
[b] CO/H_2 = 1:1
[c] Regioselectivity
[d] Zwitterionic rhodium complex $[BPh_4]^-[Rh(COD)]^+$

Table 2. Selected values of regioisomeric ratio in the hydroformylation of unsaturated substrates in the presence of unmodified rhodium precursors[a].

Substrates	T (°C)	P (bar)[b]	Reaction times (h)	b:l[c]	References
Fluoroethene	80	110	6	100/0	12b
3,3,3-Trifluoropropene	80	110	6	97/3	12b
Substituted styrenes	20	60	16	95/5-98/2	15
(Vinyl)ethyl ether	20	100	9	83/17	11d
3,3-Dimethylbutene	20	60	16	0/100	15
Vinyl acetate	20	60	16	> 99/1	15
1-Hexene	15	100	6	52/48	22b
(Allyl)ethyl ether	20	100	6	70/30	11d
2-Phenylpropene	100	100	3	< 1/99	11f
1,1-Diphenylethene	100	100	20	< 1/99	11f
2-Methylpropene	100	100	1	0/100	11f

[a] At complete substrate conversion
[b] CO/H_2 = 1:1
[c] Regioselectivity

2.2.3 Influence of temperature

Systematic studies on the influence of temperature on the regioselectivity in the hydroformylation of vinyl substrates in the presence of unmodified rhodium-based precursors have been carried out only in few cases. In particular the investigations reported in literature concern the hydroformylation of styrene, ethyl- and allyl ethers and 1-hexene, with $Rh_4(CO)_{12}$ over temperatures, ranging from 20 °C to 100 °C.

In the case of styrene a strong increase of linear aldehydic isomer with increasing temperature is observed (**b**:**l** = 98/2 at 20 °C to 64/36 at 130 °C) [12c]. For (ethyl)vinyl ether the above increase is lower, the percentage of linear aldehyde ranging from 12% at 20 °C to 24% to 100 °C [11d]. In the case of (phenyl)vinyl ether, which shows a high α-regioselectivity at 20 °C (**b**:**l** = 95/5), a negligible temperature effect is obtained [21]. In all these cases no variation of the regioisomeric ratio with increasing substrate conversion is observed.

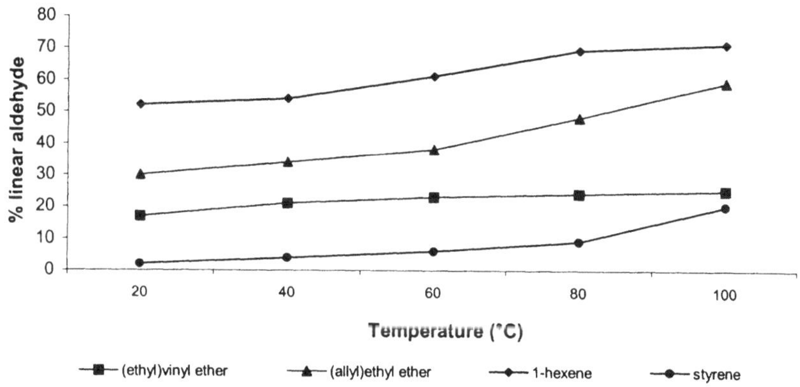

Figure 2. Influence of temperature on the hydroformylation regioselectivity of selected substrates in the presence of $Rh_4(CO)_{12}$ as catalyst precursor

The hydroformylation of allylic alkenes is characterized, at high temperatures, by competing isomerization [22]. Significant regioselectivity ratios for the hydroformylation of the sole terminal double bond have been obtained by carrying out the reaction at partial substrate conversion, when the starting terminal alkene is still present in excess relative to that resulting from isomerization [22b]. In 1-hexene hydroformylation the amount of linear isomer increases with increasing temperature, ranging from 52% at 20 °C to 72% at 100 °C. As internal alkenes are converted into aldehydes only when all the terminal alkene has reacted, it is possible to estimate, at partial substrate conversion, the chemoselectivity of the reaction, i. e. the amount of 1-hexene converted into aldehydes (**b+l**) with respect to that isomerized to internal alkene (E-2). Chemoselectivity to aldehydes decreases with increasing temperature, **(b+l)/E-2** reaching the value 60/40 at 120 °C. The effect of temperature is greater in the case of (allyl)ethyl ether, for which the percentage of linear isomer increases from 30% at 20 °C to 60 % at 100 °C. A slight increase of linear aldehyde with increasing temperature is observed also for propylene, which cannot give any internal alkenyl product.

In the case of vinylidenic alkenes the linear aldehydic isomer is obtained with a complete selectivity at any temperature [11f, 23].

2.2.4 Influence of CO and H_2 partial pressures

In the case of styrene the CO and H_2 partial pressures affect the reaction regioselectivity only when the reaction is carried out at high temperatures. In particular it has been observed that a decrease of carbon monoxide or hydrogen partial pressure causes an increase of linear aldehydic isomer, this effect being more evident at higher temperatures (100 °C). So in the case of styrene hydroformylation at 100 °C the **b:l** ratio ranges from 80/20 at 170 bar of CO/H_2 (1:1) to 56/44 at P_{H2} = 6 bar, P_{CO} = 85 bar or to 60/40 at P_{H2} = 85 bar, P_{CO} = 6 bar [12c]. In the case of 1-hexene gas pressure does not affect the regioselectivity of the reaction either at room temperature or at high temperature. By contrast the chemoselectivity to aldehydes increases with increasing temperature, **(b+l)/E-2** being 44/56 at 40 bar and 77/23 at 140 bar [22b].

2.3 Mechanism of the hydroformylation of vinyl and vinylidenic alkenes

As described above, both the nature of the substrate and the reaction conditions strongly influence the regioselectivity in the hydroformylation of vinyl substrates. The above results clearly demonstrate that, by raising the reaction temperature, and decreasing the CO and H_2 partial pressures, the amount of linear aldehydes increases. Indeed, this is a general trend in the hydroformylation of different substrates and constitutes a fundamental starting point for a rationalization of the influence of experimental parameters on the reaction selectivity.

In this context we are going to examine the above results in the light of the generally accepted mechanism for hydroformylation, taking into account the more recent findings on the behavior under reaction conditions of the main intermediate species, namely alkyl- and acyl-rhodium complexes. A simplified scheme for the hydroformylation of a typical vinyl substrate is shown in Figure 3.

The rhodium hydride tricarbonyl species easily coordinates the vinyl substrate generating the π complex (**1**), which is converted into the alkyl-rhodium intermediates (**2**) through insertion of the alkene into the Rh-H bond. Migratory insertion of the alkyl moiety on to a CO molecule coordinated to the metal center provides the acyl-rhodium species **3**, which, at the end of the catalytic cycle, interacts with hydrogen via an oxidative addition, giving rise to aldehydic products and regenerating the rhodium-hydride species.

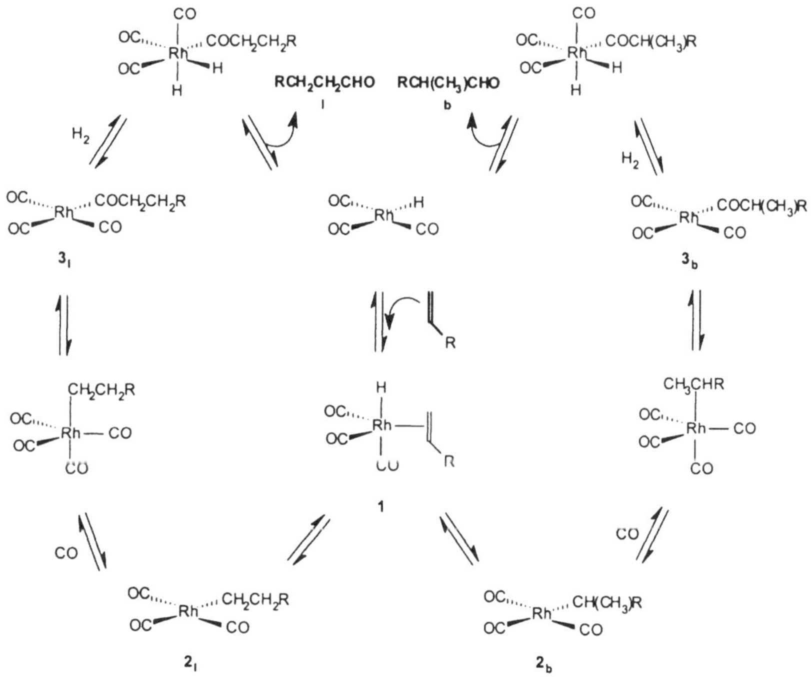

Figure 3. Generally accepted mechanism for the rhodium-catalyzed hydroformylation

2.3.1 Activation of the catalyst precursor

The mechanism of fragmentation of the rhodium cluster $Rh_4(CO)_{12}$ under oxo conditions has been extensively studied during recent years; in the course of these investigations the presence of low nuclearity species has been proposed [25]. Only since Garland and co-workers began studying the fragmentation of $Rh_4(CO)_{12}$ in the presence of alkenyl substrates via in situ IR spectroscopy the mechanism of transformation of rhodium-clusters under oxo condition become clearer. The disappearance of the typical bands due to $Rh_4(CO)_{12}$ and the appearance of those due to the acyl-rhodium intermediates were investigated under different experimental conditions in order to determine a kinetic expression for the fragmentation process [10].

In a paper by Garland [10a] concerning the hydroformylation of 3,3-dimethylbut-1-ene (33DMB) with $Rh_4(CO)_{12}$, the cluster fragmentation was investigated and a mechanism proposed, according to the following kinetic expression

$$\text{rate (I)} = k_0(\text{I})\,[Rh_4(CO)_{12}]^1[CO]^{1.8}[H_2]^{0.7}[33DMB]^{0.1}$$

This equation is consistent with i) a preequilibrium between $Rh_4(CO)_{12}$, CO and a second unstable cluster, $Rh_4(CO)_{14}$ and ii) a rate-limiting step involving the activation of the latter cluster by H_2.

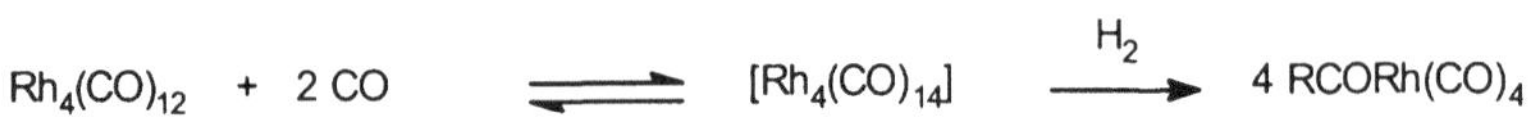

Figure 4. $Rh_4(CO)_{12}$ cluster conversion into acyl-metal intermediate

As reported in the same paper, it is likely that this unstable rhodium cluster is converted into the mononuclear rhodium-hydride species $HRh(CO)_x$ ($x = 3,4$), which are usually considered as the true catalyst system in the reaction mixture. These compounds represent extremely unstable intermediates, which would certainly recombine to form higher nuclearity rhodium species if alkene is not present in the reaction mixture. This mechanism is proposed for all the hydroformylation experiments carried out in the presence of $Rh_4(CO)_{12}$.

2.3.2 Behavior of the isomeric alkyl-metal intermediates via deuterioformylation experiments

The crucial step that determines regioselectivity is the alkene insertion into the Rh-H bond which gives rise to the alkyl-metal intermediates. This step can be reversible or irreversible, depending on the reaction conditions.

Deuterioformylation experiments carried out at partial substrate conversion has proved to be the best way to investigate the reversibility of the above step [11]. As shown in Figure 5, when a deuterated alkyl species undergoes a β-hydride elimination process, the elimination of Rh-H is favored over that of Rh-D one, because of the well documented kinetic isotope effect observed in this kind of process [26]. Thus β-hydride elimination from the linear alkyl species gives rise to an alkene deuterated at the carbon atom in position 2, whilst the analogous process for the branched alkyl intermediate generates an alkene deuterated at the terminal position of the double bond.

Examination by ^{2}H-NMR spectroscopy of the crude deuterioformylation mixture at partial substrate conversion gives direct information, both qualitative and quantitative, on the occurrence of a β-elimination process, i.e. on the reversibility of formation of the alkyl intermediates.

As a typical example, the ^{2}H-NMR spectra of a mixture resulting from deuterioformylation of (ethyl)vinyl ether at 20 °C and 100 °C and 30% substrate conversion are shown in Figure 6 [11d].

At 100 °C β-hydride elimination occurs for both the alkyl-rhodium species; this is evident from the presence of Et-O-CH=CHD (1-d) (signals at 4.10 and 3.97 ppm) derived from the branched isomer and by the presence of Et-O-CD=CH_2 (2-d) (signal at 6.44 ppm) derived from the linear one.

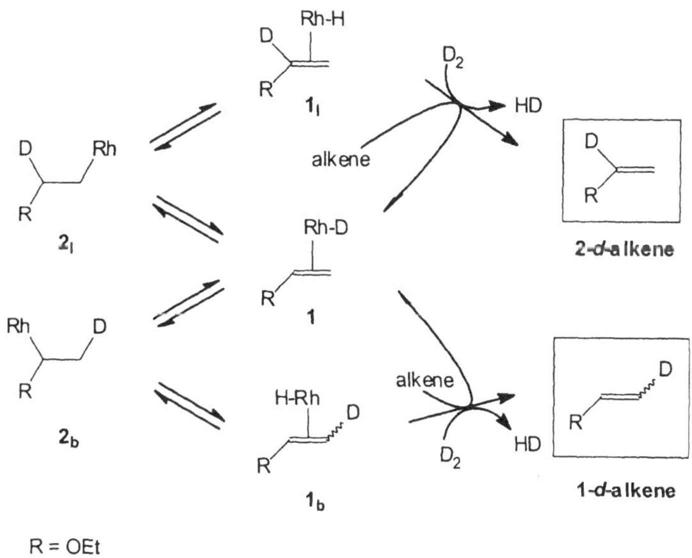

Figure 5. Rhodium-catalyzed deuterioformylation of (ethyl)vinyl ether

By contrast, at 20 °C, no signals due to the deuterated alkene are present, indicating that no β-elimination process takes place and hence the formation of the alkyl intermediates is not a reversible step. It is noteworthy that analogous experiments carried out at room temperature with styrene as well as 1-hexene [11a] and (allyl)ethyl ether [11d], clearly show that no β-hydride elimination occurs under these mild conditions. This means that the alkene insertion into the Rh-H bond is not reversible and hence the isomeric alkyl-rhodium intermediates are directly transformed into the acyl species to give isomeric aldehydes. Deuterioformylation experiments also give other interesting information. As far as hydroformylation of (ethyl)vinyl ether at 100 °C is concerned, comparing the relative intensities of alkenyl absorption and formyl group at 9.46 ppm, it clear that the ratio between the above absorptions [(Σ alkenyl deuteriums)/(formyl deuterium)] increases as the temperature is raised. It has also been observed that β-hydride elimination occurs more readily for the branched alkyl intermediate than for the linear one, as shown by the lower intensity of the signal at 6.44 ppm with respect to those at 3.97-4.14 ppm. Analogous behavior has been observed also for allylic substrates, e. g. 1-hexene and (allyl)ethyl ether. In these cases the β-hydride elimination from the branched alkyl gives mainly the isomeric internal alkene deuterated at the terminal position (Z-CH=CH-CH_2D) (Figure 5).

By contrast, in the case of styrene [12c] and other aromatic substrates [19a] β-hydride elimination occurs selectively for the branched intermediate, at least at temperatures lower than 100 °C, since the signal at low fields, typical of 2-d-alkene, is absent from the ^2H NMR spectra.

Deuterioformylation at 80-100 °C of vinylidenic alkenes containing one or two phenyl groups on the double bond [11e-f] clearly shows, in addition to the linear aldehyde, the formation of alkenes deuterated in the terminal position, arising from β-elimination from the tertiary alkyl-metal intermediates. However, there are no traces of the quaternary aldehyde deriving from migratory insertion of this alkyl group to CO and subsequent oxidative addition of H_2. In contrast, deuterioformylation experiments carried out on vinylidenic alkenes containing two alkyl groups [11f] show that the formation of a tertiary alkyl-metal intermediate occurs only to a very low extent (2-methylpropene) or that it does not occur at all (2,3,3-trimethylbutene).

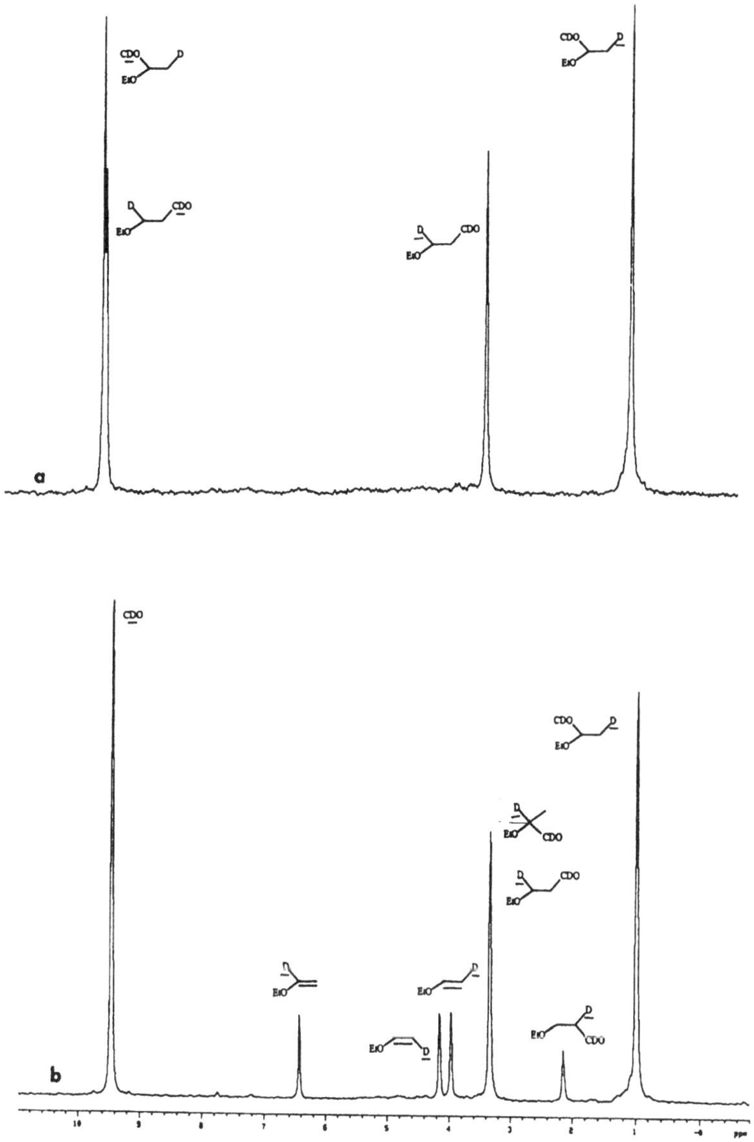

Figure 6. ^{2}H-NMR spectrum (46 MHz, 25 °C, C_6D_6 as external standard) of the crude mixture resulting from deuterioformylation of (ethyl)vinyl ether at (**a**) 20 °C and (**b**) 100 °C

2.3.3 In situ IR investigation of the formation and reactivity of acyl-rhodium intermediates

Acyl intermediates are the only detectable reaction intermediates under typical hydroformylation conditions; they can be relatively stable under CO atmosphere and at mild temperatures. Using in situ IR spectroscopy, Garland and co-workers have made an important contribution in making clear several kinetic and mechanistic aspects of the behavior of acyl intermediates [10]. The most interesting aspect of the reactivity of such species is the oxidative addition of H_2 and the following reductive elimination to provide the final aldehydic products, with the regeneration of the metal-hydride complex. The first studies were carried out with alkenes generating only an aldehydic isomer, such as 33DMB [10a] and cyclohexene [10b]. In the case of 33DMB the existence of a preequilibrium between the rather stable and detectable intermediate $RCORh(CO)_4$, CO and the unsaturated reactive complex $RCORh(CO)_3$, is observed; the latter undergoes oxidative addition of H_2 to provide the final aldehydic products, which represents the rate-limiting step of the reaction.

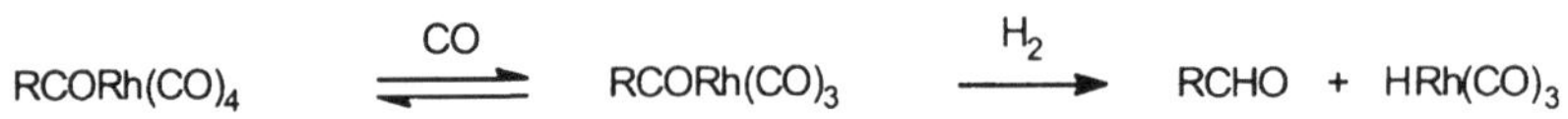

Figure 7. Hydrogenolysis of acyl-metal intermediates

The whole process can be described by the following kinetic expression

$$\text{rate (II)} = k_0(\text{II})\,[\text{RCORh(CO)}_4]^1[\text{CO}]^{-1.1}[\text{H}_2]^1[\text{33DMB}]^{0.1}$$

Investigations carried out on several alkenyl substrates indicate that in all the cases the acyl-rhodium species formed under hydroformylation conditions have a bipyramidal geometry, with C_s symmetry, the acyl group taking an axial position (Figure 8) [10e].

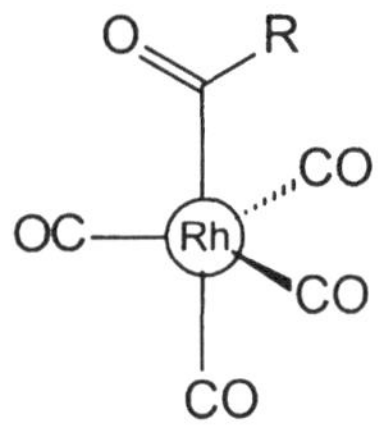

Figure 8. Structure of acyl-metal intermediates

Recently a very detailed investigation of the mechanism of the interaction between the isomeric acyl-metal species and H_2 under different reaction conditions in the case of styrene was carried out [10d]. Both the isomeric acyl-rhodium intermediates were observed, and their hydrogenolysis to give aldehydic products and the relative kinetics analyzed under different reaction conditions. The kinetic expression derived for the whole process is

$$\text{rate} = k_0{}^{\gamma}[RCORh(CO)_4]_{ss}^{1.0}[CO]^{-1.0}[H_2]^{1.0}[C_8H_8]^{0.0}$$

Of particular interest is the effect of reaction temperature on the reaction rate and regioselectivity. The most remarkable results from the experiments carried out at P(CO) = 50 bar and P(H_2) = 5 bar, in the range 298-313 °K, are summarized in Table 3.

The hydrogenolysis rate of the linear acyl-metal intermediate is higher than the one of the branched isomer, the difference being much more marked at 40 °C than at 25 °C. The regioisomeric ratio between the acyl intermediates at 40 °C is higher than the one between the corresponding aldehydes; by contrast, at 25 °C the two values are quite similar.

Table 3. Selected values of kinetic constants and regioisomeric ratios for styrene hydroformylation in the presence of $Rh_4(CO)_{12}$ as catalyst precursor, at 25 °C and 40 °C.

Values	Temperature	
	25 °C	40 °C
k_b	0.93	4.18
k_n	1.27	10.3
$\mathbf{3_b}:\mathbf{3_l}$[a]	97.5/2.5	87.5/12.5
b:l	96.7/3.3	66/34

[a] Regioisomeric ratio between branched ($\mathbf{3_b}$) and linear ($\mathbf{3_l}$) acyl-rhodium intermediates.

2.4 Origin of the regioselectivity

2.4.1 Influence of the nature of the substrate

As previously shown by deuterioformylation experiments, when the reaction is carried out at low temperature, the formation of alkyl-metal intermediates is not a reversible step. Under these conditions the regioselectivity observed for aldehydic isomers is directly determined in the step at which the alkyl metal intermediates are formed. Taking into account the structure of the linear and branched alkyl-metal intermediates it is possible to explain why the branched aldehydes are strongly favored in the case of styrene or other functionalized substrates, whereas in the case of simple 1-alkenes approximately equal amounts of aldehydic isomers are formed.

As shown in Figure 9, the metal-carbon bond in the alkyl-rhodium intermediates is polarized with a partial positive charge on the metal and a partial negative charge on the carbon atom. When this carbon atom is bonded to a strongly polarizable group (i.e. $-C_6H_5$) or to an electron withdrawing group (i.e. -F, -OR, $-CH_2OR$, $-CF_3$), the partial negative charge on the carbon atom is better delocalized owing to the inductive effect in the branched isomer $\mathbf{2_b}$ than in the linear one $\mathbf{2_l}$ [12b-c].

When R is an electron donor group (n-alkyl group), no delocalization of the partial negative charge occurs for either isomer. As a consequence the branched and linear alkyl intermediates are formed in similar amounts, hence so are the corresponding aldehydes (Figure 9).

However, when the alkyl group bonded to the vinyl moiety has a secondary or tertiary structure, steric hindrance plays in crucial role on the regioselectivity, causing the linear aldehyde to predominate.

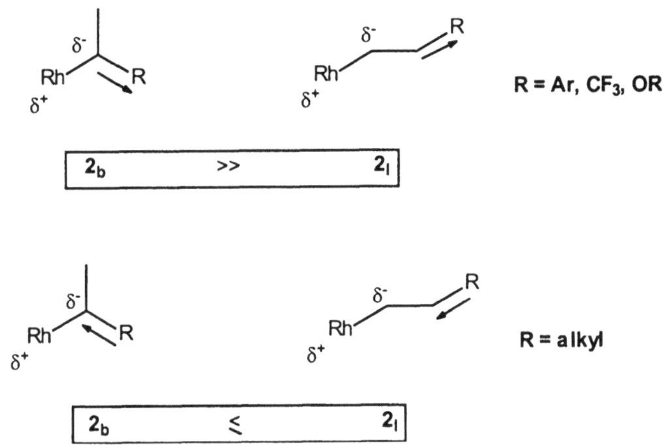

Figure 9. Stabilization of alkyl-rhodium intermediates arising from the hydroformylation of different alkenes

As far as the vinylidenic substrates are concerned, deuterioformylation of phenyl substituted vinylidenic alkenes gives interesting information about the formation of a tertiary alkyl intermediate under reaction conditions. Indeed, the formation of vinylidenic alkenes deuterated at the terminal position to a larger extent than the linear aldehyde, demonstrates that the branched alkyl predominates over the linear one. As previously mentioned, this is due to the higher stabilization induced by the two phenyl groups adjacent to the carbon-rhodium bond. However the migratory insertion on to the CO coordinated to the metal in the case of tertiary alkyls is prevented by steric reasons. Thus it seems evident that the behavior of the two isomeric alkyl-rhodium intermediates is completely different: while the primary one is converted into the linear aldehyde, the tertiary one exclusively undergoes β-hydride elimination, regenerating the starting alkene [11e].

In conclusion, the ^{2}H-NMR analysis of crude deuterioformylation products derived from vinyl or vinylidenic aromatic substrates is a direct and simple way to detect the different behavior of a primary, secondary and tertiary alkyl-metal intermediate, related to the β-hydride elimination process under typical hydroformylation conditions.

2.4.2 Influence of the reaction parameters

As far as the influence of reaction parameters, observed for vinyl and allyl substrates, is concerned, the increase of linear aldehyde with increasing temperature can be easily explained on the basis of the different behavior of the alkyl-rhodium intermediates under the reaction conditions. Thus the linear alkyl mainly undergoes the migratory insertion process and, hence, gives the linear aldehyde. In contrast, the branched one undergoes carbonylation only partially, mainly providing β-hydride elimination. It is to be noted that the π complex derived from the above elimination process regenerates both the linear and the branched alkyls. Thus the whole process brings about a partial isomerization of the branched alkyl isomer to the linear one and hence determines an increase of linear aldehyde. The different behavior of isomeric alkyl-rhodium intermediates could account also for the increase of linear aldehyde with decreasing CO and H_2 pressure. At high gas pressure both the intermediate alkyls are forced to take part in the carbonylation to provide the aldehydic products. At low pressure, the β-elimination process becomes competitive with the acyl formation and with the subsequent oxidative addition of H_2. Because the above elimination process is favored in the case of branched alkyl-metal species, the final result will be an increase of linear aldehyde.

On this basis it is possible to explain the results obtained by Garland in the hydroformylation of styrene under relatively mild reaction conditions. It is plausible that the **b:l** ratio = 66/34 obtained at 40 °C, which is lower than the one observed for the acyl-rhodium species ($\mathbf{3_b}/\mathbf{3_l}$ = 87.5/12.5), is due to the β-elimination process which is much more favorable for the branched alkyl intermediate than for the linear one.

Phosphine ligands, when employed in excess with respect to rhodium, generally block the β-elimination process, as shown by deuterioformylation experiments carried out by Casey [27] and Takaya [7a], thus accounting for the low variation of regioselectivity with temperature obtained in the presence of phosphine-modified precursors [20b].

It is to remark that in the hydroformylation of styrene, the most investigated vinyl aromatic substrate, the predominance of the branched aldehyde at room temperature is higher with unmodified rhodium precursors than with phosphine-modified ones [4, 7c, 28]. In this context, when hydroformylation of styrene with chiral phosphines occurs without asymmetric induction and with a large prevalence of the branched aldehyde (> 96%), it is likely that unmodified rhodium-catalysts are also present in the reaction mixture [14, 29, 30].

References

1 (a) Schiller, G. Ger. Pat. 965,605 **1956** (to Chem. Verwertungsges. Oberhausen); *Chem Abstr.* **1959**, *53*, 11226. (**b**) Hughes, V. **L**. Br. Pat. 801,734 **1958** (to Esso Res. Eng Comp.); *Chem Abstr.* **1959**, *53*, 7014.

2 Roelen, O. Ger. Pat. 849,548 1938; *Chem. Zentr.* **1953**, 927.

3 *Organometallic Chemistry of Transition Elements*, ed. F. P. Pruchnik, Plenum Press, New York, 1990, p 691.

4 Evans, D.; Osborn, J. A.; Wilkinson, G. *J. Chem. Soc., A* **1968**, 3133.

5 (a) *Organic Syntheses via Metal Carbonyls*, eds. I. Wender and P. Pino, Wiley-Interscience, New York, 1977, Vol. 2. (**b**) *Homogeneous Catalysis with Metal Phosphine Complexes*, ed. **L**. H. Pignolet, Plenum Press, New York, 1983. (c) *Applied Homogeneous Catalysis with Organometallic Compounds*, eds. **B**. Cornils and W. A. Herrmann, VCH, Weinheim, 1996.

6 (a) Casey, C. P.; Paulsen, E. **L**.; Beuttenmueller, E. W.; Proft, **B**. R.; Matter, **B**. A.; Powell, D. R. *J. Am. Chem. Soc.* **1999**, *121*, 63. (**b**) Casey, C. P.; Paulsen, E. **L**.; Beuttenmueller, E. W.; Proft, **B**. R.; Petrovich, **L**. M.; Matter, **B**. A.; Powell, D. R. *J. Am. Chem. Soc.* **1997**, *119*, 11817. (c) van der Van, **L**. A.; Kamer, P. C. J.; Van Leeuwen, P. W. N. M. *Organometallics* **1999**, *18*, 4765. (d) van Rooy, A.; Kamer, P. C. J.; Van Leeuwen, P. W. N. M.; Goubitz, K.; Fraanje, J.; Veldman, N.; Spek, A.; *Organometallics* **1996**, *15*, 835.

7 (a) Nozaki, K.; Sakai, N.; Nanno, T.; Higashijima, T.; Mano,, S.; Horiuchi, T.; Takaya, H. *J. Am. Chem. Soc.* **1997**, *119*, 4413. (**b**) Nozaki, K.; Nanno, T.; Takaya, H. *J. Organomet. Chem.* **1997**, *527*, 103. (c) Gladiali, S.; Bayón, J. C.; Claver, C. *Tetrahedron: Asymm.* **1995**, *6*, 1453. (d) Bayón, J. C.; Claver, C.; Masdeu-Bulto, A. M. *Coord. Chem. Rev.* **1999**, *195*, 73. (e) Miquel-Serrano, M. D.; Masdeu-Bulto, A. M.; Claver, C; Sinou, D. *J. Mol. Cat., A: Chemical* **1999**, *143*, 49.

8 (a) Thatchenko, I. *Comprehensive Organometallic Chemistry*, eds. G. Wilkinson, F. G. A. Stone and E. W. Abel, Pergamon, Oxford, 1982, Vol 8, pp. 101. (**b**) *Transition metals for Organic Synthesis*, eds. M. Beller and C. Bolm, Wiley VCH, 1999.

9 Beller, M.; Cornils, **B**.; Frohning, C. D.; Kohlpaintner, C. W. *J. Mol. Cat. A: Chem.* **1995**, *104*, 17.

10 (a) Garland, M.; Pino, P. *Organometallics* **1991**, *10*, 1693. (**b**) Garland, M. *Organometallics* **1993**, *12*, 535. (c) Fyhr, C.; Garland, M. *Organometallics* **1993**, *12*, 1753. (d) Feng, J.; Garland, M. *Organometallics* **1999**, *18*, 417. (e) Guowei, **L**.; Volken, R.; Garland, M *Organometallics* **1999**, *18*, 3429.

11 (a) Lazzaroni, R.; Uccello-Barretta, G.; Benetti, M. *Organometallics* **1989**, *8*, 2323. (**b**) Raffaelli, A.; Pucci, S.; Settambolo, R.; Uccello-Barretta, G.; Lazzaroni, R. *Organometallics* **1991**, *10*, 3892. (c) Uccello-Barretta, G.; Lazzaroni, R.; Settambolo, R.; Salvadori, P. *J. Organomet. Chem.* **1991**, *417*, 111. (d) Lazzaroni, R.; Settambolo, R.; Uccello-Barretta, G. *Organometallics* **1995**, *14*, 4644. (e) Lazzaroni, R.; Uccello-Barretta, G.; Scamuzzi, S.; Settambolo, R.; Caiazzo, A. *Organometallics* **1996**, *15*, 4657. (f) Lazzaroni, R.; Settambolo, R.; Uccello-Barretta, G.; Caiazzo, A.; Scamuzzi, S. *J. Mol. Cat., A: Chemical* **1999**, *143*, 123.

12 (a) Pino, P.; Oldani, F.; Consiglio, G. *J. Organomet. Chem.* **1983**, *250*, 491. (**b**) Ojima, I. *Chem. Rev.* **1988**, *88*, 1011. (c) Lazzaroni, R.; Raffaelli, R.; Settambolo, R.; Bertozzi, S.; Vitulli, G. *J. Mol. Cat.* **1989**, *50*, 1.

13 Botteghi, C.; Paganelli, S.; Bigini, **L**.; Marchetti, M. *J. Mol. Cat.* **1994**, *93*, 279.

14 Basoli, C.; Botteghi, C.; Cabras, M. A.; Chelucci, G.; Marchetti, M. *J. Organomet. Chem.* **1995**, *488*, C20.

15 Doyle, M. P.; Shanklin, M. S.; Zlokazov, M. V. *Synlett* **1994**, 615.

16 Amer, I.; Alper, H. *J. Am. Chem. Soc.* **1990**, *112*, 3674.

17 (a) Kalck, P.; Serein-Spiran, F. *New J. Chem.* **1989**, *13*, 515. (**b**) Lapidus, A. **L**.; Rodin, A. P.; Pruidze, I. G.; Ugrak, **B**. I. *Izv. Akad. Nauk SSSR, Ser. Khim.* **1990**, *7*, 1661.

18 Browning, A. F.; Bacon, A. D.; White, C.; Milner, D. J. *J. Mol. Catal.* **1993**, *83*, L11.

19 (a) Caiazzo, A.; Settambolo, R.; Uccello-Barretta, G.; Lazzaroni, R. *J. Organomet. Chem.* **1997**, *548*, 279. (**b**) Lazzaroni, R.; Settambolo, R.; Mariani, M.; Caiazzo, A. *J. Organomet. Chem.* **1999**, *592*, 69.

20 (a) Settambolo, R.; Scamuzzi, S.; Caiazzo, A.; Lazzaroni, R. *Organometallics* **1998**, *17*, 2127. (**b**) Caiazzo, A.; Settambolo, R.; Pontorno, **L**.; Lazzaroni, R. *J. Organomet. Chem.* **2000**, in press.

21 Lazzaroni, R.; Bertozzi, S.; Pocai, P.; Troiani, F.; Salvadori, P. *J. Organomet. Chem.* **1985**, *295*, 371.

22 (a) Hanson, **B**. E.; Davis, N. E. *J. Chem. Educ.* **1987**, *64*, 928. (**b**) Lazzaroni, R.; Pertici, P.; Bertozzi, S.; Fabrizi, G. *J. Mol. Catal.* **1990**, *58*, 75.

23 Botteghi, C.; Cazzolato, **L**.; Marchetti, M.; Paganelli, S. *J. Org. Chem.* **1995**, *60*, 6612.

24 Botteghi, C.; Marchetti, M.; Paganelli, S.; Sechi, **B**. *J. Mol. Catal. A: Chem.* **1997**, *118*, 173.

25 (a) Vidal, J. **L**.; Walker, W. E. *Inorg. Chem.* **1981**, *20*, 249 (**b**) Oldani, F.; Bor, G. *J. Organomet. Chem.* **1983**, *246*, 309.

26 Evans, J.; Schwartz, J.; Urquhart, P. W. *J. Organomet. Chem.* **1974**, *81*, C37.

27 Casey, C. P.; Petrovich, L. M. *J. Am. Chem. Soc.* **1995**, *117*, 6007.

28 Sakai, N; Mano, S.; Nozaki, K.; Takaya, H. *J. Am. Chem. Soc.* **1993**, *115*, 7033.

29 Brown, J. M.; Cook, S. J.; Khan, R. *Tetrahedron* **1986**, *42*, 5105.

30 Eckl, R. W.; Priermeier, T.; Herrmann, W. A. *J. Organomet. Chem.* **1997**, *532*, 243.

Chapter 3

Rhodium Phosphite Catalysts

Paul C. J. Kamer, Joost N. H. Reek, and Piet W. N. M. van Leeuwen

Institute of Molecular Chemistry, University of Amsterdam, Nieuwe Achtergracht 166, 1018 WV Amsterdam, The Netherlands

3.1 Introduction

The important discovery by Wilkinson [1] that rhodium afforded active and selective hydroformylation catalysts under mild conditions in the presence of triphenylphosphine as a ligand triggered a lot of research on hydroformylation, especially on ligand effects and mechanistic aspects. It is commonly accepted that the mechanism for the cobalt catalyzed hydroformylation as postulated by Heck and Breslow [2] can be applied to phosphine modified rhodium carbonyl as well. Kinetic studies of the rhodium triphenylphosphine catalyst have shown that the addition of the alkene to the hydrido rhodium complex and/or the hydride migration step is probably rate-limiting [3] (Chapter 4). In most phosphine modified systems an inverse reaction rate dependency on phosphine ligand concentration or carbon monoxide pressure is observed [4].

Since π-back bonding contributes significantly to the strength of the metal-to-ligand bond, especially for carbonyl ligands, it is not surprising that electron-withdrawing substituents on the ligands increase the reaction rate as a result of the more facile CO dissociation and stronger alkene association [5, 6]. Because phosphites are better π-acceptors than phosphines they have great potential in hydroformylation catalysis. An additional advantage of phosphites is that they can be more easily prepared than phosphines and are less sensitive to sulfur compounds and oxidizing agents. On the other hand phosphites are more sensitive to side reactions such as hydrolysis, alcoholysis, and the Arbuzov reaction. Shortly after the discovery of Wilkinson, Pruett and Smith of UCC reported the beneficial effect of

P.W.N.M. van Leeuwen and C. Claver (eds.), Rhodium Catalyzed Hydroformylation, 35–62.

phosphite ligands in the rhodium catalyzed hydroformylation of 1-octene and methyl methacrylate [7]. Their results indicated that electron-withdrawing ligands showed a higher tendency to form linear aldehyde using 1-octene as substrate. Although the relative stability of primary and secondary alkyl complexes is determined mainly by steric factors, electronic effects can also be important. Strongly electron-withdrawing ligands will create a higher positive charge on the metal, which might favor the linear metal alkyl complex formation. For diphosphine ligands electron-withdrawing substituents resulted in an increased preference for bisequatorial coordination of the phosphorus donors [6]. Since phosphites have a much higher χ value than phosphines, a high preference for bisequatorial coordination is anticipated. The effect of the coordination mode on the selectivity, however, is not always clear. For triphenylphosphine Brown and Kent concluded that coordination of two phosphines in the equatorial plane resulted in the highest linearity [13], whereas the study of structurally related diphosphines showed that the selectivity for the linear aldehyde was independent of the relative amount of bisequatorially coordinating complex [6].

In their early reports Pruett and Smith already recognized the complicated effects of ligand structure and process conditions on the product distribution and the rate of the catalytic reaction. Systematic studies of ligand effects on the hydroformylation reaction are often obscured by the presence of several catalytically active rhodium complexes in the reaction mixture (see Figure 1). These complexes containing different numbers of phosphorus ligands are in equilibrium and up to three phosphorus ligands they can all be active as hydroformylation catalyst. The composition of the equilibrium mixture is dependent on many reaction parameters, such as type of ligand, concentration, temperature and pressure.

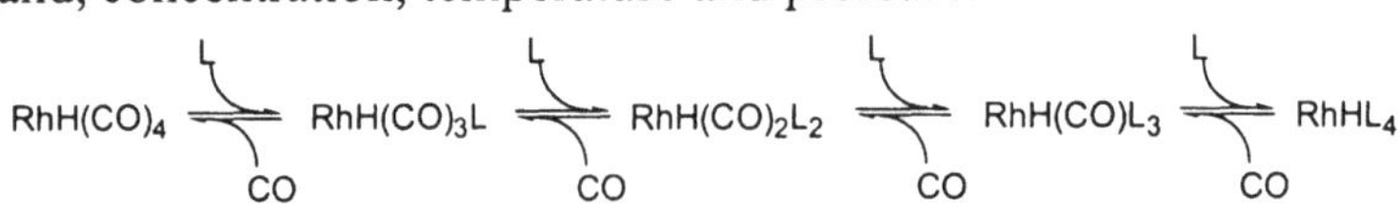

Figure 1. Actual rhodium catalyst containing various phosphorus and carbonyl ligands

The complexes of Figure 1 will show different rates and selectivities in the hydroformylation of 1-alkenes. In general, phosphorus ligands are stronger σ-donors and weaker π-acceptors than carbonyl ligands. As a consequence of the larger size, phosphines and phosphites create more steric bulk in the rhodium complex. Both stronger CO bonding and steric hindrance hamper the alkene addition. Therefore, the overall rate of the hydroformylation reaction decreases with the number of phosphorus ligands coordinating to the rhodium, whereas simultaneously the selectivity increases. For small phosphites an excess of phosphites results in the formation of an inactive rhodium complex. In laboratory experiments this can be used to quench reaction samples by addition of a large excess of a non-bulky ligand such as tributyl phosphite.

3.2 Monophosphites

3.2.1 Catalysis

As stated above, the first example of the use of phosphite ligands in rhodium catalyzed hydroformylation of 1-alkenes was reported by Pruett and Smith of Union Carbide Corporation. They studied a wide variety of phosphite and phosphine ligands [7]. The general trend that can be found in their results is an increasing selectivity for the linear aldehyde when the electron withdrawing properties of the ligand increase. The donating 4-methoxyphenyl substituent resulted in a decrease of the linear to branched ratio, whereas the 4-chloro substituted phenyl phosphite gave a relatively high l:b ratio of 13 (see Table 1). The expected higher reaction rates with the stronger π-acceptor phosphite ligands are less obvious from their results. The amounts of isomerized alkenes were not reported either. The use of ortho-substituted aryl phosphites gave lower selectivity for the linear product, whereas no remarkable effects on the rate of the reaction were reported.

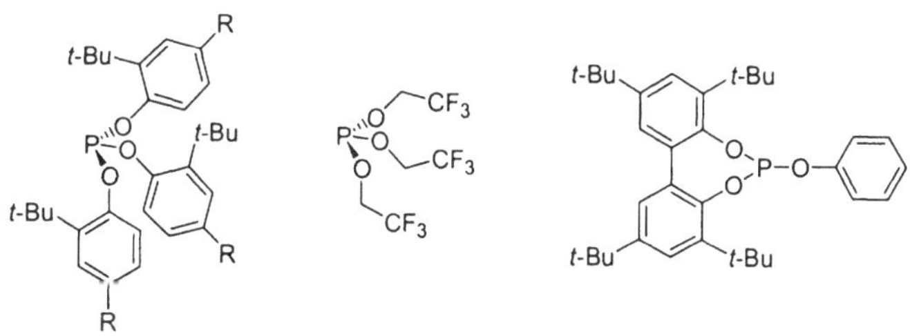

1 a R = H
b R = Me

2

3

Figure 2. Structure of bulky phosphites (**1**, **3**) and electron poor phosphite (**2**)

As one would expect that increasing steric hindrance in the catalytically active rhodium complex will result in lower reaction rates, the results of Van Leeuwen and Roobeek seemed contradictory at first. They used the very bulky tris(ortho *tert*-butylphenyl)phosphite **1a** (Figure 2) as a ligand and found high reaction rates in the rhodium catalyzed hydroformylation of otherwise unreactive alkenes like 1,2- and 2,2-dialkylalkenes (see Table 1) [8]. The high reactivity was explained by the exclusive formation of monoligated rhodium phosphite complexes, which was confirmed by *in-situ* IR and NMR studies [9].

The high catalytic activity of the rhodium bulky phosphite system was also evident in the hydroformylation of 1-alkenes [11,12]. A rhodium complex containing tris(2-*tert*-butyl-4-methylphenyl) phosphite (**1b**) as ligand gave extremely high rates up to 161,000 mol.mol $Rh^{-1}.h^{-1}$ in the hydroformylation of 1-octene with moderate selectivity for the linear aldehyde.

Table 1. Hydroformylation using rhodium bulky mono phosphite catalysts[a]

Ligand	T, °C	pCO	pH_2	Alkene	Isom. %	Rate,[b] mol. $(mol\ Rh^{-1})h^{-1}$	l:b
$P(OPh)_3$ [c]	90	3	3	1-octene	n.d.	-	6.1
$P(OC_6H_4$-*p*-$Cl)_3$ [c]	90	3	3	1-octene	n.d.	-	13.3
$P(OC_6H_4$-*p*-$OMe)_3$ [c]	90	3	3	1-octene	n.d.	-	4.9
PPh_3 [d]	80	10	10	1-octene	1.5	2200	2.8
1a [e]	90	10	10	1-heptene	n.d.	7100	3.3
1b [f]	80	10	10	1-octene	13	40,000	1.9
2 [e]	95	4.8	4.8	1-heptene	15	300	19
2 [e]	120	4.8	4.8	2-heptene	-		1.9
1a [e]	80	7	7	2-methyl-1-hexene	-	1000	>100
1b [f]	80	10	10	*c*-hexene	-	500	-
1b [f]	70	11	11	styrene	-	10,000	-
1a [e]	80	7	7	limonene	-	1700	>100
1b [g]	80	10	10	methyl oleate	-	400	-

[a] Conditions: 0.1-1 mM Rh, L/Rh = 10-20, [alkene] = 0.5-1 M in benzene or toluene. [b] Initial rate. N.d. = not determined. [c] [7]. [d] [61]. [e] [8]. [f] [9, 11, 12]. [g] [20].

Next to the high reactivity induced by sterically demanding ligands Van Leeuwen and Roobeek found remarkable reaction rates using strongly electron withdrawing ligands, even for the hydroformylation of less reactive internal alkenes. It was found that the selectivity for the linear product increased using electron withdrawing ligands. By applying tris(2,2,2-trifluoroethyl) phosphite (**2**) they obtained 96% linear aldehyde starting from 1-alkenes and 66% linearity with internal alkenes as substrate (Table 1) [8b]. To obtain a high linearity in the hydroformylation of 1-heptene a high ligand concentration was required. Probably at low ligand to rhodium ratios the phosphite ligand is partially replaced by a carbonyl resulting in higher rates but lower selectivity. The explanation for the high rates and selectivities is mainly based on electronic factors. The strongly electron withdrawing phosphite ligand induces fast replacement of a carbonyl ligand by the alkene substrate, resulting in high reaction rates. Increased isomerization to internal alkenes is a side effect of the high rates induced by electron-poor rhodium catalysts.

Ziólkowski and Trzeciak have extensively studied the use of phosphite ligands in the rhodium catalyzed hydroformylation of alkenes [16]. As a consequence of the higher χ-values compared to triphenylphosphine, an inactive hydroformylation catalyst, $RhHL_4$, is obtained using triphenyl phosphite at high ligand/rhodium ratio's (L/Rh). This is in contrast to the PPh_3 catalyst; here at high ligand concentrations when all the rhodium is maximally coordinated by the phosphines, the rate is very low and eventually becomes independent of L/Rh. Phosphites with moderate cone angles and higher χ-values give rise to acceptable l:b ratios. Good results were obtained by tri(4-chlorophenyl) phosphite, contrary to tri(2,6-dimethylphenyl) phosphite, having a cone angle of 190°. Unfortunately, the increase in regioselectivity for the former is accompanied by an additional increase of isomerization products. The high isomerization rates could be a consequence of the low pressures used; most experiments were performed at 1 bar, which can easily lead to CO depletion.

One of the commercial applications of bulky phosphites is the production of 3-methylpentane-1,5-diol by hydroformylation of 3-methylbut-3-en-1-ol by Kuraray [17]. Furthermore, they reported the use of bulky phosphite in the hydroformylation of vinyl acetate and 7-octenal, the latter providing an intermediate for the preparation of nonanediol. The high reactivity induced by bulky phosphite ligands has also led to the application of hydroformylation in the functionalization of natural product derivatives that are otherwise hardly reactive. Syntheses of important intermediates to fine-chemicals have been reported by hydroformylation of dihydrofuran [18], glucal derivatives [19] and methyl oleate [20] (see also chapter 6 for further details).

Bryant from UCC reported a very elegant application of bulky phosphite ligands. He utilized the high activity but moderate selectivity of bulky phosphites as an indicator for ligand depletion during the hydroformylation reaction using selective bulky diphosphite systems [59]. When the diphosphite ligand concentration becomes too low, as a consequence of ligand decomposition, the bulky phosphite ligand **3** will coordinate to rhodium preventing rhodium plating or cluster formation. Furthermore, the hydroformylation rate of this bulky phosphite ligand based system is orders of magnitude higher than that of the diphosphite catalyst. Because of the lower selectivity and much higher rate, diphosphite depletion will be monitored as a drop in overall product selectivity and new diphosphite can be added.

Easy recovery of the rhodium bulky phosphite catalyst could be achieved by immobilization of the catalyst [21]. A perfectly random copolymer of styrene and 2,2'-bis(4,6-di-*tert*-butylphenyl)-4-styryl phosphite was used to form rhodium complexes. The structure of the active catalyst was found to be dependent on the loading of the phosphite monomer. At very high phosphite loadings the less active rhodium bis phosphite complexes were formed whereas low phosphite loadings provided a high activity in the hydroformylation of cyclooctene, showing rates comparable to the low molecular weight analogue [21a]. Separation of the catalyst from the reaction product was facilitated by using a silica-grafted polymer bound bulky phosphite [21b]. Additionally this resulted in a very high stability of the immobilized catalyst.

3.2.2 Mechanistic and kinetic studies

The extremely large cone angle of 180° of **1a** prevents coordination of a second bulky phosphite under hydroformylation conditions [10]. The effect is twofold. First of all the overall steric hindrance at the rhodium metal is low because three relatively small carbonyl ligands are coordinated next to the bulky phosphite (structure **A**, Figure 3). Secondly the rhodium center containing only one weak phosphite donor and strongly electron withdrawing carbonyl ligands is electron poor. As a result the carbonyl ligands are very loosely bound and the fast dissociation of CO (structure **B**) and subsequent alkene addition (structure **C**) results in high reaction rates. Subsequent hydride migration results in the formation of the linear (**D**) or branched (**H**) rhodium alkyl complex.

The extreme steric bulk of the tris(ortho *tert*-butylphenyl) phosphite (**1a**) ligand not only prevents coordination of two ligands but in fact requires a high concentration of the ligand (up to 60 mM [9]) to ensure complete formation of the rhodium phosphite complex. It should be noted that the complex formation is a function of both ligand and rhodium concentration. Since the rhodium catalyst based on bulky phosphite is very active, low rhodium concentrations down to 0.1 mM Rh are often employed. Therefore, a high ligand to rhodium ratio of 50 can be necessary to prevent the formation of unsubstituted rhodium carbonyl complexes; the latter would result in low selectivity for the linear aldehyde and high isomerization rates.

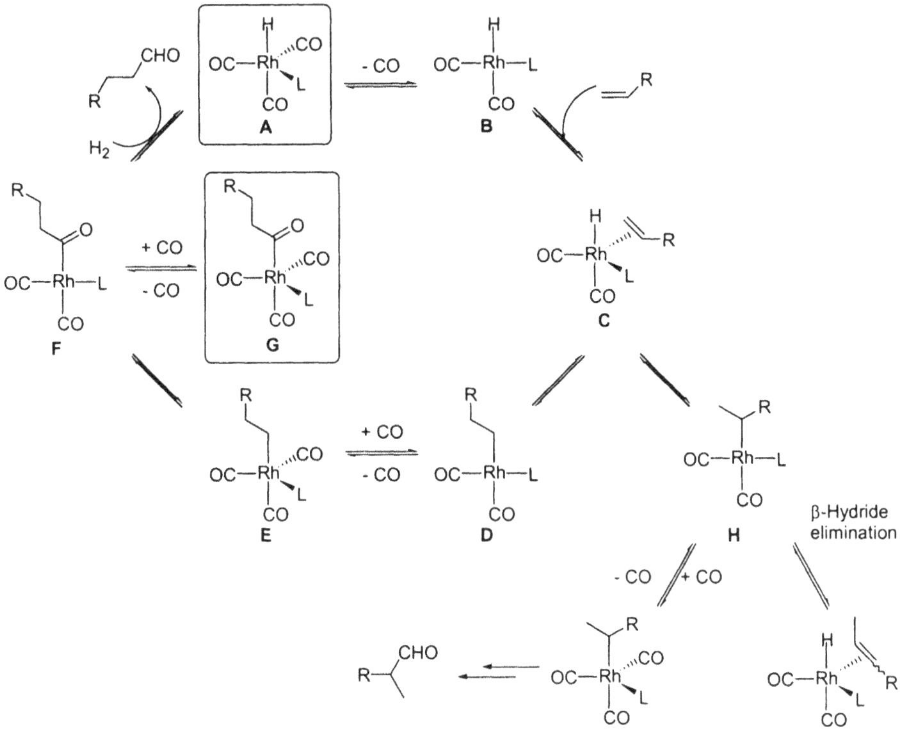

Figure 3. Mechanism of rhodium catalyzed hydroformylation using bulky monophosphites

The isomerization reaction is often ignored in catalytic studies but it is important with respect to the selectivity for the linear aldehyde. Under reaction conditions the rhodium alkyl complexes **D** and **H** (see figure 3) can give either migratory insertion forming the rhodium acyl complex or give β-hydride elimination. Starting from the primary rhodium alkyl **D**, β-hydride elimination is not productive as 1-alkene is reformed. The secondary rhodium alkyl complex **H**, however, can give both the starting 1-alkene and the internal 2-alkenes by β-hydride elimination. For both complexes the β-hydride elimination is competing with the migration reactions that lead to the product aldehydes. Since β-hydride elimination is faster for the secondary rhodium alkyl than for the primary rhodium alkyl complex high isomerization rates will reduce the formation of the branched rhodium acyl to a larger extent than the linear rhodium acyl. As a consequence the ratio of the linear to branched aldehyde will increase when the isomerization rate increases. The total selectivity for the linear aldehyde of the hydroformylation reaction decreases when the selectivity for isomerized internal alkenes increases.

Since the mechanism of alkene isomerization involves β-hydride elimination this reaction requires the creation of a vacant site [15]. Therefore, isomerization rates can be suppressed by using low temperatures, high CO pressures and high ligand concentrations, provided that the ligand is not very bulky. As stated above, a change in linear to branched ratio can often be explained, at least partially, by a change in isomerization rate. Therefore, the reported effects of pressure, temperature and ligand concentration on the selectivity for the linear aldehyde should be considered with care.

Kinetic studies showed that the catalyst based on bulky phosphites **1** differed from systems using phosphines as ligands (see chapter 4) and resembled the unmodified rhodium carbonyl catalysts (see chapter 2). Cavalieri d'Oro et al. showed that for the hydroformylation of propene with rhodium triphenylphosphine as catalyst the rate of the reaction is determined by the first steps in the catalytic cycle [3]. The kinetic studies by Van Leeuwen et al showed completely different behavior of the bulky phosphite modified rhodium catalyst [11, 12]. Using 1-octene as substrate the reaction rate has a zeroth order dependency on the alkene concentration. The rate of the hydroformylation reaction was first order in hydrogen and rhodium concentration and showed an inverse first order in carbon monoxide pressure. Increasing the H_2 : CO ratio to 7 at 80 °C resulted in extremely high reaction rates of 161,000 mol aldehyde (mol Rh)$^{-1}$ h^{-1} [11]. The reaction even has a runaway character; at low CO pressure the reaction becomes so fast that CO transfer to the solution becomes rate limiting and the reaction rate increases further until CO has been depleted. A negative order in the concentration of one of the reactants has important implications in process design (see chapter 8).

In situ IR studies showed that using the more reactive 1-octene as substrate the predominant species under reaction conditions was a rhodium acyl complex (structure **G**, figure 3), probably containing one phosphite ligand and three carbonyl ligands [14]. Before addition of the substrate under reaction conditions the rhodium hydride (**A**) containing one ligand was observed. Rapid scan IR shows that immediately after addition of the substrate a small signal at 1690 cm^{-1} is observed, while the signals of the hydride **A** in the carbonyl region are disappearing (see figure 4). The new signal at 1690 cm^{-1} is in the region of the carbonyl vibration of rhodium acyl complexes. This signal is only visible for a very short time as later it is concealed by the carbonyl band of the product aldehyde. In the metal-carbonyl region of the spectrum three new bands are observed during the reaction (see figure 4). It was concluded that during the reaction the acyl complex **G** was the only observable species. When all alkene was consumed the rhodium hydride **A** was formed back. Both the spectroscopic and the kinetic studies show that the rate limiting step for the hydroformylation of 1-alkenes using the bulky phosphite modified rhodium catalyst is the hydrogenolysis of the rhodium acyl intermediate.

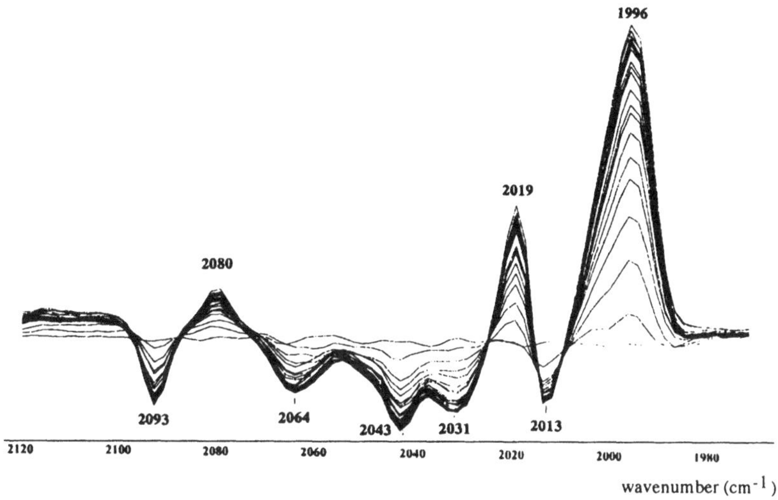

Figure 4. Metal-CO region of IR spectra recorded with rapid-scan method after addition of 1-octene to $RhH(CO)_3$**1b**

When less reactive substrates such as internal or 2,2-disubstituted alkenes are used as the substrate the kinetics of the system resemble those of the rhodium triphenylphosphine catalyst system. The alkene addition or hydride migration is rate determining again, as evidenced by kinetic studies [12] and in-situ spectroscopy [9]. NMR and IR spectroscopic studies using cyclooctene as substrate revealed that the resting state of the catalytic reaction was the monoligated hydride $HRh(CO)_3$**1a** (structure **A**, Figure 3) [9]. The structure of the catalyst is independent of the ligand concentration; even when a hundred fold excess of ligand is used, only one bulky phosphite is coordinated to rhodium.

Remarkably, Claver et al showed that in a square planar rhodium carbonyl chloride complex two bulky phosphite ligands **1b** were able to coordinate in a trans orientation [19b]. This complex was isolated from the reaction mixture after performing hydroformylation in chlorinated solvent. The steric hindrance of the bulky ligands is less in the square planar trans complex than in the trigonal bipyramidal rhodium hydride. In the absence of CO pressure one of the carbonyl ligands can be replaced by a bulky phosphite ligand. Probably, the phosphite ligand can exchange with carbon monoxide under higher CO pressure as is also observed in the formation of rhodium complexes of bulky diphosphites (see section 3.3.2).

Catalyst and ligand stability are important features in catalysis (for more details see chapter 9). The sensitivity towards hydrolytic reactions and/or solvolysis is strongly dependent on the structure of the phosphites. For instance, UCC reported that the hydrolytic stability of bulky phosphites could be improved by the use of bulky bisphenols in the ligand structure (structure **3**, figure 2). Ligand hydrolysis and other destructive side reactions like Michaelis-Arbuzov rearrangement [62] are among the problems that can be encountered when employing phosphites in catalysis (see figure 5). Many examples of metal catalyzed Arbuzov-type decomposition reactions have been reported [48]. The Arbuzov reaction is restricted to alkyl phosphites, which is probably the reason that almost exclusively aryl phosphite ligands are used in catalysis.

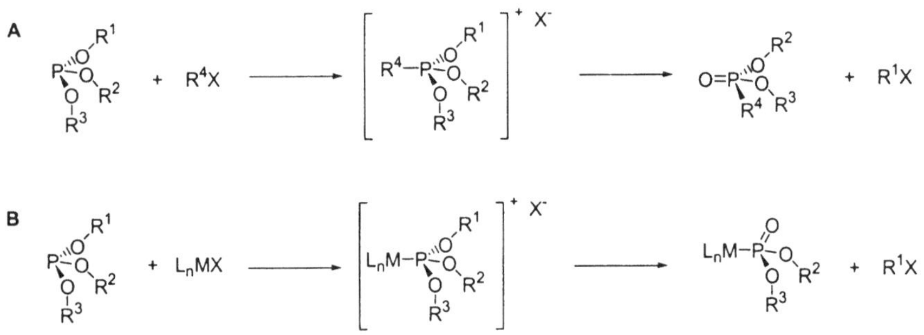

Figure 5. Classical (A) and metal catalyzed (B) Michaelis-Arbuzov reactions

3.3 Diphosphites

3.3.1 Catalysis

Phosphite ligands and especially bulky phosphites are very useful in rhodium catalyzed hydroformylation because of the higher reaction rates obtained when compared to triphenylphosphine. An important drawback, however, is the loss of selectivity. Where rhodium triphenylphosphine systems can provide selectivities of up to 92 % for the linear aldehyde, albeit at low rates, for bulky phosphite ligands the selectivity is reduced to 70%. One way to improve the selectivity was changing to diphosphite systems. It was only after the first reports from Bryant and coworkers that diphosphites were recognized as a new generation of promising ligands in rhodium catalyzed hydroformylation of alkenes [22]. This initial report was followed by numerous patents from UCC and triggered a large research effort at academia and industries.

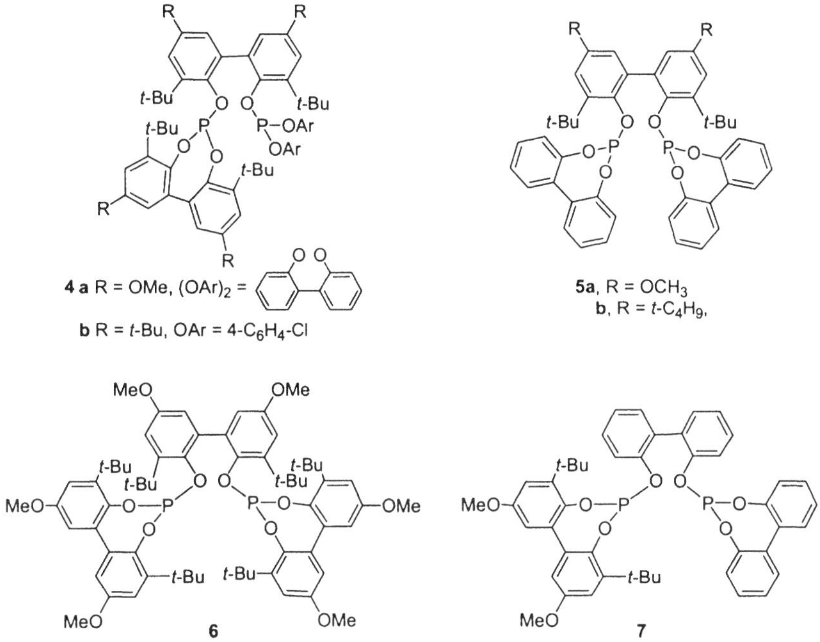

Figure 6. Bulky diphosphites developed by Bryant at UCC

The change from bulky monophosphite to bulky diphosphite ligands by using a bisphenol linker resulted, in several cases, in a tremendous increase of the selectivity for linear aldehyde in the rhodium catalyzed hydroformylation of 1-alkenes [22, 23]. The reaction rates were in general much lower than that of the bulky monophosphite system, but still relatively high compared to the triphenylphosphine based catalyst. The selectivity was found to be very dependent on the exact ligand structure (see figures 6 and 7) [24]. Selectivities higher than 95% were obtained and depending on the ligand small amounts of the branched aldehyde, isomerized alkenes or both were observed (see Table 2).

The bridge length of the diphosphites has a large influence on the selectivity for the linear aldehyde. For aliphatic bridges as in ligands **9** the optimum selectivity was found for a three-carbon bridge (**9b**), derived from 1,3-propanediol. Remarkably, the same preference for a three carbon bridge was observed for the asymmetric hydroformylation of styrene using chiral diphosphite ligands [25,26]. Although this was not recognized at first, the

Table 2. Hydroformylation using rhodium bulky diphosphite catalysts[a]

Ligand	T, °C	p CO/H_2 bar	Ratio CO/H_2	Alkene	Isom. %	Rate,[b] mol. (mol Rh)$^{-1}$. h^{-1}	l:b
4a [c]	70	2.5	1:2	1-Butene		2400	50
4a [c]	71	6.7	1:2	"		730	35
8a [c]	70	7	1:2	"		1480	3.2
6 [c]	70	7	1:2	"		160	6.3
7 [c]	70	4.3	1:1	Propene		280	1.2
6 [c]	70	4.3	1:1	"		20	2.1
5a [c]	74	4.5	1:1	"		402	53
9a [c]	90	7.1	1:1	1-Butene		1620	2.3
9b [c]	90	7.1	1:1	"		1320	3.8
9c [c]	90	7.1	1:1	"		1070	2.2
10 [c]	90	7.1	1:1	"		3660	2.0
11 [c]	90	7.1	1:1	"		1650	9.9
10 [c]	90	7.1	1:1	2-Butene		1140	0.5
11 [c]	90	7.1	1:1	2-Butene		65	2.8
12a [d]	80	20	1:1	1-octene	n.d.	11,100	1.6
12b [d]	80	20	1:1	1-octene	20	1550	2.2
5a [d]	80	20	1:1	1-octene	18	3600	>100
5b [d]	80	20	1:1	1-octene	27	6,120	51
4b [d]	80	20	1:1	1-octene	n.d.	3375	19
8b [d]	80	20	1:1	1-octene	13	520	1.2

[a] Conditions: 0.1-1 mM Rh, L/Rh = 10-20, [alkene] = 0.5-1 M in toluene. [b] Initial rate. [c] [22, 23]. [d] [24]. N.d. = not detected.

bite angle of the diphosphite ligands is probably an important parameter determining the selectivity of the hydroformylation reaction as was also found for diphosphine ligands [27-29]. The highest selectivities were, however, achieved using bisphenol bridges.

A crucial feature for obtaining high selectivities for the linear aldehyde seems to be large steric bulk at the bridging bisphenol like in ligands **4** and **5** [22-24]. Ligand **6** is probably too sterically hindered as can be concluded from the relatively low hydroformylation rates that are observed. In contrast ligand **7** is apparently too small to induce the desired high selectivity for the linear aldehyde. Interestingly, ligand **8** is structurally related to **5** and the steric hindrance is comparable to that of **4**. Nevertheless the selectivity in the hydroformylation of 1-octene is very low compared to the successful ligands **4** and **5** and even lower than obtained when using (bulky) monophosphites.

Similar to the ligands **9** containing the aliphatic bridges, the backbone of ligand **8** is lacking steric bulk. Although ligand **10** seems to have large steric bulk, the obtained selectivity is low. Probably this ligand cannot coordinate in a bidentate fashion; the distance between both phosphorus donors is estimated to be more than 7 Å. The activity is high, however, which is reminiscent of bulky monophosphites (see section 3.2.1). The behavior of ligand **10** as bulky monophosphite is even more evident in the hydroformylation of the internal substrate 2-butene. Relatively high rates are obtained albeit with moderate selectivity for the linear aldehyde.

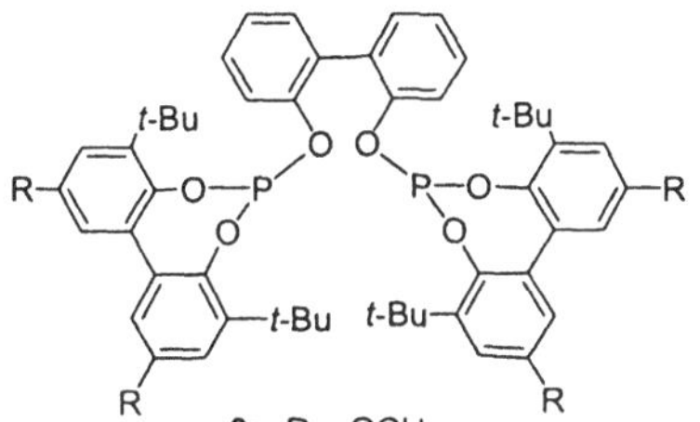

8a, R = OCH_3
b, R = $t\text{-}C_4H_9$

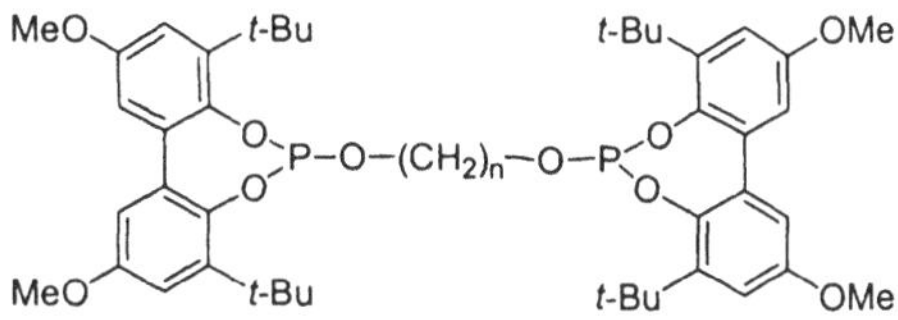

9 a, n = 2
b, n = 3
c, n = 4

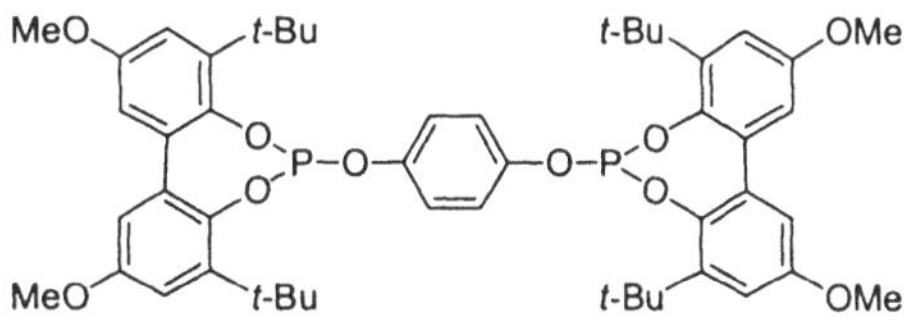

10

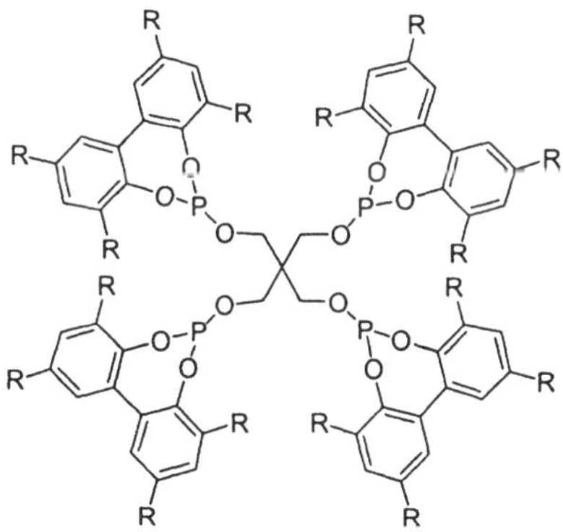

11, R = $t\text{-}C_4H_9$

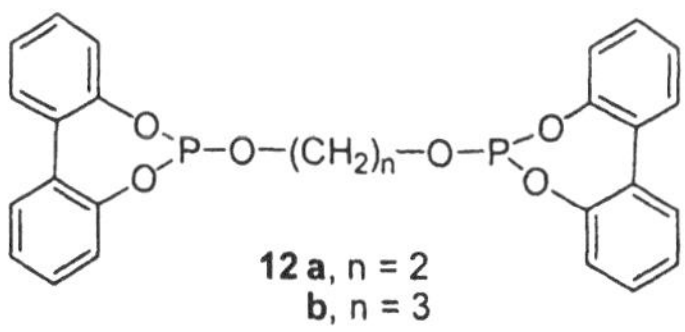

12 a, n = 2
b, n = 3

Figure 7. Bulky diphosphite ligands

In contrast to bulky monophosphite modified systems, rhodium diphosphite catalysts showed very low activity for internal alkenes like cyclohexene. The required monodentate coordination of the ligand is prevented by the chelate effect.

Bulky diphosphite based catalysts are active in the hydroformylation of styrene [24]. The rates of the reaction are generally a factor of ten lower than those for aliphatic 1-alkenes. Styrene is a substrate that has an intrinsic preference for the formation of the branched aldehyde, probably caused by stabilization of the branched rhodium alkyl by the phenyl ring [30]. The high selectivity for the formation of linear aldehydes of the bulky diphosphite catalyst was also observed with styrene as a substrate. Typically linear to branched ratios around 1 are observed while the selectivity for the branched aldehyde is generally higher than 90% when monodentate ligands are employed. The selectivity for the linear aldehyde can be enhanced by influencing the "isomerization" rate. Secondary rhodium alkyl complexes are more prone to β-hydride elimination than primary rhodium alkyls (see section 3.2.2). For styrene this effect is even more pronounced because of the formation of relatively stable branched rhodium styryl complexes. Since β-hydride elimination can only result in the reformation of the substrate, this reaction results in the preferential removal of the branched rhodium alkyl and, therefore, a relatively slow formation of the branched aldehyde. Lazzaroni et al. came to the same conclusion when they studied the deuterioformylation of styrene (see chapter 2) [57]. Employing reaction conditions that promote high isomerization rates, i.e. high temperature and low CO pressure, will result in a higher selectivity for the linear aldehyde. The high hydrogen pressure that was used resulted in the formation of a substantial amount of the hydrogenation product, ethylbenzene.

3.3.2 Mechanistic and kinetic studies

Van Leeuwen et al. studied the kinetics of the hydroformylation of 1-octene using the bulky diphosphite **4b** [24]. The reaction exhibited a first order dependency of the alkene concentration. The reaction rate was almost independent of the hydrogen pressure and showed a negative order in CO pressure. All data indicate a rate determining step early in the catalytic cycle. The kinetic data are very similar to those obtained by Cavalieri d'Oro et al. for the triphenylphosphine based catalyst (see chapter 4) [3]. Instead of a negative order in CO they found a negative order in triphenylphosphine, probably because of the more facile ligand dissociation of the phosphine from the putative resting state $HRh(PPh_3)_3CO$) compared to the $HRh(CO)_2$(diphosphite).

The selectivity to the linear product nonanal was strongly dependent on the CO pressure (see Table 2). The linear to branched ratio drops from 29 at P_{CO} = 5 bar to 4 at P_{CO} = 40 bar. Part of this selectivity change can be ascribed to enhanced isomerization at lower CO pressure (vide supra), but faster β-hydride elimination cannot account completely for the increased formation of the branched aldehyde. The reduced selectivity was attributed to partial ligand dissociation resulting in less selective rhodium monophosphites and/or ligand free complexes.

Rhodium diphosphite catalysts were studied under actual hydroformylation conditions by the groups of van Leeuwen [24] and Gladfelter [31]. The active catalyst was prepared from $Rh(CO)_2acac$ and excess diphosphite ligand. By monitoring the reaction mixture by NMR, it was shown that the first intermediate was a monoligated rhodium acac carbonyl complex [24]. Only in the absence of an excess of CO the bis-coordinated Rh(P-P)acac was observed; this complex was also reported as an intermediate by Gladfelter [31]. An X-ray crystal structure of the Rh(P-P)acac complex of ligand **4b** was reported by van Rooy [24], indicating that despite the steric bulk of the diphosphite ligand *cis*-coordination in the Rh(I) complex was feasible.

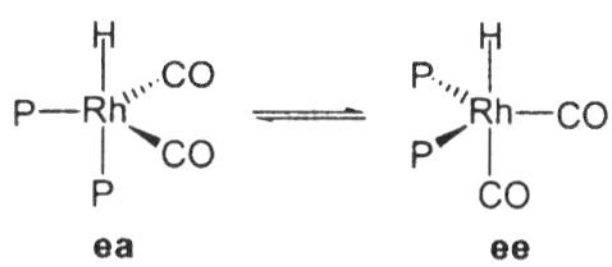

Figure 8. Equilibrium between **ee** and **ea** coordinating rhodium diphosphite complexes

Under syn gas pressure the rhodium acac precursors were converted to the catalytically active hydride complexes $HRh(CO)_2(L\text{-}L)$. The complexes are generally assumed to have a trigonal bipyramidal structure and two isomeric structures of these complexes are possible, containing the diphosphite coordinated in a bisequatorial (**ee**) or an equatorial-apical (**ea**) fashion. The structure of the complexes can be elucidated by (high pressure) IR and NMR data (Table 3). In the carbonyl region of the infrared spectrum the vibrations of the **ee** and **ea** complex can be easily distinguished. The **ee** complexes typically show absorptions around 2015 and 2075 cm^{-1} [24, 26, 31, 32], whereas the **ea** complexes exhibit carbonyl vibrations around 1990 and 2030 cm^{-1} [26,32]. The rhodium hydride vibration lies in the same region as the carbonyl ligands but is often very weak. In fact additional carbonyl signals in case of mixtures of complexes are often erroneously assigned to the rhodium hydride stretching frequencies. Van Leeuwen et al. were able to isolate some complexes as powders and the IR spectra were measured as nujol mulls (see for IR data Table 3). For complex

$HRh(CO)_2$(**4b**) it was found that three absorptions were present in the CO-stretching region instead of two observed with IR in solution. Upon measuring $DRh(CO)_2$(**4b**), only two absorptions remained (2058, 2012 cm^{-1}), which were shifted compared to $HRh(CO)_2$(**4b**) (2035, 1998, 1990 cm^{-1}). This implies that the three frequencies in the hydrido complex are a combination of two CO stretching frequencies and one rhodium hydrido stretch. The rhodium-hydride vibration disappears upon deuteration of the complex as the rhodium-deuteride vibration is situated in the fingerprint region. The large frequency shift of the highest energy absorption is indicative of a trans hydrido-CO relation [46]. In solution IR, the rhodium hydride vibration and the lowest energy CO vibration overlap, which results in two absorptions.

Additional information can be obtained from NMR data. Phosphorus donors coordinated in the equatorial plane have generally small coupling constants of phosphorus to hydrogen. The coupling constant of a trans coordinated phosphite shows a large phosphorus to hydrogen coupling constant of 180-200 Hz. The phosphorus to phosphorus coupling constant is much larger for the **ee** (around 250 Hz) than for the **ea** complex (usually <70 Hz). Some representative NMR and IR data of rhodium-diphosphite complexes are given in Table 3.

Table 3. NMR and IR data for HRh(P-P)(CO)2 complexes[a]

Ligand	$\delta\ ^{31}P$	$\delta\ ^{1}H$	$^{1}J_{Rh\text{-}P}$	$^{1}J_{Rh\text{-}H}$	$^{2}J_{P\text{-}H}$	$^{1}J_{P\text{-}P}$	$\nu_{Rh\text{-}CO}$
4b	160.5	-10.4	237, 226	3.5	-19, 70	170	2049
	159.8						1966
5a	174.5	-10.8	239	3.5	4		2074
							2013
5b	173.0	-10.3	234	3	9		
13	167.4	-9.60	238, 154	n.r.	190	58	2035
	156.9						1994
14	168.1	-10.11	244, 225	<2	<2	247	
	165.5						

[a] Data taken from references 24 and 26.

When C_2-symmetric ligands are employed both phosphorus donor atoms become inequivalent no matter if the **ee** or **ea** complex is formed. Buisman prepared complexes of ligands **13** and **14** (figure 9) that exhibited exclusive **ea** or **ee** coordination respectively. These complexes showed fluxional behavior at room temperature [32]. Both the **ee** and the **ea** complexes have inequivalent phosphorus donor atoms which are in fast exchange on the NMR time scale. For the **ea** complex of **13** an averaged phosphorus to hydrogen coupling constant of 80 Hz was observed. At low temperature the exchange process was halted and the large trans coupling constant of the phosphorus to the hydride was observed, whereas the cis coupling constant remained very small and unresolved [32]. By variable temperature NMR studies the thermodynamic activation parameters could be determined by comparison of the calculated and experimental NMR spectra (see figure 10). The $\Delta H^{\ddagger}$, calculated from the Eyring equation, was 35 $kJ.mol^{-1}$ and the $\Delta S^{\ddagger}$ was 6 $J.K^{-1}.mol^{-1}$. Both $\Delta H^{\ddagger}$ and $\Delta S^{\ddagger}$ were significantly higher for the **ee** rhodium complex derived from ligand **14**, 62 $kJ.mol^{-1}$ and 46 $J.K^{-1}.mol^{-1}$ respectively. From the relatively low entropy of activation it was concluded that the exchange process was an intramolecular process and did not involve ligand dissociation. Remarkably, it was observed that the exchange rate of the **ea** complex was an order of magnitude higher than that of the **ee** complex.

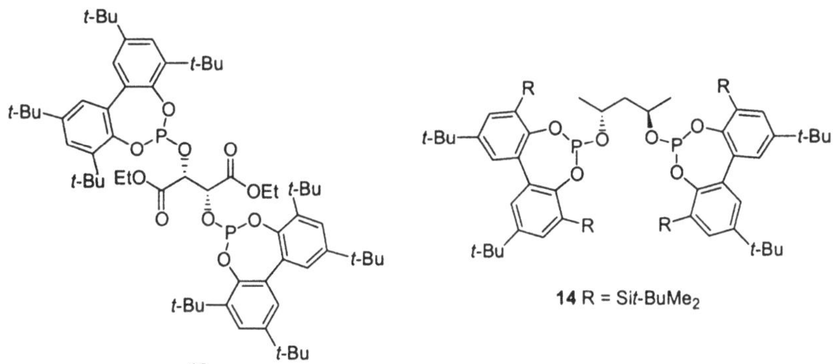

Figure 9. Chiral diphosphites forming exclusively **ea** (**13**) or **ee** (**14**) rhodium complexes

The exchange process of the phosphorus donor atoms of the **ea** and **ea** complexes was suggested to proceed by the low energy rearrangement mechanism, described by Meakin [47] (see figure 11). A simultaneous bending motion of the hydride and carbonyl ligands takes place in the hydridorhodium diphosphite dicarbonyl complexes containing **ea** coordinating diphosphites (figure 11A). For **ee** coordinating ligands, however, a motion of the hydride the equatorial phosphite functions is responsible for the exchange (figure 11b). The latter process is expected to be more difficult than the former, which explains the higher fluxionality of the **ea** coordinating complexes.

For ligand **4b** an X-ray crystal structure of the hydrido rhodium dicarbonyl complex could be obtained [24]. The P-Rh-P angle in $HRh(CO)_2$(**4b**) is 116° and the complex is indeed a trigonal bipyramid with the diphosphite in the equatorial plane. The structure is somewhat distorted; the larger ligands in the equatorial plane bend toward the small hydride. This explains why the coupling constants between phosphorus and hydrogen are sometimes larger (up to 70 Hz, see table 3) than might be expected in a complex showing a pure cis-relationship between hydride and phosphorus ligand (< 10 Hz). This crystal structure confirms the NMR analysis, but so far it is the only example of an X-ray crystal structure of a diphosphite rhodium catalyst.

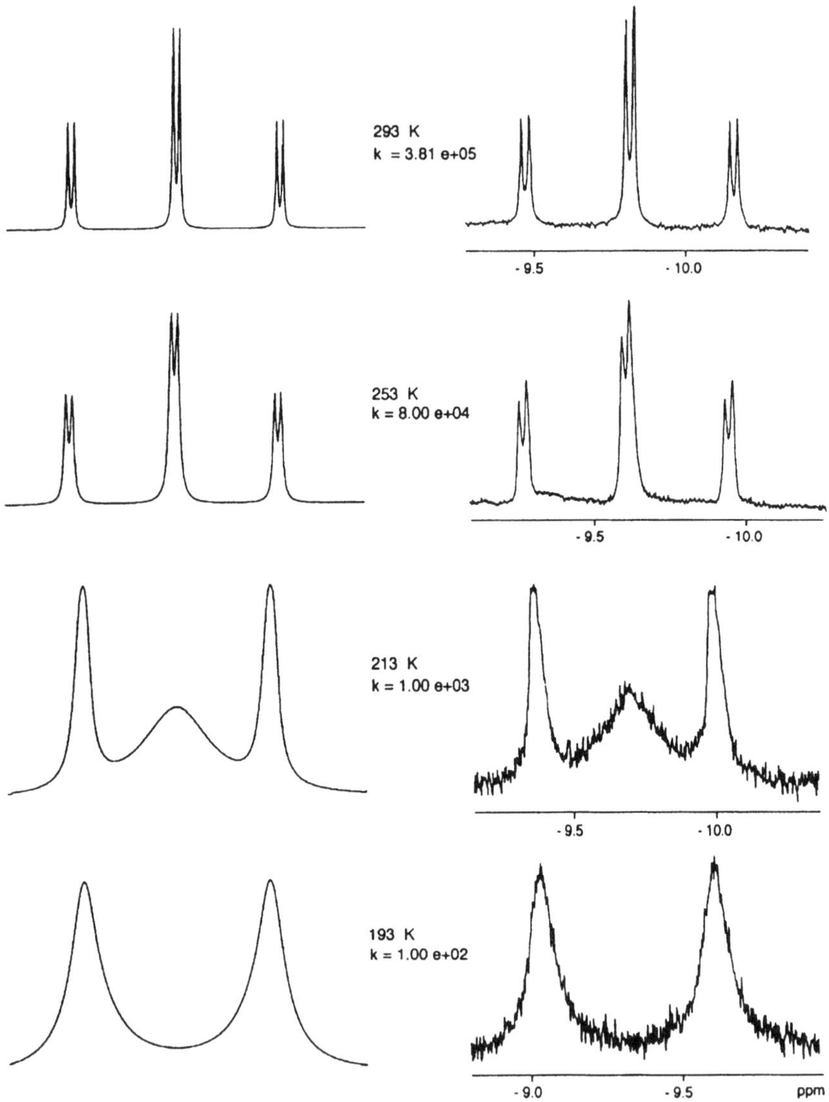

Figure 10. Calculated and observed variable-temperature ^1H (300 MHz) NMR spectra for $HRh(CO)_2$**13**

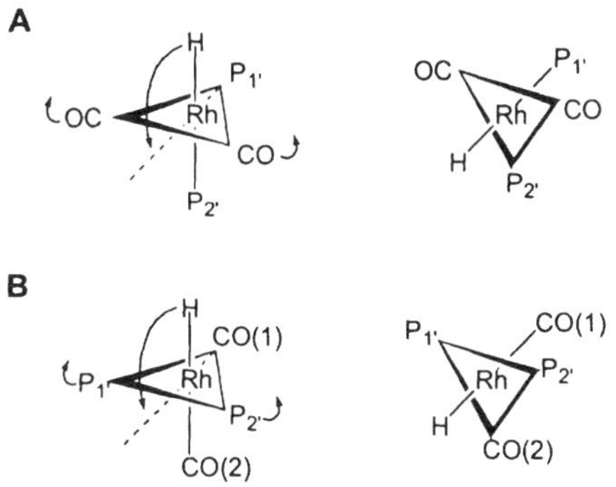

Figure 11. A: Equatorial-apical phosphorus exchange. B: Equatorial-equatorial phosphorus exchange

Summarizing, the relationship between ligand and complex structure and selectivity of the hydroformylation reaction is not always straightforward. In general bisequatorial coordination of the ligand is required for obtaining a high preference for the formation of linear aldehyde. Bisequatorial coordination does not *always* lead to high selectivity, as the preference for the linear aldehyde is very dependent on the exact ligand structure. Best results are obtained with bridging backbones based on bisphenol; aliphatic bridges often give poor results. An important feature of the ligands is that they must create sufficient steric bulk around the rhodium center. The steric bulk must be located on the bridging bisphenol and at most one of the terminal end-groups. Introduction of electron donating and withdrawing substituents on the dissymmetric bulky bisphenols both induced higher selectivity for the linear aldehyde. This shows that the effect of electronic variations of the system on the catalytic performance is not very straightforward either [24]. When aliphatic bridges are introduced instead of bisphenol, the obtained selectivity is dependent on the length of the bridge; the optimum is found when a three carbon linker is applied. A similar feature was observed in the asymmetric hydroformylation of styrene using chiral diphosphites [26] (see also chapter 5). The obtained linearity is generally much lower than for the successful bisphenol bridged bidentate phosphites.

In many instances the formation of inactive dimers next to active, monomeric catalyst species is observed during catalysis. When weak or instable ligands are used even larger rhodium carbonyl clusters like $Rh_4(CO)_{12}$ and $Rh_6(CO)_{16}$ can be observed [49, 50, 51]. The formation of dimers is often an equilibrium which is reversible. This only leads to a reduction of the amount of catalyst available and it does not kill the catalyst. One of the first examples is the formation of the so-called orange dimer from $HRh(PPh_3)_3CO$, already reported by Wilkinson [1c] and characterized by

Chan (see figure 12) [58]. They also observed the formation of red dimers after losing two carbon monoxide molecules.

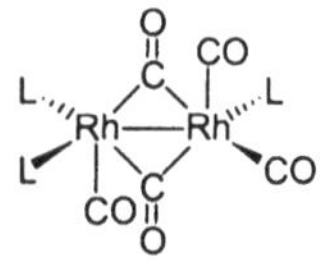

Figure 12. Rhodium dimer characterized by Chan (L=PPh_3)

Since the hydroformylation reaction for most substrates shows a first order dependence in the concentration of rhodium hydride, the reaction becomes slower when considerable amounts of rhodium are tied up in dimers. This will occur at low pressures of hydrogen and high rhodium concentrations (see chapter 4). Dimer formation was mainly reported for phosphine ligands [1c, 49, 52], but similar dimeric rhodium complexes from monophosphites [53] and diphosphites [26,32] have been reported. The orange side product obtained from HRh**14**$(CO)_2$ was characterized as the carbonyl bridged dimeric rhodium species Rh_2(**14**)$_2$$(CO)_2$ [32].

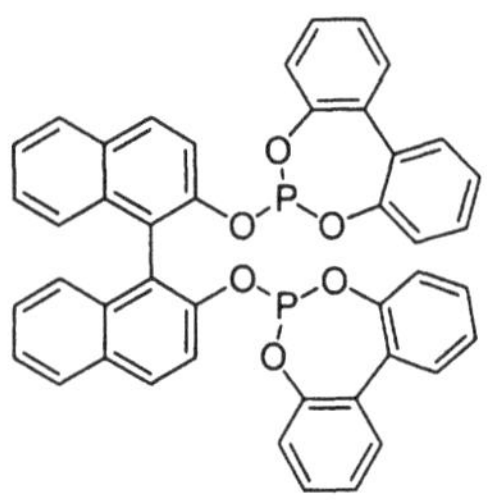

15

Figure 13. Diphosphite ligand studied by Gladfelter

Gladfelter et al. reported the presence of dimeric complexes during the rhodium catalyzed hydroformylation of 1-octene using ligand **5a** [31]. The appearance of the dimers was mainly observed at the final stage of the reaction at low hydrogen pressure. No effect of the formation of dimers on the actual catalytic reaction was reported. By reducing the steric bulk of the ligand using ligand **15** they even isolated a tetranuclear rhodium complex, illustrating the importance of steric hindrance in the ligand structure [54].

In addition to dimerization and clustering to multinuclear rhodium carbonyl complexes, active catalyst can also be removed from the reaction mixture by orthometallation. Several reports have appeared on orthometallation of rhodium triphenyl phosphite complexes in literature but structural proof for the orthometallated species has not always been convincing. Parshall reported the ortho-deuterium incorporation in phosphite ligands [56]. In agreement with this, Coolen showed that the hydrogenation catalyst $HRh[P(OPh)_3]_4$ provided ortho-deuterated product when stirred under deuterium atmosphere [55]. Both researchers suggested that the deuterium exchange proceeded via orthometallation. The intermediate orthometallated complex was characterized by ^{31}P NMR [55].

3.4 Hydroformylation of internal alkenes.

3.4.1 Hydroformylation of less reactive internal and functionalized alkenes.

The great advantage of phosphite ligands compared to phosphines is the higher reactivity of the corresponding rhodium catalysts. As a result, a much larger variety of substrates can be converted at acceptable rates. An additional advantage of the phosphite modified rhodium catalysts is that the hydroformylation can be performed under very mild conditions. Because of these mild reaction conditions many functionalized alkenes can be used as substrate; the functional group compatability is excellent, which renders the hydroformylation reaction also suitable for natural product synthesis [18-20, 38-41] (see chapter 6).

Bulky monophosphite ligands proved to be very useful for the functionalization of very unreactive substrates. Already in their first study van Leeuwen and Roobeek obtained relatively high rates for the hydroformylation of substrates such as cyclohexene and limonene. [8]. Van Rooy performed a systematic study to the rhodium catalyzed hydroformylation of substituted alkenes and compared the reaction rates with the triphenylphosphine system [42]. The bulky monophosphite derived catalyst was up to two orders of magnitude faster and gave acceptable rates using substrates for which the Wilkinson hydroformylation catalyst gave hardly any activity.

Fatty acid derivatives containing an additional functional group are important high-value-added oleochemicals. Therefore, hydroformylation of unsaturated fatty acids and their esters can be a very profitable process. Since the substrates are again internal alkenes and generally will contain impurities that are known to retard the catalytic reaction, a very active catalyst is required. Muilwijk studied the hydroformylation of pure and technical grade methyl oleate using the rhodium bulky phosphite catalyst [20]. The double bond of methyl oleate of 99% purity was hydroformylated without any problem at relatively high rates. Next to formation of the desired product methyl formylstearate (as a mixture of isomers) the rapid formation of the trans isomer of methyl oleate, methyl elaidate, was observed. As expected the hydroformylation of the trans isomer was slower than that of the cis isomer (see figure 14). Comparison of the bulky phosphite modified system with the rhodium triphenylphosphine catalyst revealed that the reaction rates of the former were much higher, as expected for internal alkene substrates.

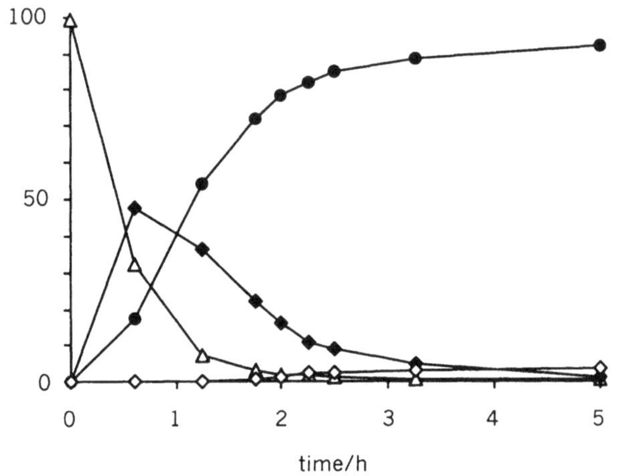

Figure 14. Hydroformylation of pure methyl oleate (3.64 mmol), $P_{initially}$ = 20 bar, CO/H_2 = 1:1, T=100 °C, Rh=4.10^{-3} mmol, Ligand:Rh =25:1, Δ methyl oleate, ◆ methyl elaidate, ● methyl formylstearate, ◊ methyl formylstearate (isomers)

Technical grade methyl oleate, however, contains about 14% of the diene methyl linoleate that easily isomerizes to a conjugated diene. These conjugated dienes strongly retard the hydroformylation reaction, probably by the formation of stable π-allyl type rhodium complexes [43]. The initial rate of the hydroformylation reaction is much lower than for the pure methyl oleate substrate that does not contain significant amounts of dienes. Only after the methyl linoleate has been converted to other hydrogenation and hydroformylation products the rate of the reaction increases. Again the reaction mixture shows isomerization of methyl oleate to the trans product methyl elaidate during the hydroformylation process (see figure 15). Since hydrogenation of the diene methyl linoleate to mono alkenes was required prior to hydroformylation, the reaction rate increased significantly when higher hydrogen pressures were applied.

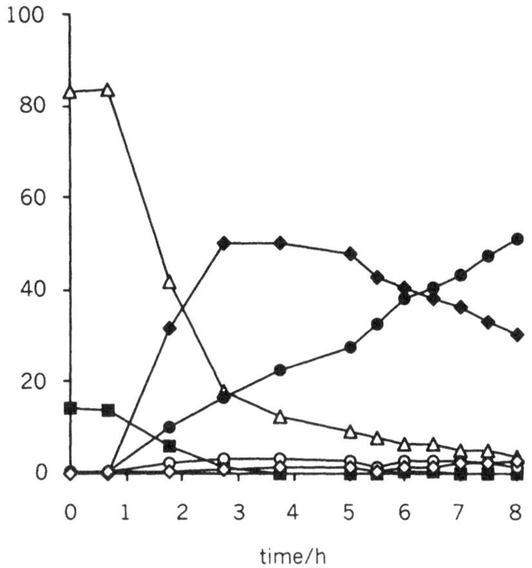

Figure 15. Hydroformylation of technical grade methyl oleate (3.64 mmol), $P_{initially}$ = 20 bar, CO/H_2 = 1:1, T=100 °C, Rh = 4.10^{-3} mmol, Ligand : Rh =25:1, Δ methyl oleate, ◆ methyl elaidate, ■ methyl linoleate, o methyl linoleate (isomers), ● methyl formylstearate, ◊ methyl formylstearate (isomers)

3.4.2 Formation of linear aldehydes starting from internal alkenes.

The selective formation of linear aldehydes starting from internal alkenes is still one of the greater challenges in hydroformylation chemistry. To achieve this goal a catalyst is required that obviously has a high isomerization activity since the thermodynamically less stable terminal alkene has to be formed prior to hydroformylation. Furthermore, the selectivity for the formation of the linear rhodium alkyl complex has to be high. Both isomerization and hydroformylation products are formed from the same rhodium alkyl intermediate. Therefore, a very active catalysts is required to obtain addition of the unreactive internal alkene and subsequent hydride migration. Moreover the branched rhodium alkyl complex must have a large preference for β-hydride elimination compared to migration to a carbonyl ligand, which would result in the formation of branched aldehydes. If isomerization rates are very high compared to hydroformylation rates, the catalyst can maintain the small concentration of 1-alkenes in the alkene mixture governed by thermodynamics. Once the terminal alkene has been formed it will react with the rhodium complex because the alkene addition is much faster than for internal alkenes. If the catalyst is very selective and gives a high linear to branched ratio in combination with fast isomerization, only the linear rhodium alkyl intermediate will show preferential carbonylation. As a consequence the linear aldehyde will be the main product, even using internal alkenes as starting material. So far only few catalysts are known that show these properties.

The strongly electron withdrawing monophosphite **2** was one of the first examples that provided rhodium catalyst that gave good selectivity for the formation of linear aldehyde starting from internal alkenes (see table 1) [8]. The high χ-value of the phosphite induces enhanced selectivity for the formation of the rhodium alkyl complex and high isomerization rates, resulting in a high overall selectivity for the linear aldehyde.

For lower alkenes such as 2-butenes UCC has achieved high contents of linear products (see Table 2). Bryant reported 74% selectivity for the formation of linear pentanal by hydroformylation of 2-butene using the rhodium bulky diphosphite catalyst [22, 23].

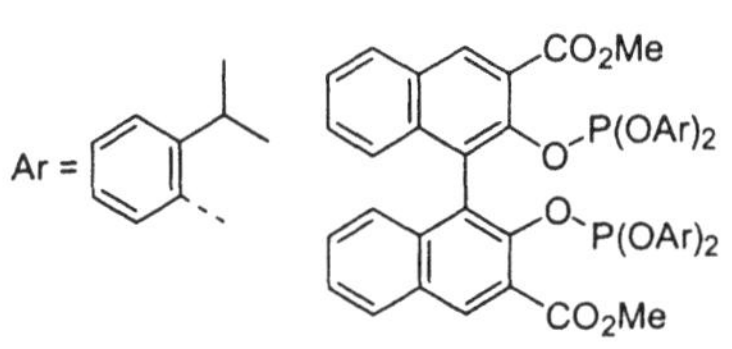

16

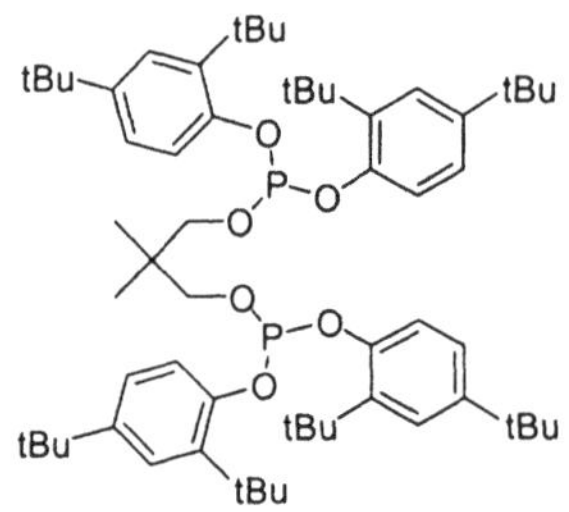

17

Figure 16. DSM and Mitsubishi diphosphite ligands

Du Pont and DSM have patented the use of ligand **16** (Figure 16) for the hydroformylation of methyl 3-pentenoate to linear product [33]. Instead of *t*-butyl groups on the bridge they use electron-withdrawing ester groups, while the remaining substituents are monophenols [34, 35], rather than diols [36, 37] or bisphenols. The necessary isomerization to the linear alkene prior to hydroformylation reaction is probably promoted by the electron withdrawing ester groups. They were able to obtain high selectivity for the linear aldehyde, providing a useful intermediate to nylon-6 feedstocks. Comparable selectivity up to 97% was reported for the hydroformylation of 2-hexene. The hydroformylation of butadiene to the linear 1,6-hexanedial proved to proceed less satisfactory; the main products were pentanal and 3-pentenal and only small amounts of the desired 1,6-hexanedial were formed.

Similar results were obtained using the UCC systems; in several patents the reported yields for dialdehyde in the hydroformylation of butadiene were relatively low [60]. Moderate yields for the linear aldehyde were generally obtained when methyl 3-pentenoate was hydroformylated with the bulky diphosphite modified UCC catalyst [60]

The efficacy of monophenols containing bulky substituents is described in the patents from Mitsubishi Chemical Corporation [34,35] (**17**, Figure 16). They report high yields of aldehyde (> 90%) for the hydroformylation of 1-alkenes with high linearities. The l:b ratios are generally above twenty using this ligand.

3.5 Calixarene based phosphites.

Recently several reports appeared of phosphite ligands based on well-defined supramolecular backbones like calixarenes and diphenylglycoluril. Although ligands based on these types of structures have been known for quite some time very few examples of actual applications in catalysis have appeared. One of the first examples of an application of a calix[4]arene based diphosphite stems from BASF [44]. They used a calix[4]arene substituted at the 1 and 3 position with bulky phosphite substituents in the rhodium catalyzed hydroformylation of 1-alkenes (Figure 17). Although they were able to obtain high linear to branched ratios the catalytic reaction gave large amounts of isomerized internal alkenes.

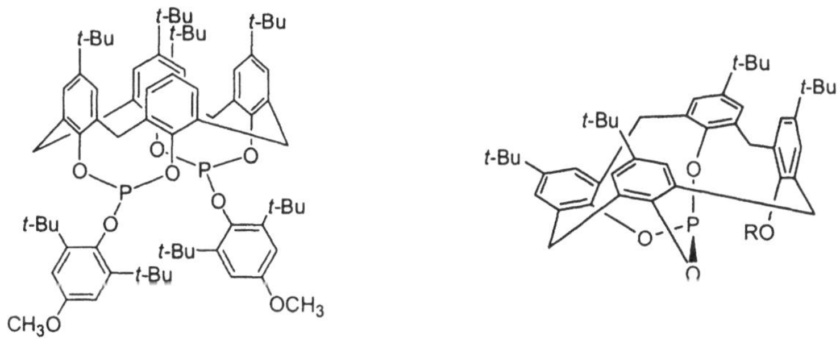

Figure 17. Calix[4]arene based diphosphite by BASF [44] and a monophosphite employed in hydroformylation [45]

Parlevliet has shown that calix[4]arene based monophosphites can exist in different conformations [45]. By using the different conformations as ligands in the rhodium catalyzed hydroformylation of 1-octene he showed that the exact conformation influenced the performance of the catalyst. Some of the conformations behaved more like triphenyl phosphite, whereas others showed catalytic results like bulky monophosphites, giving high rates with moderate selectivity for the linear aldehyde.

References

1 Young, J. F.; Osborn, J. A.; Jardine, F. A.; Wilkinson, G. *J. Chem. Soc., Chem. Comm.* **1965**, 131. (b) Evans, D.; Osborn, J. A.; Wilkinson, G. *J. Chem. Soc. (A)*, **1968**, 3133. (c) Evans, D.; Yagupsky, G.; Wilkinson, G. *J. Chem. Soc. A* **1968**, 2660. (d) Brown, C. K.; Wilkinson, G. *J. Chem. Soc. A* **1970**, 2753.

2 Heck, R. F; Breslow, D. S. *J. Am. Chem. Soc.* **1961**, *83*, 4023.

3 Cavalieri d'Oro, P.; Raimondt, L.; Pagani, G.; Montrasi, G.; Gregorio, G.; Oliveri del Castillo, G. F.; Andreeta, A. *Symposium on rhodium in homogeneous catalysis*, Veszprem, **1978**, pp 76-83; b) Gregorio, G.; Montrasi, G.; Tampieri, M.; Cavalieri d'Oro, P.; Pagani, G.; Andreetta, A. *Chim. Ind.* **1980**, *62*(5), 389; Cavalieri d'Oro, P.; Raimondi, L.; Pagani, G.; Montrasi, G.; Gregorio, G.; Andreetta, A. *Chim. Ind.* **1980**, *62*, 572

4 Van Leeuwen, P. W. N. M.; Van Koten, G. *Catalysis. An Integrated approach to Homogeneous, Heterogeneous and Industrial Catalysis*, Moulijn, J. A.; Van Leeuwen, P. W. N. M.; Van Santen, R. A. Eds., Elsevier: Amsterdam, London, New York, Tokyo, 2nd ed. **1995**, pp 199-222.

5 Unruh, J. D.; Christenson, J. R. *J. Mol. Catal.* **1982**, *14*, 19.

6 L. A. van der Veen, M. D. K. Boele, F. R. Bregman, P. C. J. Kamer, P. W. N. M. van Leeuwen, K. Goubitz, J. Fraanje, H. Schenk and C. Bo, *J. Am. Chem. Soc.*, 1998, **120**, 11616.

7 Pruett, R. L.; Smith, J. A. *J. Org. Chem.* **1969**, *34*, 327. (b) Pruett, R. L.; Smith, J. A. S. African Pat. 6804937, 1968 (to Union Carbide Coorporation); *Chem. Abstr.* **1969**, *71*, 90819.

8 (a) Van Leeuwen, P. W. N. M.; Roobeek, C. F. *J. Organomet. Chem.* **1983**, *258*, 343. (b) Van Leeuwen, P. W. N. M.; Roobeek, C. F. Brit. Pat. 2 068 377, **1980** (to Shell); *Chem. Abstr.* **1984**, *101*, 191142.

9 Jongsma, T.; Challa, G.; van Leeuwen, P. W. N. M. *J. Organometal. Chem.* **1991**, *421*, 121.

10 Tolman. C. A. *Chem. Rev.* **1977**, *77*, 313.

11 Van Rooy, A.; Orij, E. N., Kamer, P. C. J.; Van den Aardweg, F.; Van Leeuwen, P. W. N. M. *J. Chem. Soc., Chem. Commun.* **1991**, 1096.

12 Van Rooy, A.; Orij, E. N.; Kamer, P. C. J.; Van Leeuwen, P. W. N. M. *Organometallics,* **1995**, *14*, 34.

13 Brown, J. M.; Kent, A. G. *J. Chem. Soc. Perkin Trans. II* **1987**, 1597.

14 Van Rooy, A. PhD thesis, University of Amsterdam 1995.

15 Collman, J. P.; Hegedus, L. S.; Norton, J. R.; Finke, R. G. *Principles and Applications of Organotransition Metal Chemistry* University Science Books: Mill Valley, CA 2nd ed. **1987**.

16 Trzeciak, A. M.; Ziólkowski, J. J.; Aygen, S.; Van Eldik, R. *J. Mol. Catal.* **1986**, *34*, 337. (b) Trzeciak, A. M.; Ziólkowski, J. J. *Trans. Met. Chem.* **1987**, *12*, 408. (c) Trzeciak, A. M.; Ziólkowski, J. J. *Inorg. Chim. Acta Lett.* **1982**, *64*, L267. (d) Trzeciak, A. M.; Ziólkowski, J. J. *J. Mol. Catal.* **1986**, *34*, 213. (e) Janecko, H.; Trzeciak, A. M.; Ziólkowski, J. J. *J. Mol. Catal.* **1984**, *26*, 355. (f) Trzeciak, A. M.; Ziólkowski, J. J. *J. Mol. Catal.* **1987**, *43*, 13.

17 Yoshinura, N.; Tokito, Y. Eur. Pat. 223 103, 1987 (to Kuraray); *Chem. Abstr.* **1987**, *107*, 154896. (b) Omatsu, T. Eur. Pat. 0 303 060, 1989 (to Kuraray); *Chem. Abstr.* **1989**, *111*, 38870.

18 Polo, A.; Real, J.; Claver, C.; Castillón, S.; Bayón, J. C. *J. Chem. Soc. Chem. Commun.* **1990**, 600.

19 (a) Polo, A.; Fernandez, E.; Claver, C.; Castillón, S. *J. Chem. Soc. Chem. Commun.* **1992**, 639. (b) Fernandez, E.; Claver, C.; Castillón, S.; Polo, A.; Piniella, J. F.; Alvarez-Larena, A. *Organometallics* **1998**, *17*, 2857.
20 Muilwijk, K. F.; Kamer, P. C. J.; van Leeuwen, P. W. N. M. *J. Am. Oil Chem. Soc.* **1997**, *74*, 223.
21 Jongsma, T.; Fossen, M.; Challa, G.; van Leeuwen, P. W. N. M. *J. Mol. Catal.* **1993**, *83*, 17. (b) Jongsma, T.; van Aert, H; Fossen, M.; Challa, G.; van Leeuwen, P. W. N. M. *J. Mol. Catal.* **1993**, *83*, 37.
22 Billig, E.; Abatjoglou, A. G.;. Bryant, D. R. U.S. Pat. 4,668,651; Eur. Pat. Appl. 213,639, 1987 (to Union Carbide); *Chem. Abstr.* **1987**, *107*, 7392.
23 Billig, E.; Abatjoglou, A. G.; Bryant, D. R.; Murray, R. E.; Maher, J. M. U.S. Pat. 4,599,206, 1986 (to Union carbide); *Chem. Abstr.* **1988**, *109*, 233177.
24 van Rooy, A.; Kamer, P. C. J.; van Leeuwen, P. W. N. M.; Goubitz, K.; Fraanje, J.; Veldman, N.; Spek, A. L. *Organometallics,* **1996**, *15*, 835.
25 Babin, J. E.; Whiteker, G. T. W.O. 93 03839, U.S. Pat. 911 518, 1992 (to Union Carbide Corporation); *Chem. Abstr.* **1993**, *119*, 159872.
26 Buisman, G. J. H.; Vos, E.; Kamer, P. C. J.; Van Leeuwen, P. W. N. M. *J. Chem. Soc., Dalton Trans.*, **1995**, 409.
27 Casey, C. P.; Whiteker, G. T.; Melville, M. G.; Petrovich, L. M.; Gavney Jr., J. A.; Powell, D. R. *J. Am. Chem. Soc.* **1992**, *114*, 5535.
28 Kranenburg, M.; van der Burgt, Y. E. M.; Kamer, P. C. J.; van Leeuwen, P. W. N. M. *Organometallics,* **1995**, *14*, 3081.
29 van der Veen, L. A.; Keeven, P. H.; Schoemaker, G. C.; Reek, J. N. H.; Kamer, P. C. J.; van Leeuwen, P. W. N. M.; Lutz, M.; Spek, A. L. *Organometallics,* **2000**, *19*, 872.
30 Tolman, C. A.; Faller, J. W. *Homogeneous Catalysis with Metal Phosphine Complexes*, Pignolet, L. H. Ed., Plenum Press: New York and London, **1983**; Chapter 2, pp 88-89.
31 Moasser, B.; Gladfelter, W. L.; Roe, C. D. *Organometallics* **1995**, *14*, 3832.
32 Buisman, G. J. H.; van der Veen, L. A.; Kamer, P. C. J.; van Leeuwen, P. W. N. M. *Organometallics* **1997**, *16*, 5681.
33 Burke, P. M.; Garner, J. M.; Tam, W.; Kreutzer, K. A.; Teunissen, A. J. J. M. WO 97/33854, 1997, (to DSM/Du Pont); *Chem. Abstr.* **1997**, *127*, 294939.
34 Sato, K.; Karawagi, J.; Tanihari, Y. Jpn. Kokai Tokkyo Koho JP 07,278,040 (to Mitsubishi); *Chem. Abstr.* **1996**, *124*, 231851.
35 Sato, K.; Kawaragi, Y.; Takai, M.; Ookoshi, T. US Pat. 5,235,113, EP 518241 (to Mitsubishi); *Chem. Abstr.* **1993**, *118*, 191183.
36 Röper, M.; Lorz, P. M.; Koetter, D. Ger. Offen. DE 4,204,808 (to BASF); *Chem. Abstr.* **1994**, *120*, 133862.
37 Lorz, P. M.; Bertleff, W.; Röper, M.; Koeffer, D. EP 472,071 (to BASF); *Chem. Abstr.* **1992**, *117*, 34513.
38 Cuny, G. D.; Buchwald, S. L. *J. Am. Chem. Soc.* **1993**, *115*, 2066.
39 Johnson, J. R.; Cuny, G. D.; Buchwald, S. L. *Angew. Chem. Int. Ed. Engl.* **1995**, *34*, 1760.
40 Van den Hoven, B. G.; Alper, H. *J. Org. Chem.* **1999**, *64*, 3964.
41 Van den Hoven, B. G.; Alper, H. *J. Org. Chem.* **1999**, *64*, 9640.
42 van Rooy, A.; de Bruijn, J. N. H.; Roobeek, K. F.; Kamer, P. C. J.; van Leeuwen , P. W. N. M. *J. Organomet. Chem.* **1995**, *507*, 69.
43 (a) Van Leeuwen, P. W. N. M.; Roobeek, C. F. Brit. Patent Appl. 33,554, **1981** (to Shell Resarch); *Chem. Abstr.* **1982**, *96*, 6174. (b) Van Leeuwen, P. W. N. M.; Roobeek, C. F. *J. Mol. Catal.*, **1985**, *31*, 345.

44 Paciello, R.; Röper, M.; Kneuper, H.-J.; Langguth, E.; Peter, M. *DE 4321194*, 1995, (BASF AG); *Chem. Abstr.* **1995**, *122*, 160937. (b) Paciello, R.; Siggel, L.;Röper, M. *Angew. Chem. Int. Ed. Engl.* **1999**, *38*, 1920. (c) Paciello, R.; Siggel, L.; Walker, N.; Röper, M. *J. Mol. Catal. A* **1999**, *143*, 85.

45 Parlevliet, F. J.; Kiener, C.; Fraanje, J.; Goubitz, K.; Lutz, M.; Spek, A. L.; Kamer, P. C. J.; van Leeuwen, P. W. N. M. *J. Chem. Soc. Dalton Trans.* **2000**, 1113.

46 Vaska, L. *J. Am. Chem. Soc.* **1966**, *88*, 4100.

47 Meakin, P.; Jesson, J. P.; Tebbe, F. N.; Muetterties, E. L., *J. Am. Chem. Soc.*, **1971**, *93*, 1797. (b) Meakin, P.; Muetterties, E. L.; Jesson, J. P., *J. Am. Chem. Soc.*, **1972**, *94*, 5271

48 Brill, T. B.; Landon, S. J. *Chem. Rev.* **1984**, *84*, 577. (b) Werener, H.; Feser, R. *Z. Anorg. Allg. Chem.* **1979**, *458*, 301.

49 Whyman, R., *J. Chem. Soc., Chem. Commun.*, 1970, 230-231.

50 Booth, B. L.; Else, M. J.; Fields, R.; Haszeldine, R. N., *J. Organomet. Chem.*, 1971, **27**, 119-131.

51 Cattermole, P. E.; Osborne, A. G., *Inorg. Synth.*, **1977**, *17*, 115.

52 Castellanos-Páez, A.; Castillón, S.; Claver, C.; van Leeuwen, P. W. N. M.; de Lange, W. G. J. *Organometallics,* **1998**, *17*, 2543.

53 James, B. R.; Mahajan, D.; Rettig, S.J .; Williams, G. M. *Organometallics,* **1983**, *2*, 1452

54 Moasser, B.; Gladfelter, W. L. *Inorg. Chim. Acta* **1996**, *242*, 125.

55 Coolen, H. K. A. C.; van Leeuwen, P. W. N. M.; Nolte, R. J. M. *J. Organomet. Chem.* **1995**, *496*, 159.

56 Parshall, G. W.; Knoth, W. H.; Schunn, R. A. *J. Am. Chem. Soc.* **1969**, *91*, 4990.

57 Lazzaroni, R.; Rafaelli, A.; Settambolo, R.; Bertozzi, S.; Vitulli, G. *J. Mol. Catal.*, **1989**, *50*, 1. (b) Lazzaroni, R.; Settambolo, R.; Rafaelli, A.; Pucci, S.; Vitulli, G. *J. Organomet. Chem.* **1988**, *339*, 357.

58 Chan, A. S. C.; Shieh, H. S.; Hill, J. R. *J. Chem. Soc., Chem. Commun.* **1983**, 688.

59 Bryant, D. R; Leung, T. W. US patent 5,741,943, **1998**, (to UCC); *Chem. Abstr.***1998**, *128,* 283881.

60 Packett, D. L.; Briggs, J. R.; Bryant, D. R.; Phillips, A. G. U.S Patent 5,886,237 (to UCC); *Chem. Abstr.***1998**, *128,* 14354.

61 Buhling, A.; Kamer, P. C. J.; van Leeuwen, P. W. N. M. *J. Mol. Catal. A: Chemical*, **1995**, *98*, 69.

62 Bhattacharya, A. K.; Thyagarjan, G. *Chem. Rev.* **1981**, *81*, 415.

Chapter 4

Phosphines as ligands

Bite angle effects for diphosphines

Piet W. N. M. van Leeuwen[†], Charles P. Casey[‡], Gregory T. Whiteker[#]

[†]Institute of Molecular Chemistry, University of Amsterdam, Nieuwe Achtergracht 166, 1018 WV, Amsterdam, the Netherlands; [‡] Department of Chemistry, University of Wisconsin-Madison, 1101 University Avenue, Madison, WI 53706, U.S.A.; [#] Union Carbide Corporation, P.O. Box 8361, South Charleston WV 25303, U.S.A.

4.1 Monophosphines as ligands

4.1.1 Introduction

The most famous rhodium catalyst precursor for hydroformylation is undoubtedly $RhH(PPh_3)_3CO$. It was first reported by Vaska in 1963 [1], but its activity for hydroformylation was discovered by Wilkinson and co-workers a few years later [2, 3]. Initially the activity was assigned to chloro-rhodium species [4]. The catalysis and the organometallic chemistry involved were studied in great detail, such as kinetics, substrate activity, metal-ligand equilibria, equilibria between monomeric and dimeric complexes, reaction mechanisms, spectroscopic properties, etc. Soon thereafter the first commercial applications were reported [Chapter 1]. The chemistry reported in the late sixties and early seventies is still relevant for the hydroformylation studies of today. The chemistry and catalysis of rhodium-PPh_3 has been reviewed many times [3, 5]. By way of introduction for the catalysts based on diphosphines a few salient features of the PPh_3-based catalysts will be summarized. These will include mechanistic studies, in situ spectroscopic studies, kinetics, and phosphine ligand effects. Special consideration will be given to "*the rate-determining step*" under common circumstances (*vide infra*); we will see that it is not the oxidative addition of hydrogen as most textbooks erroneously say [6].

P.W.N.M. van Leeuwen and C. Claver (eds.), Rhodium Catalyzed Hydroformylation, 63–105.

4.1.2 The mechanism

In Figure 1 the well-known mechanism, as first proposed by Heck [7], has been depicted. It corresponds to Wilkinson's so-called dissociative mechanism [2]. The associative mechanism involving 20-electron intermediates will not be considered. For PPh_3 as the ligand a common starting complex is $RhH(PPh_3)_3CO$, complex **1**, which under 1 bar of carbon monoxide forms the complexes **2ee** and **2ae**, containing the phosphine ligands in equatorial positions (denoted **ee** throughout the scheme) or one in an apical position and the other ligand in an equatorial position (complexes denoted **ae**). Brown [8] found a preference for the **ee** isomer in the hydride. Dissociation of either equatorial L or equatorial CO from **1** or **2** leads to the square-planar intermediates **3c** and **3t**, which have phosphines in *cis* or *trans* configurations respectively. Preferential dissociation of equatorial ligands from trigonal bipyramids is normally observed. Complexes **3** associate with ethene to give complexes **4**, again in two

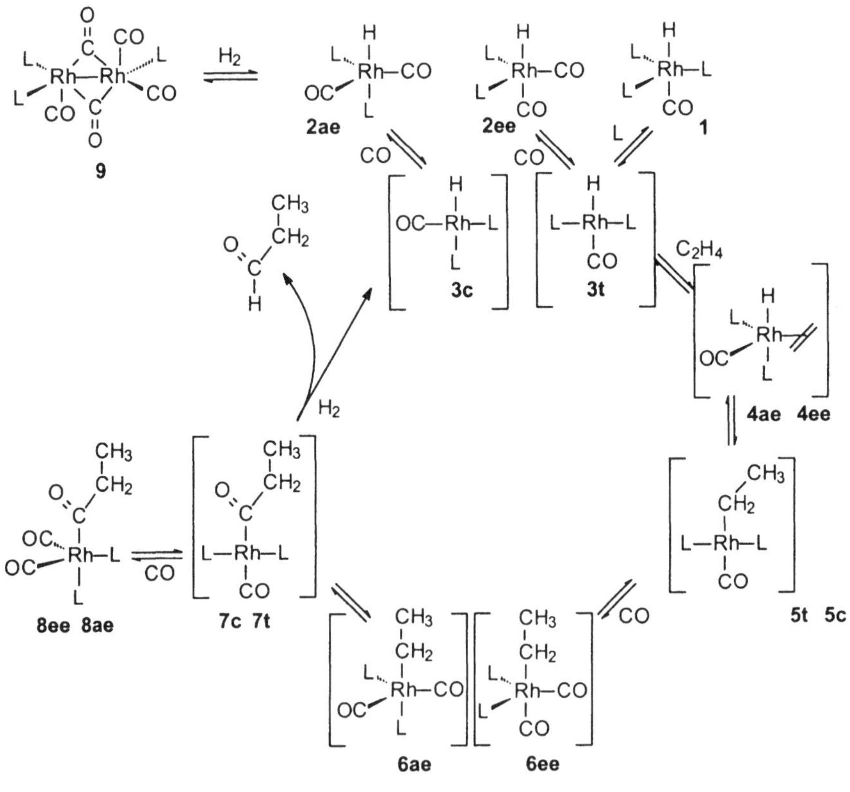

Figure 1. Simplified mechanism for hydroformylation of ethene (L = PPh_3)

isomeric forms **ae** and **ee**, having a hydride in an apical position and ethene coordinating in the equatorial plane. While four-coordinate species rhodium **3c** (*cis*) have never been observed, the analogous four-coordinate complex for **3t** *trans*-$Rh(PCy_3)_2(CO)H$ has been isolated and structurally characterized [56]. In addition, magnetization transfer experiments indicate that both **1** and **2** are in equilibrium with free PPh_3 [8]. These results are highly suggestive of the formation of the four-coordinate complex **3** by phosphine dissociation from **1** and **2**. The complexes never observed experimentally are shown in brackets in Figure 1.

Although the complexes shown in Figure 1 contain at least two coordinated PPh_3 ligands, a large body of indirect evidence suggests the influence of equilibria involving catalytically active species containing a single coordinated phosphine ligand. Such monophosphine species were initially invoked to explain the positive effect on hydroformylation regioselectivity that is typically seen with increasing P:Rh ratios (see 4.1.6). The ^{31}P NMR magnetization transfer experiments described by Brown [8] also indicate that PPh_3 dissociation from the RhL_2 complex **2** occurs, albeit at a significantly slower rate than the corresponding PPh_3 dissociation from tris-phosphine complex, **1**. In addition, IR spectra have revealed the presence of $Rh[P(OAr)_3](CO)_3H$ with a very bulky coordinated phosphite ligand [48].

It has not been established experimentally whether alkene complexation is reversible or not; in the scheme we have drawn all steps but the hydrogenolysis at the end as reversible. Brown observed that complex **1** catalyzes the isomerization of *cis*-1,2-dideuteriostyrene under nitrogen [8]. Although under these conditions the isomerization of this labeled alkene is slower than the rate of hydroformylation using 1, this result suggests that alkene coordination and insertion into the Rh-H bond can be reversible processes. Complex **4** undergoes a migratory insertion to give square-planar alkyl complexes **5c** and **5t**, which are respectively *cis* or *trans*. In the absence of CO, the 16-electron complexes of this type can be isolated as mixtures of *cis* and *trans* isomers for aryl groups or alkyl groups not containing ß-hydrogen atoms, [9]. Complex **5** can undergo ß-hydride elimination, thus leading to isomerization when higher alkenes are used, or it can react with CO to form trigonal bipyramidal complexes **6**. Thus, under low pressure of CO more isomerization may be expected. At low temperatures (< 70 °C) and a sufficiently high pressure of CO (> 10 bar) the insertion reaction is usually irreversible and thus at this point also the regioselectivity of hydroformylation of α-alkenes is determined.

Complexes **6** undergo the second migratory insertion in this scheme to form the acyl complexes **7**. Complexes **7** can react either with CO to give the saturated acyl intermediates **8**, which have been observed spectroscopically, or with H_2 to give the aldehyde product and the unsaturated intermediates **3**. The reaction with H_2 involves presumably oxidative addition and reductive elimination, but for rhodium no trivalent intermediates have been observed. For iridium the trivalent intermediate acyl dihydrides have been observed [10]. The Rh-acyl intermediates **8** have also been observed by Brown [8] and due to the influence of the more bulky acyl group, as compared to the hydride atom in **2e** and **2a**, isomer **8ae** is the most abundant species.

At low hydrogen and high rhodium concentrations, formation of dirhodium species such as **9** becomes significant. Since several of the studies of the Wilkinson group were carried out under such conditions, formation of di-rhodium species was very relevant to their chemistry. Not only did they report the formation of the orange dirhodium species of type **9**, they also observed the formation of red dirhodium species containing two carbon monoxide molecules less than **9**. A dirhodium complex **9** containing three molecules of PPh_3 was fully characterized by Chan [11]. Regeneration of rhodium hydrides from dormant rhodium species formed by impurities, is another cause for the positive rate response to raising the H_2 pressure. Note that in Figure 1 not all reagents have been depicted in the scheme for the sake of simplicity.

Usually, the full mechanistic scheme will be more complicated than that shown in Figure 1:

1. Species containing one to four molecules of L will occur, depending on the concentrations of L and CO, and the nature of L (*vide infra)*,
2. When alkenes such as propene and higher are used the regio-isomers of linear and branched alkyl and acyl groups should be included,
3. Alkyl species **5** containing 2-butyl species or higher may lead to internal alkenes via ß-hydride elimination.

The scheme shown in Figure 1 is typical of NMR studies using PPh_3, carried out at room temperature, ~1 bar of CO (and H_2), and a complex concentration of 10^{-1} to 10^{-2} M [8]. It is also characteristic of "catalytic" conditions, having as typical values 70-120 °C, 10-30 bar, rhodium concentration of $\approx 10^{-3}$ M, and a hundred-fold excess of aryl phosphines. In situ IR spectroscopy will show whether RhL_1 or RhL_2 or RhL_3 ($L = PPh_3$) species are present. NMR studies at high concentrations and low pressures and temperatures will prove the formation of dimers **9**. Quantitative data for monomer-dimer equilibria have been obtained for bidentate phosphines [12].

4.1.3 Ligand effects

Steric and electronic properties of a ligand can drastically influence the rate and selectivity of the hydroformylation reaction (see Chapter 3, Phosphites as ligands). A systematic study though is absent for several reasons. For monodentate ligands a systematic study is impeded by the variety of complexes that may be involved at the various stages of the catalytic cycle. Secondly, $RhH(PPh_3)_3CO$ was used as the precursor in several studies, while aiming at the study of the effect of ligands added. Use of this procedure without removal of the liberated PPh_3 leads to possible mixed phosphine complexes, which complicate interpretations. Thirdly, complexes of type **1** were often used without the addition of free ligand, which is needed to ensure a high equilibrium concentration of **1** or **2**. Other complications arise from the use of different experimental conditions. As we shall see in section 4.1.5. changes in temperature, concentrations and pressure can significantly influence rate and regioselectivity. For these reasons, studies which are performed under conditions outside those typically used for hydroformylation (10-30 bar, temperature 70-120 °C, [Rh] = $\approx 10^{-3}$ M, [alkene] = 0.1-2 M, [L] depending on complex stability) can result in conclusions which are not particularly relevant. Impurities in one of the feedstocks often led to oxidation of the phosphine ligand. Another source of erroneous results is the use of rhodium chloride (and the like) as the precursor, while even in the late sixties it was known that this gave only partial conversion to hydrido species [2]. When discussing selectivities isomerization has often not been taken into account. In addition, many interesting effects have been published only in patents and these results are less accessible in common literature searches.

Thus, we are facing an impossible task to summarize the "ligand effect" for monodentate phosphines (and phosphorus ligands) and to give credit to the numerous contributions. For the sake of didactics we will present a few rules of thumb, which at this point will not be fully exemplified by literature data, but which will be supported when diphosphines are discussed in section 4.2.

Electronic effects. Electron donating ligands, such as alkylphosphines, lead to slower catalysts and as result they require higher temperatures [13]. In recent years it has been shown that triethylphosphine gives turnover numbers of 700 $mol.mol^{-1}.h^{-1}$ for 1-hexene and very high turnover numbers for ethene at 120 °C and 40 bar. As a secondary solvent-dependent reaction hydrogenation to alcohols takes place [13c]. Arylphosphines containing electron-withdrawing substituents give a faster catalytic reaction than triphenylphosphine [14-16]. Phosphites can give faster catalysts than phosphines [17, 18], but this is certainly not true for all cases. A recent study in supercritical CO_2 also shows that electron withdrawing arylphosphines form more active catalysts [19], but the low concentrations of $RhHL_3(CO)$, in the absence of extra L added, and the relatively high concentration of CO in the supercritical medium may lead to excessive dissociation of the phosphine ligand. Dibenzophospholes are more electron-withdrawing than diphenylphosphino groups and without exception the former lead to faster catalysts [8, 20, 43b, 59]. In general, ligand effect studies are hard to interpret, because for a particular ligand the rates may differ by at least an order of magnitude, depending on the concentration of rhodium, ligand and CO.

Most of the authors cited agree on the explanation; electron-withdrawing ligands lead to a decrease of the back-donation to carbon monoxide and thus to a weaker binding of the carbonyls. This will affect the formation of species **3** and **7**, such that their rate of formation, or their equilibrium concentration increases. Alkene complexation, giving complexes **4**, may also be accelerated or become more favored thermodynamically. Migratory insertions are not particularly sensitive to electronic properties of the ligand [21 and references therein], but it is important to remember that oxidative addition will slow down when electron-withdrawing ligands are used.

Steric effects. More sterically demanding ligands will favor the formation of species containing a low number of ligands L and therefore more CO ligands. A high proportion of CO ligands also leads to electron poor rhodium species and thus to enhanced dissociation of CO. For phosphites this effect has been clearly observed [22] as was discussed in Chapter 3.

4.1.4 In situ studies

Moser has studied monophosphine systems of substituted arylphosphines by *in situ* IR reflectance spectroscopy [14]. During 1-hexene hydroformylation, at 70 °C, 11-24 bar of syn gas, [Rh]=10 mM, L/Rh = 6, the species observed in most instances was $RhH(CO)_2L_2$,**2**. The three bands observed were assigned to two carbonyl stretching frequencies and a rhodium-hydride stretching frequency. Claver and Van Leeuwen studied the system containing PPh_3 as the ligand during the hydroformylation of 1-octene (60-100 °C, 5-20 bar, [Rh] = 1 mM, PPh_3/Rh = 5) using *in situ* IR transmission spectroscopy [23]. They observed four bands (2042, 1992, 1981, 1947 cm^{-1}), which were assigned to the mixture of $RhH(CO)_2(PPh_3)_2$ isomers **2ae** and **ee** as reported in the NMR studies by Brown [8], without any other species being present. The rhodium-hydride bands are very weak and usually only the four carbonyl stretches are observed. The spectral assignments are derived from the spectra of the diphosphine complexes (section 4.2). The concentration ranges applied in the IR and catalysis studies are similar. Thus, both studies infer that under the conditions of "practical catalysis" species **2ae** and **2ee** are the resting state of the catalyst.

NMR spectroscopy is less useful for *in situ* studies as the concentrations of metal and ligand are often two orders of magnitude higher, which influences the positions of the equilibria considerably. Secondly, in most

NMR devices replenishment of the consumed gases would be slow compared to the rate of reaction, which would lead to the formation of species that differ from the ones under actual catalytic conditions.

Using the same ligand to metal ratios (4-9:1) as in the IR studies above [23] and pressures up to 4 bar but significantly higher ligand concentrations of ≈ 10 mM, Oswald [13b] found in NMR studies that the predominant species is now the tris-triphenylphosphine complex **1**. Dissolving **1** in toluene at 21 mM at ambient conditions (1 bar of CO, ratio 3:1) gives a mixture of free PPh_3, **1** and **2**, the latter being the major species as was reported by Brown [8]. His studies show that ligand exchange is much faster than the catalytic processes. These examples clearly illustrate the importance reporting concentrations rather than metal to ligand ratios, a common but erroneous way of describing catalyst composition.

Upon addition of alkene to solutions containing **1** and **2** under a CO atmosphere, Brown observed the formation of five-coordinate acyl complexes, which were characterized by NMR spectroscopy.

4.1.5 Kinetics

In spite of the industrial importance of the rhodium-PPh_3 catalyst, very few data have been published on the kinetics of the hydroformylation reaction and the data known are often contradictory. Early mechanistic work by Wilkinson (at < 1 bar) demonstrated the inhibitory effect of increased CO and PPh_3 concentrations on 1-hexene hydroformylation rate [2c]. Increased Rh and alkene concentrations were observed to lead to higher catalytic rates. Wilkinson also observed an accelerated hydroformylation rate caused by increasing H_2 pressure, which is most likely, an artifact of inactive dirhodium species formed under "non-standard", low-pressure conditions. We will argue that under "standard" catalytic conditions the reaction is first order in the concentration of alkene, first order in the rhodium concentration, zero order in hydrogen, and the reaction shows a negative order in ligand concentration (phosphine or carbon monoxide, or both). The most detailed, early study is the report by Cavalieri d'Oro et al.[24] from Montedison. They found the following expression for the rate:

$$v = k\,[C_3H_6]^{0.6}\,[PPh_3]^{-0.7}\,[CO]^{-0.1}\,[Rh]^{1}\,[H_2]^{0}$$

(conditions 90-110 °C, p(CO) = 1-25 bar, $p(H_2)$ = 1-45 bar, $[PPh_3]$ = 0.05-5 M, [Rh] = (0.5-7).10^{-3} M, PPh_3/Rh = 300:1 to 7:1, $[propene]_{t=0}$ = 2-7 M)

They reported an overall *activation energy* of 84 kJ.mol^{-1} for the process. The important features of this kinetic study are the zero order dependence in dihydrogen, the negative order in PPh_3 ligand (and CO), and the positive order in alkene concentration. Under "standard" conditions, i.e. the "industrial operating" conditions, chosen by many workers in the field, this rate equation seems a good starting point for arylphosphine modified catalysts. These "*standard" conditions*# (T = 70-120 °C, CO = 5-25 bar, H_2 = 5-25 bar, Rh ≈ 1 m<u>M</u>, alkene = 0.1 - 2 <u>M</u>) are assumed in later sections unless otherwise specified.

The observed *order in propene* concentration is less than one, which might point to saturation kinetics. Indeed, high concentrations were used, but perhaps the non-ideal behavior of propene (critical temperature 94 °C) plays a role in this. Under similar conditions for 1-hexene and 1-octene a neat first order behavior in alkene has been observed using Rh-PPh_3 catalysts [18, 25a].

At *high PPh_3 concentrations*, where the catalyst resting state is $(PPh_3)_3Rh(CO)H$, phosphine dissociation must occur to form the coordinatively unsaturated intermediates **3c** and **3t**. This dissociation is suppressed by increased PPh_3 concentration, which serves to reduce the concentration of active Rh species in the catalytic cycle.

At *lower PPh_3 concentrations* where the predominant resting state observed by in situ studies is $(PPh_3)_2Rh(CO)_2H$, species **3c** and **3t** are formed by CO dissociation, which is likewise inhibited by increased CO concentration. Consistent with this mechanism is the recent determination that dissociation/association of CO is reversible and faster than hydroformylation for arylphosphines (see 4.2). An inverse order in CO pressure was reported by Moser [14]. Strohmeier [25a] also noted that the rate dropped by 30 % when the total pressure was raised from 10 to 15 bar, as did Wilkinson [2a] when the pressure was raised from 50 to 80 bar. Both observations are in accord with a negative order in CO pressure and a zero order dependency on H_2 pressure. At very low pressures of CO or H_2 (a few tenths of a bar at 25 °C) the reaction will slow down as a sufficient concentration of the reagents is controlled by mass transfer limitations.

Hydroformylation reactions are extremely sensitive to experimental conditions. As detailed in the following sections, various equilibria exist which influence catalytic rates. These equilibria are controlled by concentrations of Rh, CO, H_2, alkene and added phosphine ligand. For a given catalyst, hydroformylation rate and regioselectivity are, therefore, often dependent upon the concentrations of these reactants. In addition, the concentrations of gaseous reactants can become controlled by mass transfer limitations and result in increased amounts of hydrogenation and isomerization products. Variations in added phosphine concentration can also dramatically impact catalyst performance. Under typical industrial conditions, additional phosphine ligand is added, however some studies utilize a low P/Rh ratio of ≤3, especially when chelating diphosphines are used. Unfortunately, experimental conditions reported in the literature for Rh-phosphine catalysts vary widely and make comparisons difficult. For this reason, we will define a window of "standard" conditions that cover the ranges typically utilized in industrial applications.

Under "standard" conditions, we propose that the best starting point for the kinetics is an equation of the type:

$$Rate(type-I) = \frac{A[alkene][Rh]}{B+[L]}$$

(the constants A, B do not refer to specific rate constants;
[L] is proportional to [PPh_3] and [CO])

Two possible scenarios exist which are consistent with this rate equation. Rate determining *alkene coordination* by **3c** and **3t** followed by rapid alkene insertion into the Rh-H bond is one possibility. An equally valid alternative explanation for the observed kinetics involves rate determining migratory *insertion of alkene* into Rh-H preceded by fast, reversible alkene coordination. In both cases, the concentrations of coordinatively unsaturated **3c** and **3t** are influenced by PPh_3 and CO concentrations.

The kinetics at moderate PPh_3 concentrations are in agreement with a catalyst resting state of $RhH(PPh_3)_2(CO)_2$ and a transition state of composition $RhH(PPh_3)_2(CO)$(alkene). At very high PPh_3 concentrations, the likely resting state is $RhH(PPh_3)_3(CO)$ and a transition state of composition $RhH(PPh_3)_2(CO)$(alkene) is again likely. The set of rate-limiting reactions are dissociation of CO (or PPh_3), complexation of alkene, and migratory insertion. Dissociation/association of CO is reversible and faster than hydroformylation for arylphosphines (see 4.2). Complexation of alkene is most likely reversible, although there are no experimental data on this process. Theoretical studies [26] indicate that complexation of CO or ethene to species **3** (Figure 1) has a low barrier. For migratory insertion of ethene into the metal-hydride an early transition state was found, involving a re-organization of the complex. This means that steric hindrance will play a crucial role and especially the rotation of the alkene from the in-plane coordination to a perpendicular coordination mode contributes to the barrier.

There is a strong body of evidence in the literature in disagreement with the kinetics presented above [2-4, 27, 28]. Ever since the first statements of the Wilkinson group [2a, not repeated though in 2c!] the oxidative addition of hydrogen has been considered as the rate-determining step [5, 6]. The first suggestion was based on an analogy with the hydrogenation reaction employing Wilkinson's catalyst $RhCl(PPh_3)_3$. In the textbooks [6] and reviews [5] oxidative addition of dihydrogen is still mentioned as the most likely "rate-determining" step. The theoretical work cited shows the highest barriers for the migration reactions [21, 26], but yet supports the common point of view!

The scheme shown in Figure 1 can explain why in many instances a different rate equation has been obtained, having a positive order in dihydrogen concentration. Without experimental effort we cannot re-interpret data of specific publications and we will confine ourselves to general statements. Clearly, at low pressures a different rate equation should be obtained, even though rhodium and ligand concentrations are within our "standard" regime. At low dihydrogen pressures, high rhodium concentrations, and low temperatures di-rhodium species of the type **9** will form, which will lead to a positive order in the pressure of dihydrogen due to the H_2 requirement to cleave these di-rhodium species and re-enter the catalytic cycle [12, 27b].

These basic findings were reported by Wilkinson, who observed at pressures below 1 bar the presence of dimers, the dependency of the rate on hydrogen pressure, and yet the positive order in alkene concentration. If the oxidative addition were rate-determining, the rate of reaction would be zero order in alkene (unless rapid pre-equilibration would occur all the way from **1** through **8**).

As has been discussed in Chapters 2 and 3, for carbonyl systems and bulky phosphite systems, the resting state of these catalysts (for 1-alkenes) is the acylrhodium species of type **8** and thus, the reaction with dihydrogen is rate limiting. At low Rh concentration, using $RhH(PPh_3)_3CO$ as the precursor without addition of additional PPh_3, substantial dissociation will occur, giving rise to the kinetics of mono-phosphine or phosphine-free catalysts, i.e. a type II rate equation. Under these conditions, indeed the acylrhodium complex **8** is the resting state and oxidative addition is rate-determining [29]. We would like to stress that Type-II kinetics is the exception rather than the rule.

$$Rate(type - II) = \frac{C[H_2][Rh]}{D + [CO]}$$

4.1.6 Regioselectivity

The regioselectivity of the catalysts based on PPh_3 has been extensively studied (see reviews [5]). The regioselectivity of 1-alkene hydroformylation varies from 70 to 92 % for the linear aldehyde. The highest selectivity is obtained at high concentrations of PPh_3 [5, 25], or even liquid PPh_3 [2, 30], and low pressures of CO. Isomerization should be considered as well, as this forms an "escape" route for the branched alkyl intermediate leading to internal alkene isomers instead of branched aldehyde. Since the activity of most phosphine catalysts for internal alkenes is low, the apparent linearity of the aldehyde product is high, but the overall selectivity to linear aldehyde may be low. This will be discussed further in 4.2.

Clearly, the number of phosphines coordinated to rhodium, along with the stereochemistry at Rh, determines the regioselectivity. At high PPh_3 concentrations, the resting state of the catalyst is **1** (Figure 1, 2), which undergoes PPh_3 dissociation to form selectively 4-coordinate intermediate **3t**. At lower PPh_3 concentrations, the resting state is an equilibrating mixture of **2ae** and **2ee** which undergo either CO dissociation to form **3c** and **3t** or PPh_3 dissociation to form **10c** and **10t**. According to NMR studies, conversion between axial and equatorial phosphines in **2ae** and **2ee** is many orders of magnitude faster than the rate of hydroformylation [8]. The rate of hydroformylation is only two orders of magnitude slower than that of carbon monoxide exchange of species of type **2** [31]. The relative concentrations of these four-coordinate diphosphine (**3**) and monophosphine (**10**) intermediates are controlled by the PPh_3 and CO concentrations. Wilkinson [2] and Andreetta [24b] suggested that species **3**, formed at high PPh_3

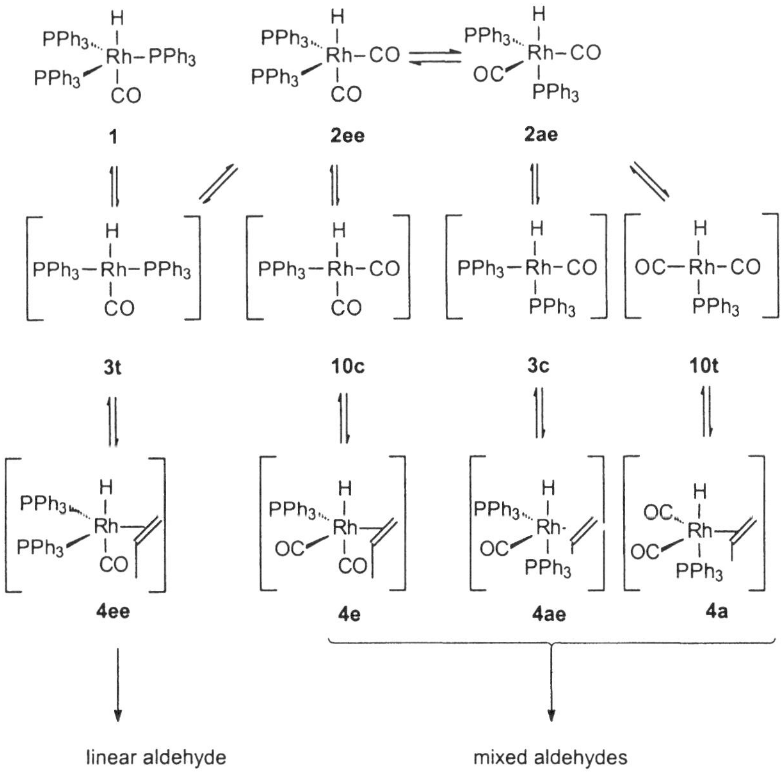

Figure 2. Intermediates leading to linear and branched products

concentration, lead to higher linear aldehyde selectivity (l:b = 20:1), and that species **10**, containing only one phosphine, lead to a lower selectivity for linear aldehyde (l:b = 4:1) (see Figure 2).

Dissociation of PPh_3 from **1** gives **3t**, dissociation of CO from **2ee** also gives **3t**, while dissociation of CO from **2ae** will give **3c**. Dissociation of PPh_3 from **2** gives the isomers **10t** and **10c**. The use of diphosphines such as dppe and dppp gives modest linear-branched ratios and their putative intermediates **3** must have *cis* structure **3c**. Therefore, **3c** is not a likely intermediate for the selective hydroformylation. For linear-branched ratios of 4:1 we don't have to invoke mono-phosphine species of type **10**, as already structure **3c** suffices for the production of moderate l:b ratios (3:1).

Hydroformylation regioselectivity can be established by either four-coordinate or five-coordinate intermediates. In one possible mechanism, the regioselectivity is controlled by the relative concentrations of four-coordinate intermediates, **3c**, **3t**, **10c** and **10t,** which undergo irreversible *formation of alkene hydride complexes*, **4**. Accordingly, this mechanism requires that interconversion of isomeric pairs **4ee/4ae** and **4e/4a** be slow relative to irreversible migratory insertion of alkene.

Alternatively, the regioselectivity can be determined during migratory *insertion of alkene* into the Rh-H bond in five-coordinate intermediates, **4**. The concentrations of these five-coordinate alkene hydride complexes are controlled by the relative concentrations of PPh_3 and CO. In addition, these stereoisomers (**4ee/4ae**, **4e/4a**), which differ in the nature of the apical ligand, presumably undergo rapid interconversion via a pseudorotation mechanism, as has been observed for **2ee/2ae** [8].

In both scenarios, the dependence of l:b on PPh_3 concentration is attributed to changes in the extent of phosphine ligation, and the stereochemistry at Rh is also proposed to influence regioselectivity. Studies using diphosphine ligands with well-defined coordination number and geometry (section 4.2) have provided much insight into these factors that control hydroformylation regioselectivity.

Substituted alkenes will be discussed in Chapters 5 and 6. From a mechanistic viewpoint the results obtained with substituted 1-alkenes are worth mentioning as they stress the importance of steric interactions. Simple 1-alkenes carrying methyl substituents at the carbon atoms 3 or 4 give progressively more linear product for a higher degree of substitution closer to the alkene bond, without losing much activity as is shown in Table 1 [18]. The table shows that in most instances the rate of hydroformylation to the terminal product is retained and that a remarkably high linearity can be obtained when steric hindrance is provided at carbon-3. Substitution at carbon-2 under these conditions gives rates of only 100 $mol.mol^{-1}.h^{-1}$, and

Table 1. Hydroformylation of methyl-substituted 1-alkenes

Alkene	Rate, $mol.mol^{-1}.h^{-1}$	Linearity, %
1-pentene	11,300	78.4
4-Me-1-pentene	9,300	78.0
4,4-Me_2-1-pentene	5,300	85.0
3-Me-1-pentene	9,600	91.0
3,3-Me_2-1-pentene	7,600	99.0

Conditions: 90 °C, $p(CO/H_2)$ = 20 bar, [Rh]=0.5 mM, [PPh_3]=5 mM, [alkene]=0.5 M, initial rates at <20% conversion, no isomerization was observed [18].

only linear product is formed. Internal alkenes also show rates that are one to two orders of magnitude lower than those of 1-alkenes [2].

Cis-2-alkenes react a few times faster than *trans*-2-alkenes. Steric influences, both from the complex part and the alkene, play a dominant role in determining the fate of hydroformylation of alkenes not containing functional groups. For alkenes such as styrene, 3,3,3-trifluoropropene, acrylates, and vinyl esters, the electronic properties of the alkene determine to a large extent the regioselectivity.

4.1.7 Conclusion

Monodentate ligands, be they phosphines, phosphites, or phosphoramidites, are very versatile ligands for rhodium catalyzed hydroformylation. Their performance as ligands though is strongly obscured by the equilibria in which they are involved. Literature articles, reviews and textbooks disagree on the rate-determining step, whether this is the insertion (or complexation) of the alkene (type-I) or the oxidative addition of dihydrogen (type-II). We have argued that under "standard" conditions type-I kinetics prevail. We have mentioned a number of pitfalls that might obscure the kinetic analysis and we will return to that in Chapters 8 and 9. The differences in the rates of the respective steps are small, and it is not surprising that different rate equations have been postulated due to differences in experimental conditions. Steps eligible as "rate-determining" are complexation of the alkene, hydride migration, migration of the alkyl group forming the acyl group, and oxidative addition of dihydrogen. We propose that *grosso modo* in phosphine catalyst systems migratory insertion of the alkene into Rh-H is the rate-determining step under standard, industrial process conditions. The concentrations of the reagents, metal and ligands may disturb the equilibria, causing deviations from our rule of thumb. As we will see in section 4.2 the uncertainties arising from the use of monodentate ligands can be reduced by the use of bidentate phosphines.

4.2 Diphosphines as ligands

4.2.1 Introduction

The key role of intermediate species **3** containing two triphenylphosphines for obtaining a high selectivity to linear products has been recognized since 1970 [2]. The way to guarantee that these intermediates are formed in excess with respect to the less selective monophosphine intermediates **10**, was the use of a high excess of PPh_3. The first commercializations (1974, Celanese, 1976, UCC) utilize this finding and linearities greater than 90 % can be achieved. It would seem obvious, in hindsight, that the use of suitable diphosphines might greatly facilitate the formation of relatively stable L_2Rh complexes. Other benefits later derived from bidentate diphosphines include control of metal geometry and efficient asymmetric induction. Favorable effects of diphosphines in catalysis were reported as early as 1966 by Iwamoto [32], who found that dppp was the best ligand for the codimerization of butadiene and ethene using iron catalysts. During the seventies the advantages of the use of bidentates were especially exploited in the field of asymmetric hydrogenation using chiral diphosphines such as DIOP and DIPAMP [33], but it would take many years before suitable bidentates were evaluated for selective hydroformylation to linear aldehydes.

Sanger [34] studied the effect that the addition of common diphosphines exerts on catalyst **1** (2 mM in benzene) at room temperature and 1 bar. While the amount of ligand added strongly influenced the rate, the l:b ratio of the aldehyde product ranged from 3.2 to 3.9 for all diphosphines (dppe, dppp, dppb, DIOP). With the exception of dppb, which forms also dimeric species, it is expected that these diphosphines will form chelated intermediates of the structures **2ae** and **3c**, due to the small bite angle of the phosphines. The behavior in such mixtures of ligands (PPh_3 and diphosphine) remains hard to disentangle.

Pittman [37b] also studied the effect of added diphosphines (dppe, dppp, and dppb) to catalyst **1** (10 mM) using higher temperatures and pressures (60-120 °C, 7 bar) on the hydroformylation of 1-pentene. He found low l:b ratios of only 1. He concluded that apparently a *cis* coordination as in **3c** gives low l:b ratios. Thus, under these "standard" conditions the l:b ratio is much lower than that found by Sanger at ambient conditions. The presence of PPh_3 in both sets of experiments would make an explanation speculative.

Later studies have shown that dppp and other ligands containing a propane bridge smoothly form complexes **11a** [18, 35]. It was noted by Brown that dppe formed less clean solutions. The small bite angle of 84° is apparently not attractive for the formation of such complexes. Bayón and coworkers observed that constrained 1,2-diphosphines form dimeric complexes in addition to the expected monomeric ones [36].

Studies using dppe or dppp under "standard" conditions, in the absence of PPh_3, have not been published in the seventies, but it would seem likely that several industries have conducted research in this field. As described in 4.1.1, two explanations for the low regioselectivities observed with such diphosphines can be brought forward. Resting state **11a** can lead to **3c** by dissociation of CO. Since the small bite angle imposed by chelates such as dppe and dppp cannot accommodate the *trans* orientation in **3t**, the resultant *cis* complex **3c** leads to low regioselectivity. Alternatively, **3c** can reversibly add alkene to form the five-coordinate alkene hydride complex, which contains an aical-equatorial diphosphine chelate. This **ae** isomer, analogously to **4ae**, leads to low regioselectivity. In addition to these two possibilities, **11a** may also undergo dissociation of one phosphine donor giving **12t**, the analog of unselective **10t**. The kinetics would be different for the two possibilities, but unfortunately these have not been reported.

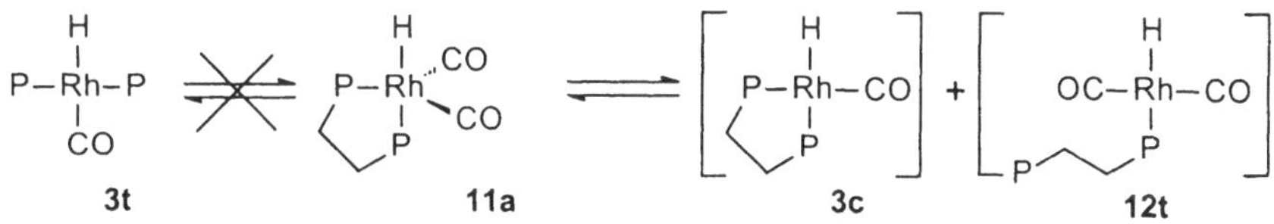

Figure 3. Reaction scheme *cis*-bidentates

In the late seventies enhanced rates for platinum hydroformylation catalysts were reported for diphosphines containing a bridge that was longer than the common two or three carbon atoms [38]. The selectivities to linear product are very high using these platinum catalysts (>>95%). One rhodium system has been reported giving linearities > 99% using dppe as the ligand [39], but this uses 1,3-butadiene as the substrate and most likely the mechanism involves allylic species and does not concern our discussion of the mechanism in Figures 1-3.

DIOP was first used as a bidentate ligand added to **1** as early as 1973 by Consiglio while aiming at asymmetric catalysis. For 1-alkenes they observed high l:b ratios of 13 at 25 °C and 1 bar [37a].

In 1981 Hughes and Unruh (Celanese) were the first to publish on diphosphines, which unlike the more common dppe and dppp led to higher l:b ratios under standard conditions (see Figure 4) [40] (100 °C, 8 bar, [Rh] = 2 mM). High l:b (5-8) ratios were obtained in systems consisting of **1** and added DIOP or *trans*-dppm-cyb, both having rigid *trans* conformations. Low l:b ratios (≈ 1-3) were obtained for systems containing *trans*-dppm-cyh, *cis*-

dppm-cyb, and non-rigid diphosphines dppe, dppp, and dppb. According to a later report [15c] the use of *trans*-2,3-dppm-nor (Figure 4) (containing trifluoromethyl substituents) also gave a high ratio (l:b=7) at a high rate.

Steric bulk and the chelate effect are brought forward as an explanation. Pre-empting our later discussion, we might propose that ligands with small chelation angles such as dppe and dppp give intermediates **2c** which lead to low l:b ratios, while ligands with wider angles behave differently.

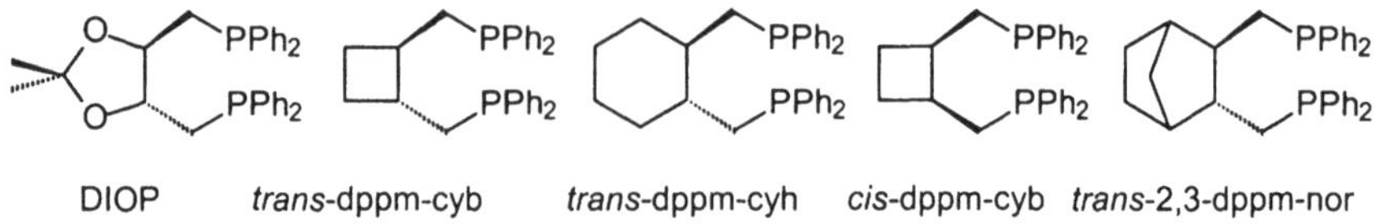

Figure 4. Diphosphine ligands used by Unruh

In the following we will present the results obtained with ferrocene based ligands and a variety of bidentate phosphine ligands containing bite angles larger than 100°. Since the catalysis for these ligands has been carried out at somewhat different conditions we will present them separately and discuss the results in three parts. Finally we will speculate about a general view in an attempt to find a "global" explanation.

4.2.2 Ferrocene based diphosphine ligands

Unruh and Christenson [15a,b] studied the use of ferrocene diphosphine ligands in 1982. Their study was interesting in that the results were different from the Wilkinson work in several ways. We will quote their own interpretations, but we cannot avoid adding more recent viewpoints that may be relevant. Dppf is a flexible ligand and we now know from crystal structures that the phosphorus-metal-phosphorus angle in complexes of dppf can vary from 92° up to 120°, with a strong preference for values around 97±2° [46]. Rotation of the cyclopentadienyl fragments with respect to one another should help to accommodate a variety of structures. From this point of view, dppf is an interesting choice. In addition to dppf they studied three electronically modified ferrocenyl-based ligands, each containing electron withdrawing groups, see Table 2 and Figure 5.

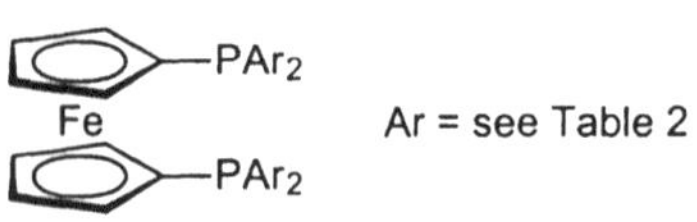

Figure 5. Unruh's substituted dppf ligands

Table 2. Hydroformylation using substituted dppf ligands

Ar =	χ_i-value	l:b ratio	Relative rate	2-hexene, %
Ph	4.3	5.4	7.2	3.7
4-ClC_6H_4	5.6	6.8	9.3	5.0
3-FC_6H_4	6.0	8.0	13.7	5.0
4-$CF_3C_6H_4$	6.4	11.4	13.8	6.4

Conditions: 110 °C, p(CO/H_2) = 7.9 bar, [Rh] = 0.6 mM, ligand/Rh = 3, for ligand structure see Figure 5. χ_i-value, see Chapter 1.

The table shows that with increasing accepting ability, i.e. increasing χ-value (the authors used Hammett constants), the rate of the reaction increases and so does the l:b ratio. For phosphites a similar trend had been observed, see Chapter 3. The rate increase can be easily understood, as dissociation of CO will be more facile when π-back-donation decreases. Assuming that steric properties are the same coordination of alkene may also be enhanced by a more electrophilic rhodium center. Rates of migration reactions are usually not affected by electronic changes according to MO studies [21], thus we observe an overall rate increase with electron withdrawing phosphines. Steric properties of the ligands are the same, except that the electronic changes do have an influence on the mode of coordination as we shall see later, but this was not yet known at the time. Substituting trifluoromethyl groups for hydrogens in other arylphosphines (*trans*-2,3-dppm-nor) also has a favorable effect on rate and selectivity [15c].

The l:b ratio of the product aldehyde increases even though electron-withdrawing phosphines give rise to increasing amounts of alkene isomerization. The total preference for linear alkyl formation still increases when the 2-hexene is included in the total amount of intermediary 2-hexylrhodium that must have been formed. Thus, we can say that the intrinsic preference for 1-alkyl formation increases with increasing χ-values of the ligand. Figure 6 explains the influence of 2-hexene formation on the observed l:b ratio in the aldehyde product in a system in which the internal alkene is not converted into (branched) aldehyde. The higher l:b ratio for more positively charged metal centers was supposed to originate from a higher preference for 1-alkyl formation for such centers. An increasing rate of isomerization is a general observation, also for phosphite catalysts. Note that none of the species in Figure 6 has been observed experimentally.

The authors note that under their conditions, involving relatively high temperatures and low pressures, many steps of the cycle have similar rates, as the kinetics are rather complex. The best results were obtained when an excess of phosphine was employed. A ligand to metal ratio of 1.5 was preferred, as at higher ratios the selectivity did not improve further.

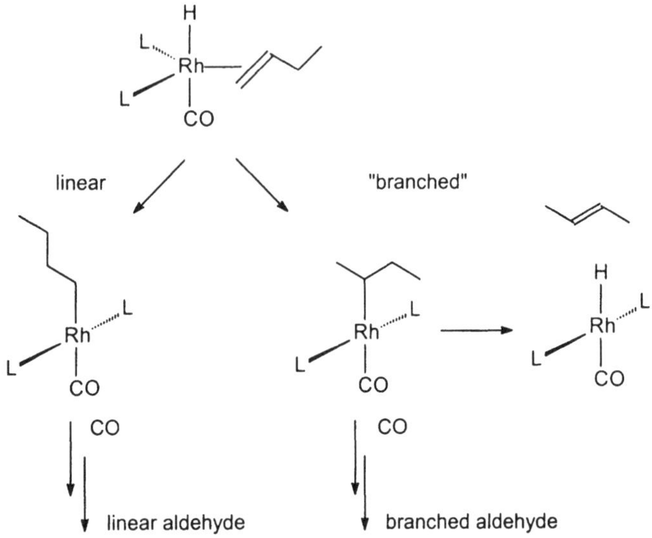

Figure 6. Influence of 2-hexene formation on observed l:b ratio

This led Unruh and Christenson to the conclusion that the most selective species was one containing more than two phosphorus ligands, while the current view was, Wilkinson's view, that complexes containing two phosphines were the ones leading to high l:b ratios of 20. They propose that the most selective species contains three phosphorus atoms, one dppf as a bidentate on one metal and one dppf bridging between two metals. The two metals are to function independently in the catalysis.

Complexes of this type have indeed been identified [41] in ^{31}P NMR studies. A whole range of diphosphines was studied including those of Figure 4, Figure 5, and the common diphosphines dppe and dppp. The study involved the addition of various amounts of the diphosphine to complex **1**. The spectra were all interpreted in terms of the species shown in Figure 7.

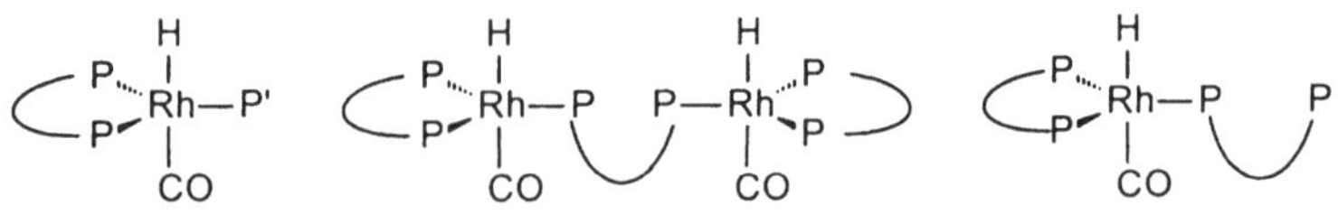

13 **14** **15**

Figure 7. Hughes's phosphine complexes **13**, **14**, and **15**

Unfortunately no 1H NMR data for the hydride signals were reported. The $^{2}J_{P\text{-}P}$ coupling constants of the ^{31}P atoms in the equatorial positions, typically ranging from 100 to 130 Hz, clearly show that in most cases tris-phosphine complexes are present having all phosphorus atoms in the equatorial plane. In a few instances the coupling constants are smaller (50 – 100 Hz) indicating that the phosphines occupy in part an apical position, probably involving a fast exchange. These results support the catalytic results suggesting that three phosphine ligands are coordinated to the catalyst in its resting state. However, it should be borne in mind that the concentrations for the catalysis experiments and the NMR experiments differ considerably. In catalysis the free ligand concentration is at most 1.2 mM ([Rh] = 0.6 mM) and p(CO) = 4 bar, while in the NMR experiments the concentrations are one order of magnitude, maybe two, higher (27 mM for [Rh] and 130 mM for phosphine in one example) and no CO is added. Thus one might wonder whether indeed tris-phosphine species are present under the reaction conditions of the catalytic experiments.

Re-examination of the dppf-Rh system has shown that the coordination behavior under syn-gas is different [42]. Both NMR spectra and *in situ* IR ([Rh] = 8 mM, [dppf] = 17 mM, p(CO) = 20 bar, 40 °C) spectra show that the only species under these conditions is $RhH(dppf)(CO)_2$ being a mixture of rapidly equilibrating **2ae** (approximately 80 %) and **2ee** (20 %). The l:b

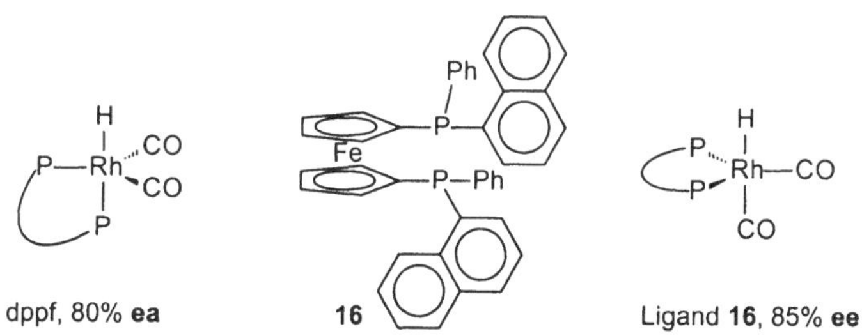

Figure 8. Complexes for dppf and **16**

ratio for 1-octene hydroformylation is 5.4, in accord with the results by Unruh. Interestingly, replacement of two phenyl groups by more bulky 1-naphtyl groups, rendering a S,S-chiral diphosphine **16** (Figure 8), gives a much higher proportion of a structure of type **2ee**, *viz.* 85 %, but the linearity only slightly improves as the l:b ratio is 7.3. Thus, by increasing the steric bulk at phosphorus, and perhaps by introducing the C_2 chirality, the ferrocene ligand can adopt a larger (averaged) bite angle. Ferrocene remains an interesting backbone and perhaps still better results may be achieved.

4.2.3 BISBI ligands and the *natural bite angle*

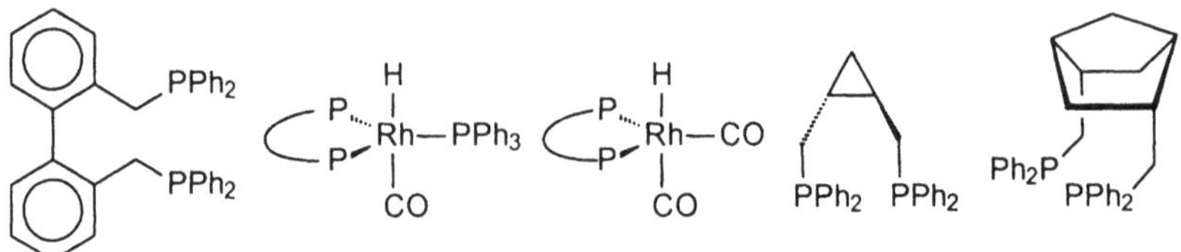

Figure 9. BISBI, **17**, and its complexes **18**, and **19**, and ligands **20** and **21**

In 1987 Devon and co-workers from Texas Eastman reported on a new diphosphine ligand, BISBI **17** (see Figure 9), that shows a very high regioselectivity for the formation of linear aldehydes [43a]. Propene was hydroformylated using BISBI under "standard" conditions in a vapor stripped reactor. The aldehyde product shows a high linearity (l:b = 30). For comparison, other ligands known hitherto gave much lower linearities as shown in Table 3, taken from their work.

Table 3. Hydroformylation results using BISBI and other ligands [43a]

Ligand	Rate ($mol.mol^{-1}.h^{-1}$)	Ratio l:b
BISBI	3650	25.1
trans-dppm-cyb	3200	4.4
trans-2,3-dppm- nor	2550	4.3
dppf	3800	3.6
DIOP	3250	4.0
dppp	600	0.8
PPh_3 [a]	6000	2.4

Conditions: syn gas pressure 16 bar, [Rh] = 1.5 mM, ligand/Rh = 2.4, 95-125 °C, solvent 2,2,4-trimethyl pentane-1,3-diolmonoisobutyrate (Texanol), p(propene) 5 bar.
[a] [Rh] = 0.7 mM, ligand/Rh = 124.

More recently improved variations were described in a later patent [43b]. Instead of diphenylphosphino groups dibenzophosphole groups were used, which often leads to an increase in rate and selectivity. In addition, BISBI was substituted with tertiary butyl groups to increase the solubility of the ligand. Rates were in the same order of magnitude as those mentioned in Table 3. It was also mentioned that the rate of reaction was accelerated by high propene pressures, hinting at the common rate equation for hydroformylation under "standard" conditions. Under optimized conditions (i.a. low CO pressure) l:b ratios as high as 288 were obtained for propene.

Examination of molecular models of metal complexes of BISBI indicates that the chelate bite angle is much greater than 90°, the most common bite angle for bidentate ligands used in coordination chemistry and organometallic chemistry [44]. The work quoted so far, and the results shown in Table 3 led to the speculation that ligands with wider bite angles may perhaps have a fortuitous effect on the l:b ratio in the hydroformylation of 1-alkenes. However, at this time few chelating diphosphines had been proven by X-ray crystallography to be capable of adopting large bite angles. To identify ligands with wide bite angles Casey and Whiteker [45] introduced the concept of the "natural bite angle", which was estimated using molecular mechanics. It is defined as the preferred chelation angle of the bidentate ligand, as determined only by ligand backbone constraints and not by metal valence angles. The neglect of the electronic preferences of the metal for particular P-M-P angles facilitates the calculation, as the force fields for the organic part provide a high degree of accuracy. The natural bite angle is calculated by minimizing the strain energy of the diphosphine-M fragment with a P-M-P bending force constant of 0 $kcal.mol^{-1}.rad^{-2}$, for certain M-P distances. In addition, by fixing the bite angle at various angles by using a large bending force constant and then calculating the strain energy, potential diagrams can be constructed, which allow an estimation of the flexibility of the ligand. The outcome will be different for different programs used, even if the same program and ligand parameters were the same, since dihedral force constants and stretching modes involving the metal may be treated differently. Comparisons with crystal structures are useful, but one should not expect the same outcome [46].

The two wide bite angle ligands studied in the publication of 1992 were BISBI and *trans*-bisdiphenylphosphinocyclopropane (*trans*-dppm-cyp, **20**). Using these ligands the complexes **18** and **19** were synthesized and structurally characterized by X-ray crystallography. This study demonstrated for the first time the unequivocal bis-equatorial coordination of a bidentate diphosphine. The measured bite angle for BISBI in **18** turned out to be 124°, compared to the calculated natural bite angle in this study of 112°. When complexes **18** are subjected to 1 bar of CO the ligand PPh_3 is replaced by CO and **19** is formed. NMR and IR spectroscopy proved that the structure of complexes **19** indeed contained the two phosphorus atoms in the equatorial plane and hydrogen in one of the apical positions.

Yamamoto and Casey separately reported that 2,5-dppm-nor **21** (Figure 9) which has a wide natural bite angle (126°) [47a,b], gave an unexpectedly low l:b ratio of only 2.6, which was attributed to its inability to form stable chelates. Indeed, no monomeric metal complexes of **21** have been reported. Molecular mechanics calculations show that the Rh **21** chelate has 12 kcal more strain than the free ligand; in contrast, BISBI and DPPE chelates are less than 3 kcal more strained than the free ligands. In designing new wide bite angle ligands, it is suggested that the extra strain imposed by chelation be calculated to determine whether the new ligand will from a stable chelate.

Hydroformylation of 1-hexene under mild conditions (34 °C, 6 bar of syn gas, 4mM solution of **19**)gave a l:b ratio of 66.5 for BISBI and for *trans*-dppm-cyp it was 12. DIOP and dppe afforded values of 8.5 and 2.1. No isomerization was observed. It was concluded that apparently a bis-equatorial coordination of a diphosphine lead to high l:b ratios, although the reason for this was not understood in detail.

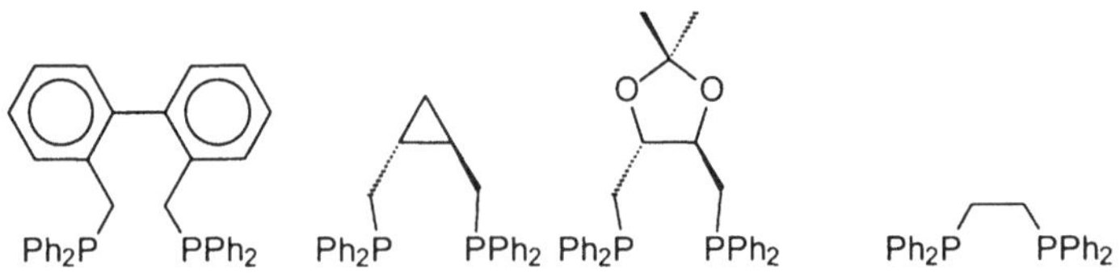

bite angle	113°	107°	102°	85°
l:b	66	12	8.5	2.1

Figure 10. Chart showing bite angles and selectivities

The correlation between natural bite angle and l:b ratio is very interesting and was an important finding for future work. Thus, complexes **2ee** (Figure 1) lead to higher l:b values than complexes having structure **2ae**, or alternatively ligands with larger bite angles lead to more effective steric bulk of the diphosphine and a relative destabilization of branched alkyl intermediates.

Before developing detailed explanations of the control of regiochemistry based on steric or electronic differences between bis-equatorial and equatorial-apical diphosphine complexes it was considered crucial to test whether alkene insertion to produce 1-alkyl and 2-alkyl rhodium species was irreversible [47c]. For instance, previous work on the deuterioformylation of cyclohexene using bulky phosphite had shown that H-D exchange was at least ten times faster than formylation (40-80 °C, < 10 bar) [48], but for 1-alkenes insertion is irreversible, provided that the CO concentration is maintained at a sufficiently high level [49]. Similar results have been reported for unmodified carbonyl catalysts (see Chapter 2). In the alternative case of fast alkene exchange, the regioselectivity would depend upon the equilibrium of 1- and 2-alkylrhodium species and their respective rates of CO insertion. Deuterioformylation of 1-hexene at 6 bar and relatively low temperature (32 °C) has shown that insertion of the 1-alkene is an irreversible process as very little deuterium was incorporated in recovered alkenes. The small amounts of secondary alkyl chains that are formed lead mainly to (deuterated) 1- or 2-hexene and only 25 % proceeds to branched aldehyde. Little deuteration α to the carbonyl is observed indicating that alkene dissociation occurs after β-hydrogen elimination. These results establish that the regiochemistry of aldehyde formation is set largely by the rhodium hydride addition to coordinated 1-alkene. Plausible steric explanations were considered and a whole range of intermediates for diequatorial and equatorial-apical resting states were discussed, but molecular mechanics calculations failed to support such steric explanations. We will return to this in section 4.2.5.

Since steric arguments failed to give a satisfactory explanation for the much higher l:b aldehyde ratios observed from diequatorial chelates like BISBI compared with apical-equatorial chelates like dppe, it was considered whether the electronic differences between diequatorial chelates and apical-equatorial chelates are responsible for the differences in regioselectivity. Because the electronic interaction between two apical ligands or between two equatorial ligands is stronger than the interaction between apical and equatorial ligands, electronic differences can be expected between diequatorial chelated complexes and apical-equatorial chelated complexes. For example, back bonding from rhodium to the alkene ligand in the equatorial plane would be expected to be stronger for the BISBI complex with two strong donor phosphines in the equatorial plane than for the dppe complex with only a single donor phosphine in the equatorial plane. Also, the apical hydride of the BISBI complex is trans to a CO ligand and would be expected to be more acidic than the hydride of the dppe complex which is trans to a phosphine. Previously, Unruh (see 4.2.2) reported favorable effects of CF_3 substitution at the aryl rings on phosphorus.

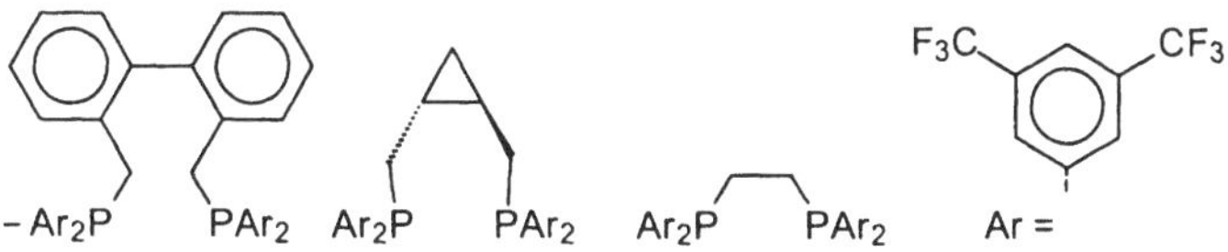

22 **23** **24**

Figure 11. Ligands having electron withdrawing substituents

To investigate electronic effects of equatorial and apical phosphines, analogs of chelating diphosphines of BISBI, *trans*-dppm-cyp, and dppe having electron-withdrawing substituents on the aryl rings were synthesized and used in hydroformylation [50] (see Figure 11, **22-24**). The introduction of electron withdrawing substituents on the aryl rings of the diequatorial chelate **22** leads to an *increase* in linear aldehyde selectivity as well as rate. In contrast, introduction of electron withdrawing substituents on the aryl rings of the apical-equatorial chelate **24** resulted in a *decrease* in linear aldehyde selectivity when compared with the phenyl-substituted dppe.

Table 4. Regiochemistry of Hydroformylation of 1-Hexene with Rhodium Diphosphine Catalysts

diphosphine	ee:ae[a]	l:b[b]	Rate[c]
BISBI	100:0	66.5	12.4
22	100:0	123	61.9
Trans-dppm-cyp	37:63	12.1	3.7
23	90:10	17.7	13.7
dppe	0:100	2.6	3.5
24	0:100	1.3	4.3

[a] Ratio of diequatorial:apical-equatorial chelates of (diphosphine)$Ir(CO)_2H$ at room temperature.[b] Mol heptanal:mol 2-methylhexanal.[c] Turnover rate = [mol aldehyde] × [mol $Rh]^{-1}$ h^{-1}.

Interestingly, electron-withdrawing groups in equatorial and apical positions had opposite effects on l:b regioselectivity. Hydroformylation results with the electron withdrawing *diequatorial* coordinated ligands **22** and **23** showed an increase in the n:i ratio, while *apical-equatorial* coordinated ligand **24** showed a decrease in the l:b ratio. Thus, electron withdrawing aryl substituents in the equatorial position lead to higher l:b regioselectivity and in the apical position to lower regioselectivity.

It was hypothesized that a dissymmetric dppe derivative ($Ar_2PCH_2CH_2PAr'_2$) could lead to greater l:b selectivity than either of the related symmetric dppe derivatives. A chelate designed to have an electron withdrawing phosphine in the equatorial position and a electron-donating phosphine in the apical position should give higher l:b regioselectivity than either of the parent symmetric dppe compounds.

The new ligand **25** was synthesized and both its coordination behavior and selectivity in catalysis were studied [51]. Ligand **25** shows a l:b ratio of 4.2, higher than either of the symmetric ligands dppe or **24**, supporting our hypothesis that creating an electronically dissymmetric environment about our hydroformylation catalysts enhances linear aldehyde selectivity. An electron-withdrawing substituent on an equatorial phosphine increases the l:b ratio while an electron-withdrawing substituent on an apical phosphine decreases the l:b ratio. Therefore, the best dppe catalysts place an electron-

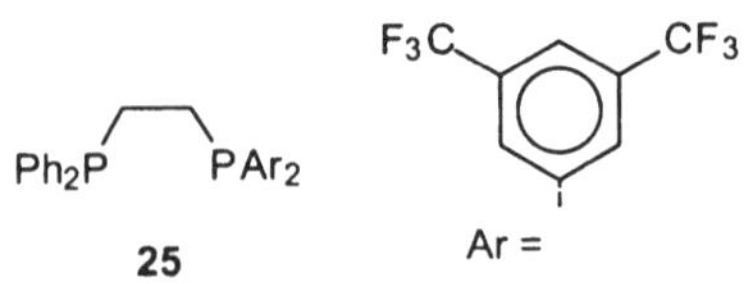

Figure 12. Dissymmetric dppe derivative

withdrawing substituted phosphine in the equatorial position and an electron-donating substituted phosphine in the apical position. Since ligand **24** gave a lower l/b than dppe, the effect of the electron-withdrawing phosphine in the apical position to decrease l/b must override the increase in l/b caused by the electron-withdrawing phosphine in the equatorial position.

However, an inconsistency emerges if the macroscopic differences between BISBI and dppe catalysts are considered. The macroscopic changes in going from H(alkene)Rh(CO)(BISBI) to H(alkene)Rh(CO)(dppe) involve replacement of an equatorial phosphine (one equatorial BISBI arm) with a strongly electron accepting CO ligand and replacement of an apical CO ligand by a strongly electron donating phosphine ligand (apical dppe arm). The second arm of each diphosphine, in the equatorial position, remains unchanged. Overall, the replacement of BISBI by dppe places a better acceptor ligand in the equatorial position and a better donor ligand in the apical position. By analogy to the effect of electron donor aryl substituents, both of these changes should have resulted in higher regioselectivity not in the observed lower regioselectivity seen for dppe. Clearly, the effects on regioselectivity for equatorial electron-withdrawing phosphines and π-accepting CO ligands are not identical.

4.2.4 Xantphos ligands: tunable bite angles

In 1990 a program was started at the University of Amsterdam aiming at the synthesis of ligands having bite angles outside the common range of 75 to 99°. It was thought that ligands having bite angles wider than this might induce interesting catalysis by stabilizing or destabilizing initial states, transition states or final states of catalytic reactions catalyzed by square-planar, tetrahedral, or trigonal-bipyramidal metal-ligand complexes. Molecular mechanics methods, as developed in Shell [52], were used for the design of the ligands, but after the publication by Casey and Whiteker [45] the latter method was adopted. A variety of backbones have been designed supporting wider bite angles, but often the synthesis was tedious.

Readily available backbones turned out to be those based on xanthene and related compounds (Figure 13). The diphosphines derived from these cover a range of angles from 102 to 121°. Key feature is the oxygen ether atom in the bridge that prevents metallation and that in many instances will not participate in the coordination to the metal. By changing the fragment in the back of the molecule (C_1, C_2, N, S) the bite angle can be easily tuned to other values. The calculated natural bite angles are shown in Figure 13 and Table 5. In cationic metal complexes the oxygen will often participate in the coordination to the metal, which leads to P-M-P bite angles around 165°.

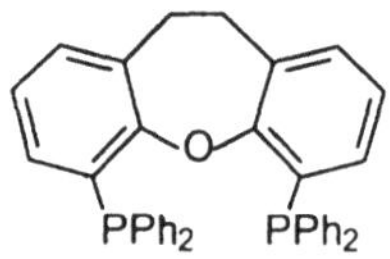

Homoxantphos (**26**) 102.0°

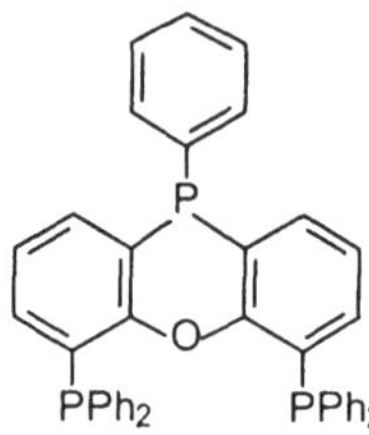

Phosxantphos (**27**) 107.9°

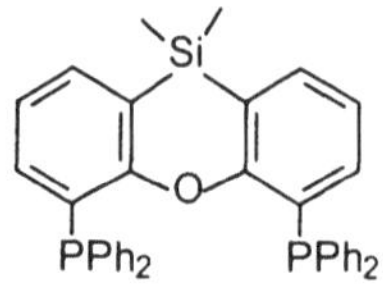

Sixantphos (**28**) 108.5°

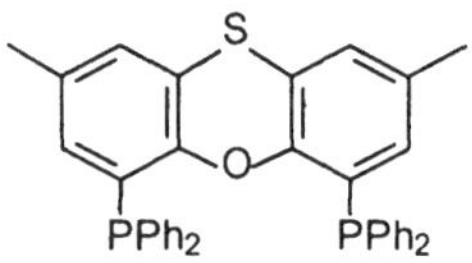

Thixantphos (**29**) 109.6°

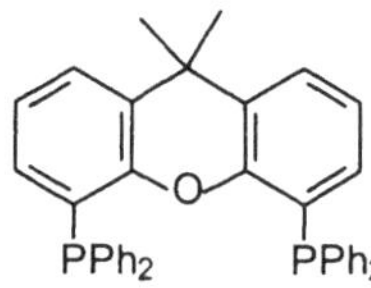

Xantphos (**30**) 111.4°

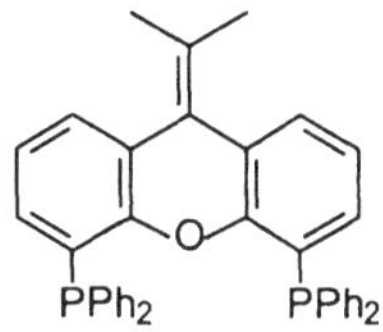

Isopropxantphos (**31**) 113.2°

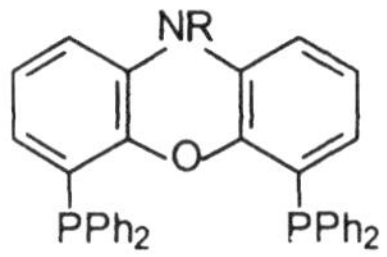

R = H, Nixantphos (**32**) 114.1°
R = Bn, Benzylnixantphos (**33**) 114.2

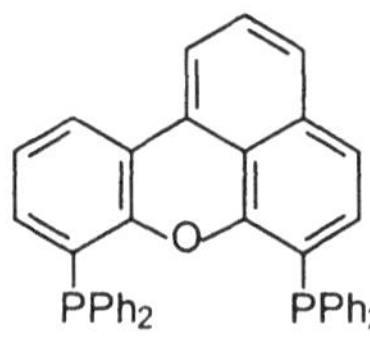

Benzoxantphos (**34**) 120.6°

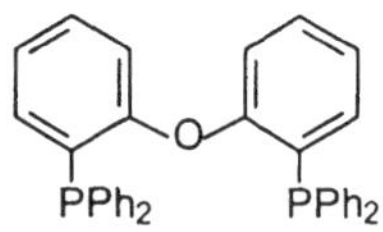

DPEphos (**35**) 102°

Figure 13. Xantphos type ligands and their calculated β_n values

Hydroformylation. Changing the bite angle does usually affect the catalytic reaction, be it the rate or the selectivity, and in several cases a favorable effect was reported [46] for Xantphos ligands. For rhodium catalyzed hydroformylation the first results were published in 1995 by Kranenburg et al. [53]. In Table 5 we have combined his results (**26-34**) with those of Van der Veen to complete the list of Xantphos ligands tested to date [54]. From the table one can see that the selectivity for linear aldehyde increases with wider bite angles. For ligands **26-30** the l:b ratio increases, but at higher bite angles the correlation breaks down. The rate of reaction increases with increasing bite angle (with the exception of ligands **31-33**) as one might expect since two phosphines with a small bite angle are a better donor pair than two phosphines having a wider bite angle. Ligand **34**, which will adopt a flattened confirmation compared with the other ligands, gives an interesting result in that a low isomerization rate is observed. Ligand **26** assumes a local C_2 type symmetry, in contrast to the other ligands all having C_s symmetry. Ligand **35**, the diphosphine made from diphenyl ether, gives a moderate linear branched ratio of 6.7 but no isomerization [53] at a pressure of 10 bar.

Table 5. Hydroformylation of 1-octene using Xantphos ligands[a]

Ligand	β_n, deg	ee:ae[d]	l:b ratio[b]	% lin. ald.[b]	% isomer.[b]	tof[b,c]
26	102.0	3:7	8.5	88.2	1.4	37
27	107.9	7:3	14.6	89.7	4.2	74
28	108.5	6:4	34.6	94.3	3.0	81
29	109.6	7:3	50.0	93.2	4.9	110
30	111.4	7:3	52.2	94.5	3.6	187
31	113.2	8:2	49.8	94.3	3.8	162
32	114.1	7:3	50.6	94.3	3.9	154
33	114.2	8:2	69.4	94.9	3.7	160
34	120.6	6:4	50.2	96.5	1.6	343

[a] Conditions: $CO/H_2 = 1$, $P(CO/H_2) = 20$ bar, ligand/Rh = 5, substrate/Rh = 637, [Rh] = 1.00 mM, number of experiments = 3. In none of the experiments was hydrogenation observed. [b] Linear over branched ratio, percentage linear aldehyde, percent isomerization to 2-octene, and turnover frequency were determined at 20% alkene conversion. [c] Turnover frequency = (mol of aldehyde) (mol of Rh)$^{-1}$ h^{-1}. [d] Calculated from $J_{P\text{-}H}$ NMR data.

The X-ray crystal structures of (**33**)Rh(CO)H(PPh_3) and (**34**)Rh(CO)H(PPh_3) were determined and both structures reveal distorted trigonal bipyramidal complex geometries with all the phosphines occupying equatorial sites. The observed P-Rh-P bite angles for ligands **33** and **34** in the crystal structures are 110.2° and 109.2°, respectively. The observed bite angles are within the flexibility ranges calculated for the ligands and close to the bite angle observed in the complex containing ligand **29** (111.7°) [55], but especially the angle for **34** is smaller than the calculated one (120.6°). The bite angle of approximately 110° found in all three crystal structures, is probably the result of stereoelectronic interactions between the diphosphine and the large triphenylphosphine ligand. When this PPh_3 ligand is replaced by CO and subsequently by alkene during the catalytic cycle, the P-Rh-P angle presumably can increase as a result of decreased steric congestion. Furthermore, the crystal structures show several possibilities for π-π-stacking interactions between the different phenyl moieties, which could also influence the final complex geometry.

Electronic effects. To study the nature of the electronic effect in the rhodium diphosphine catalyzed hydroformylation, a series of thixantphos **29** ligands with varying basicity was synthesized **36-41** (Figure 14). In this series of ligands, steric differences are minimal so purely electronic effects could be investigated. To investigate the influence of phosphine basicity on chelation behavior, the solution structures of the (diphosphine)Rh(CO)H(PPh_3) complexes and the (diphosphine)Rh(H)$(CO)_2$ complexes were studied in great detail for all ligands. Displacement of PPh_3 in $(PPh_3)_3$Rh(CO)H by the diphosphines **29** and **36-41** resulted in formation of the (diphosphine)Rh(CO)H(PPh_3) complexes. Crystals of the (diphosphine)Rh(CO)H(PPh_3) complexes suitable for X-ray structure determination were obtained for ligands **29** and **37**.

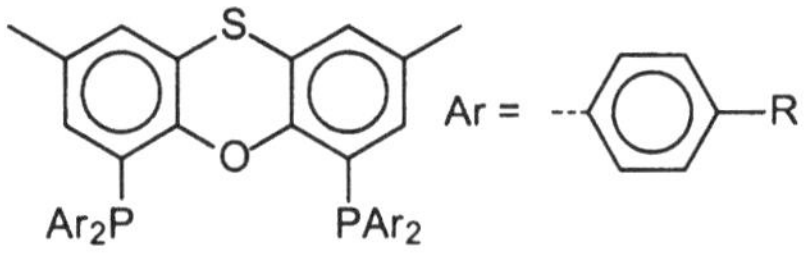

R	
R = $N(CH_3)_2$	**36**
OCH_3	**37**
CH_3	**38**
H	**29**
F	**39**
Cl	**40**
CF_3	**41**

Figure 14. Substituted thixantphos ligands

The X-ray crystal structures of (**29**)Rh(CO)H(PPh_3) and (**37**)Rh(CO)H(PPh_3) are very similar. The P-Rh-P bite angles are 111.7° for the thixantphos ligand **29** and 109.3° for the *p*-CH_3O-thixantphos ligand **37**. The small difference in observed bite angles confirms the assumption based on molecular mechanics that *p*-substitution on the aryl rings has only minor steric effects. The better donor prefers a smaller bite angle. A noticeable difference between comparable bond lengths in the complexes is found in the Rh-CO distance. The decreased Rh-CO bond length for the *p*-CH_3O-thixantphos complex (1.86 Å) compared to the thixantphos complex (1.89 Å) is a result of increased back-bonding of rhodium to the carbonyl ligand.

In situ IR spectra. Unambiguous evidence for the presence of **ee** and **ae** complex isomers was supplied by HP-IR spectroscopy. The spectra of the (diphosphine)Rh$(CO)_2$H complexes all showed four absorption bands in the carbonyl region (Figure 15). In an effort to assign the bands to **ee** and **ae** isomers the (thixantphos)Rh$(CO)_2$D complex was measured for comparison. Upon H/D exchange, only ν_1 and ν_3 shift to lower wavenumbers (respectively 18 and 14 cm^{-1}), and therefore, these two bands are assigned to the carbonyls of the **ee** complex. The two bands that do not shift, ν_2 and ν_4, belong to the **ae** complex. From the disappearance of a low-frequency shoulder upon H/D exchange, it can be concluded that one of the rhodium hydride vibrations is partly hidden under ν_4. DFT calculations were fully in line with the observed shifts and wavenumbers, even for the Rh-H stretching frequencies. The shift in isomer composition found in the NMR spectra (*vide infra*) was confirmed when comparing the IR spectra of the series of complexes (Figure 15).

During hydroformylation the IR spectra of the rhodium species (diphosphine)Rh$(CO)_2$H for ligands **37**, **29**, and **41** do not change. The respective **ee**:**ae** equilibria are not influenced, and no other carbonyl complexes are observed.

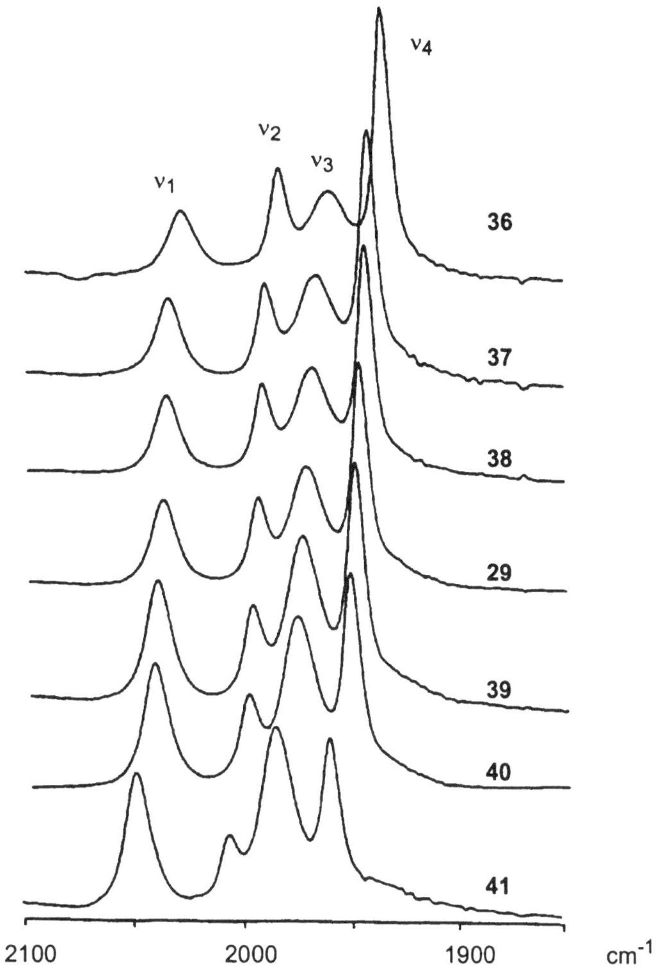

Figure 15. IR spectra of complexes RhH(ligand)CO (ligand = **36-41, 29**)

NMR spectra. For all diphosphine ligands, the formation of the (diphosphine)Rh$(CO)_2$H complexes was evidenced by ^{31}P NMR spectra and an apparent triplet of doublets for the rhodium hydride in the ^{1}H NMR spectra. Decreasing phosphine basicity gives decreasing rhodium-proton (^{1}J(Rh,H)) and phosphorus-proton (^{2}J(P,H)) coupling constants but increasing rhodium-phosphorus (^{1}J(Rh,P)) coupling constants. The rhodium-proton and phosphorus-proton coupling constants and their dependence on phosphine basicity also show that these complexes exist as mixtures of **ee** and **ae** isomers that are in rapid, dynamic equilibrium.

DFT calculations. The enhanced preference of the basic phosphines for apical coordination and the preference of the nonbasic phosphines for equatorial coordination was demonstrated by DFT calculations on the **ee** and

ae isomers of complexes $(PH_3)_2Rh(CO)_2H$ and $(PF_3)_2Rh(CO)_2H$. The energies of the **ee** and **ae** isomer of the complex modified with PH_3 as model ligand for a basic phosphine were the same within 0.1 kcal mol^{-1}. Uisng PF_3 as a model ligand for a nonbasic phosphine the **ee** isomer was found to be favored by 2.4 kcal mol^{-1}. This corresponds to an **ee**:**ae** equilibrium ratio of approximately 50:1 at room temperature. This trend is consistent with the experiments.

Hydroformylation results in Table 6 show that, with the exception of ligands **39** and **40**, the rate of the reaction increases with decreasing phosphine basicity. An explanation for the deviant behavior of **39** and **40** can be incomplete catalyst formation or deactivation of the catalyst. Decreasing phosphine basicity facilitates CO dissociation from the (diphosphine)$Rh(CO)_2H$ complex and enhances alkene coordination to form the (diphosphine)Rh(CO)H(alkene) complex, and therefore, the reaction rate increases.

Table 6. Electronic effects in hydroformylation using thixantphos derivatives

ligand	χ_i	ee:ae	l:b	2-octene, %	nonanal, %	t.o.f.
36	1.7	47:53	44.6	4.8	93.1	28
37	3.4	59:41	36.9	5.3	92.1	45
38	3.5	66:34	44.4	4.7	93.2	78
29	4.3	72:28	50.0	4.9	93.2	110
39	5.0	79:21	51.5	5.7	92.5	75
40	5.6	85:15	67.5	6.9	91.7	66
41	6.4	92:8	86.5	6.8	92.1	158

1-octene, 80 °C, 20 bar of 1:1 CO/H_2, 1.0 mM rhodium (diphosphine/Rh = 5)

While the l:b ratio increases with the χ-value, no electronic effect of the ligands on the selectivity for linear aldehyde is observed. The selectivities for linear aldehyde are all between 92 and 93%. The increase in l:b ratio with decreasing phosphine basicity can be attributed completely to an increased tendency of the branched alkyl rhodium species to form 2-octene instead of branched aldehyde (Figure 6). The increasing electrophilicity of the rhodium center leads to a higher reactivity of the rhodium alkyl species toward CO dissociation and β-hydrogen elimination [15a]. The total amount of all other products (branched aldehyde and 2-octene) is 7-8% for all ligands.

The constant selectivity for linear aldehyde in the hydroformylation of 1-octene implies that for the basic ligands the l:b ratio reflects the regioselectivity of the formation of the rhodium alkyl species. For the less basic ligands the increase in l:b ratio results from the different behavior of branched and linear rhodium alkyls toward -hydrogen elimination, as was already reported by Lazzaroni and co-workers for the deuterioformylation of 1-hexene [57]. The linear alkyl is mainly converted to linear aldehyde, while the branched alkyl partially generates 2-octene. Since 2-octene is far less reactive in the hydroformylation, its formation is irreversible in these experiments. We conclude that for this catalytic system the ratio of linear to branched rhodium alkyl species formed is determined not by phosphine basicity but by steric constraints only.

Kinetics of ^{13}CO dissociation from (Diphosphine)Rh(^{13}CO)$_2$H Complexes. The rate determining step in the hydroformylation of 1-octene with xantphos-type ligands is in an early stage in the catalytic cycle. Dissociation of CO, alkene coordination, and migratory insertion are relevant for the determination of the rate. So far, only the first step has been studied separately [54]. An experimental set-up was introduced enabling the measurement of the rate constants for CO dissociation from the (diphosphine)Rh(CO)$_2$H complexes. By exchanging the ^{12}CO ligands for isotopically pure ^{13}CO ligands, the carbonyl absorptions in the IR spectra of the (diphosphine)Rh(CO)$_2$H complexes shift 30-40 cm^{-1} to lower energy.

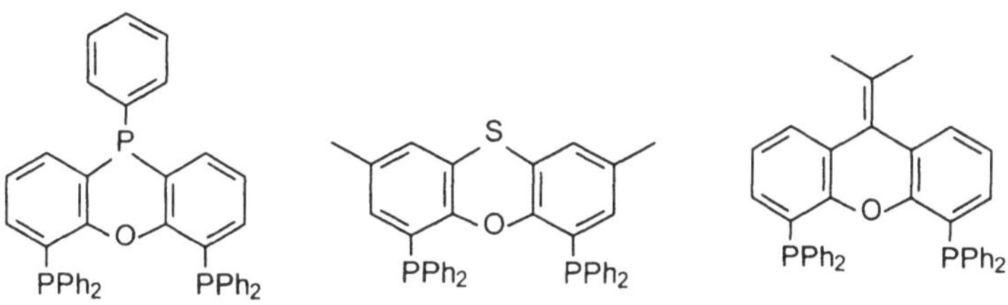

bite angle 107.9 | 109.6 | 113.2

27 | **29** | **31**

Figure 16. Diphosphines used for CO dissociation studies

The rate constants k_1 for the dissociation of ^{13}CO from the (diphosphine)Rh(^{13}CO)$_2$H complexes were determined by monitoring the exchange of ^{13}CO for ^{12}CO by exposing the ^{13}CO labeled complexes to a large excess of ^{12}CO, so any dissociated ^{13}CO will be replaced quantitatively by ^{12}CO. The ligands are shown in Figure 16.

To investigate the possible effect of the natural bite angle on the rate of CO dissociation from the (diphosphine)Rh(CO)$_2$H complexes ^{13}CO exchange was measured for complexes containing ligands **27**, **29** and **31**. The exchange of ^{13}CO by ^{12}CO in the (diphosphine)Rh(^{13}CO)$_2$H complexes was monitored by rapid-scan HP IR spectroscopy at 40 °C. The $^{13}CO/^{12}CO$ exchange was initiated by adding a large excess of ^{12}CO. The difference between the reactivity of the **ee** and **ae** isomers of the (diphosphine)Rh(^{13}CO)$_2$H complexes cannot be determined, since the exchange between **ee** and **ae** isomers is several orders of magnitude faster than the rates of ^{13}CO dissociation, as is evident from the NMR spectra.

Table 7. Kinetics of ^{13}CO dissociation in (diphosphine)Rh($^{13}CO)_2$H complexes[a]

Ligand	Bite angle, deg.	P(^{12}CO), bar	k_1, h^{-1}	Rate, t.o.f[b]
27	107.9	25	177	74
29	109.6	25	206	110
29[c]		25	200	
31	113.2	20	182	162
31		25	185	
31		30	181	
43[d]	123.1	25	1200	1560

[a] Reaction conditions: [(diphosphine)Rh($^{13}CO)_2$H] = 2.00 mM in cyclohexane, $P(^{13}CO)$ = 1 bar, $P(H_2)$ = 4 bar, T = 40 °C, diphosphine/Rh = 5. [b] In mol.mol^{-1}.h^{-1} at 80 °C and 20 bar CO/H_2 [c] [(diphosphine)Rh($^{13}CO)_2$H] = 3.00 mM. [d] See Figure 17 below.

The observed rate constants, k_1, are listed in Table 7. It is commonly accepted that CO exchange in (diphosphine)Rh$(CO)_2$H complexes proceeds via the dissociative pathway. The decay of the carbonyl resonances of the (diphosphine)Rh($^{13}CO)_2$H complexes indeed followed simple first-order kinetics. The experiments with ligand **31** at different ^{12}CO partial pressure show that the rate of CO displacement is independent of the CO pressure. Furthermore, the rate is also independent of the (diphosphine)Rh($^{13}CO)_2$H complex concentration, as demonstrated by the experiments with ligand **29**. It can therefore be concluded that CO dissociation for these complexes obeys a first-order rate-law and proceeds by a purely dissociative mechanism.

Table 7 shows that there is no correlation between the rate of CO dissociation and the natural bite angle. The rate constants k_1 for ligands **27**, **29**, and **31** do not differ significantly and cannot explain the trend in observed hydroformylation activities. The comparison of the k_1 values for ligand **27**, **29**, and **31** with the turnover frequencies depicted in Table 5, reveal that the rates of CO dissociation, measured at 40 °C, are higher than the hydroformylation rates at 80 °C. Since reaction rates increase approximately an order of magnitude with a temperature rise of 20 degrees, the CO dissociation rate at 80 °C is about 100 times as fast as the hydroformylation reaction.

In this study no influence of the natural bite angle on the rate of formation of the square planar (diphosphine)Rh(CO)H intermediate (k_1) was found, implying that the activation energy for the formation of these complexes is not affected significantly. Therefore, the increase in hydroformylation rate with increasing bite angle might originate either from an increase in the concentration of these four-coordinate complexes, or from a decrease in the activation energy for alkene coordination (if this step were rate-determining), or a faster migratory insertion of the alkene.

It is not likely that increasing the bite angle would lower the activation energy for alkene coordination. Increasing the bite angle results in increased steric congestion around the rhodium center and consequently in more steric hindrance for the alkene entering the coordination sphere. What kind of electronic effect the widening of the bite angle has on the activation energy for alkene coordination is unclear, since it depends on the bonding mode of the alkene. Rhodium to alkene back-donation is promoted by narrow bite angles, while alkene to rhodium donation is enhanced by wide bite angles [62].

Widening the bite angle of a *cis* bidentate in a square planar complex would certainly accelerate a migration reaction, but is is not clear how this would work out in trigonal bipyramid having the diphosphine as a bis-equatorial ligand. We will return to this in section 4.2.5.

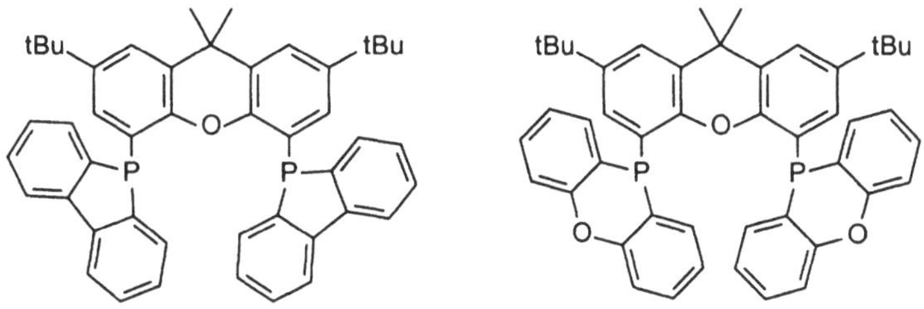

bite angle 121.4 123.1

42 **43**

Figure 17. New phosphole and phenoxaphosphino xantphos ligands

Internal alkenes. Dibenzophosphole- and phenoxaphosphino substituted xantphos ligands **42** and **43** were reported by van der Veen et al. [58, 59] (Figure 17). They show a high activity and selectivity in the rhodium catalyzed linear hydroformylation of 1-octene (l:b = > 60). More importantly, ligands **42** and **43** exhibit an unprecedented high activity and selectivity in the hydroformylation of trans 2- and 4-octene to linear nonanal.

Table 8. Hydroformylation of 2- and 4-octene[a,d]

Ligand	Substrate	l:b ratio[b]	% linear ald	t.o.f[c].
PPh_3	2-octene	0.9	46	39
42		9.5	90	65
43		9.2	90	112
PPh_3	4-octene	0.3	23	2
42		6.1	86	15
43		4.4	81	20

[a] [Rh] = 1.0 mM, Rh:P:octene = 1:10:673. [b] l:b Ratio includes all branched aldehydes. [c] Turnover frequencies were calculated as (mol aldehyde)(mol catalyst)$^{-1}$(h)$^{-1}$ at 20-30% conversion. [d] 120 °C and p = 2 bar.

They constitute the first rhodium phosphine modified catalysts for such a selective linear hydroformylation of internal alkenes. The extraordinary high activity of **43** even places it among the most active diphosphines known. Since large steric differences in the catalyst complexes of these two ligands are not anticipated, the higher activity of **43** compared to **42** might be ascribed to very subtle bite angle effects or electronic characteristics of the phosphorus heterocycles.

The high l:b ratios of ligands **42** and **43** are in good agreement with earlier observations that diphosphines with natural bite angles close to 120° give very selective catalysts. The calculated bite angles are largely due to the interference of the rigid substituent rings at phosphorus. Table 8 shows the acitivities and selectivities achieved for internal alkenes. To enhance the rate of isomerization and to prevent hydroformylation of the internal alkenes, a relatively high temperature and low pressure were used. The dissociation rate of CO from the resting state (**43**)$RhH(CO)_2$ was very fast, several times faster than that found for diphenylphosphino Xantphos ligands (see Table 7). Thus larger (natural) bite angles promote the formation of four-coordinate species, as may be expected. For the same reason they will enhance isomerization, since this involves a competition between CO association and β-hydrogen abstraction by the vacant site.

4.2.5 The mechanism, regioselectivity, and the bite angle Concluding remarks

Summary of ligand effects. A wider bite angle leads to higher l:b ratios and to a higher proportion of bis-equatorial hydride resting states. Above a certain value of β_n or the **ee:ae** ratio the effect on the l:b ratio levels off. It might be that in a region of 104 to 113° a change in mechanism occurs.

When the bite angle of the diphosphine is changed with large increments from 78° (dppe), to 107° (*trans*-dppm-cyp), to 112° (BISBI or Xantphos), the trend is clear; a higher l:b ratio is found. The same applies to electronic effects in symmetrically substituted diphosphines; a higher χ-value leads to higher l:b ratios. However, the latter was not true for the Xantphos series shown in Figure 14 and Table 6 when the selectivity is considered instead of the l:b ratio. Wider bite angles and higher χ-values both lead to faster catalysts and a relatively higher rate of isomerization. For propene the isomerization is immaterial and as a consequence very high linearities can be obtained using ligands displaying these properties. When isomerization becomes sufficiently fast, even internal alkenes can be converted to linear products, albeit at moderate rates (ligands **42** and **43**).

Bis-equatorial coordination as in **2ee** is not a prerequisite for high linearity, as selective catalysts show a substantial amount of **2ae**. On the other hand, ligands showing purely equatorial-apical coordination give the lowest l:b ratios, even lower than monophosphines as PPh_3. Monodentate phosphines can always give trans-like intermediates, while ligands having a small angle cannot (intermediates **4ae** and **5c** in Figure 19).

Ligands having high χ-values prefer equatorial sites and lead to higher l:b ratios. Dissymmetrically substituted dppe ligands give, indeed, the acceptor arm of the bidentate in the equatorial plane together with an electron-accepting CO ligand, but the unexpected result in hydroformylation is a lower l:b ratio.

After dissociation of carbon monoxide (or a phosphine) the unsaturated intermediates **3c**, **3t**, **10c**, and **10t** may form, which will subsequently undergo association with alkene to give species **4.**

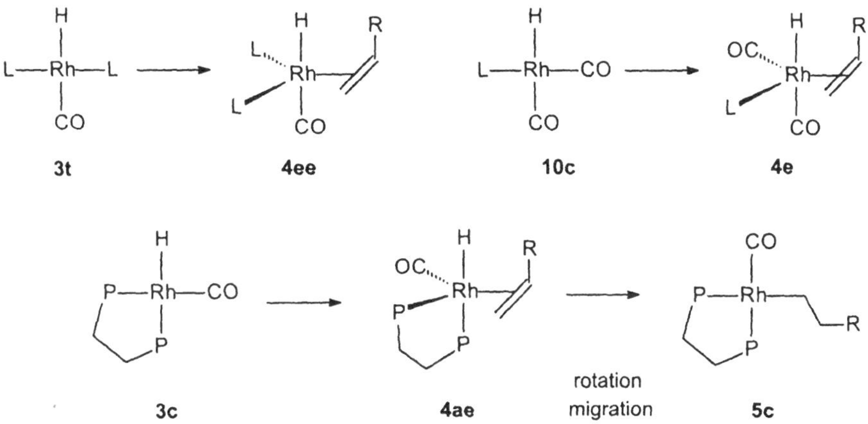

Figure 18. Migratory insertion in tbp-like structures for mono and bidentate dppp

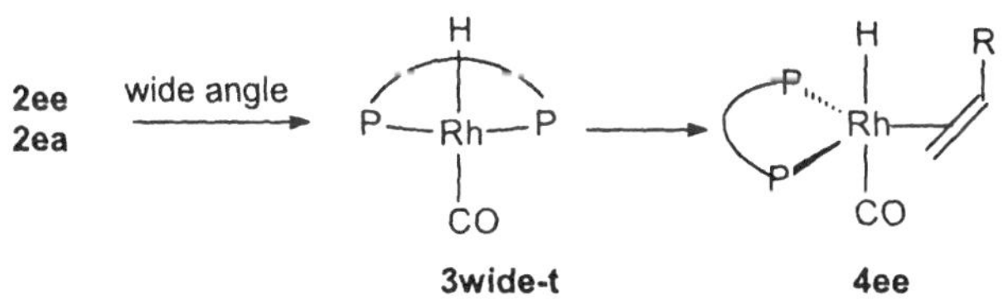

Figure 19. Structure of key intermediate **3wide-t**

Bidentate ligands having wide bite angles will lead to species similar to **3t** and **4ee**. Ligands with wide bite, such as BISBI and Xantphos, should impose little deformation of the *trans* P-Rh-P angle of **4t**. Indeed, in $RhH(CO)(PCy_3)_2$, the P-Rh-P angle is 164° [56], which seems well within reach of the ligands having wide bite angles such as BISBI and Xantphos [45, 46]. This species **3wide-t** associates with the alkene to give **4ee**. The overall large steric hindrance in such a complex could be sufficient explanation for the selective formation of a primary rhodium alkyl species. Ligands with bite angles in excess of 125°, such as **21** [58] and **44** [57] (see Figure 20), lead to low linearities [53]. They do not form the desired species **2ee** [47].

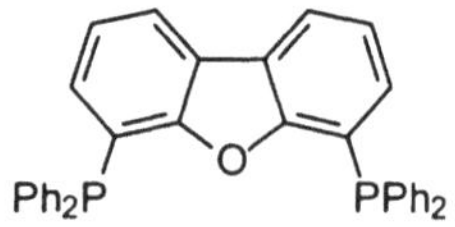

44

Figure 20. Structure of dibenzofuran based ligand with $\beta_n = 131°$

Transition state geometry. As we have seen above, a better knowledge of the geometry of the transition state is of the utmost importance before we can arrive at a better understanding of the "bite angle" effect. Two theoretical studies have been devoted to the mechanism of the hydroformylation reaction, one by Morokuma [60] and the other by Gleich [61]. Using DFT methods Morokuma finds that the trigonal bipyramid containing an apical hydride and ethene in the equatorial plane has to rearrange to a square pyramid having both hydride and ethene in the basal plane prior to migration. This leads to a structure that has an orbital of dx^2-y^2 character as the empty orbital, a situation very similar to insertion processes in palladium(II) complexes.

In Figure 21 we have depicted the square pyramidal intermediate containing the diphosphine in the basal plane, together with the hydride and ethene ligand. A carbonyl is drawn at the apical position, as was found to be the most stable geometry. The directing effect of the bidentate is now very similar to that in square planar complexes. The alkyl substituent of the alkene will experience steric hindrance from the diphosphine, in the basal plane and the most favorable position for the alkyl group will be away from the ligand, leading to linear alkyl formation when migration takes place. During migration the carbonyl moves to the vacancy left by the hydride. The wider the bite angle, the stronger the steric hindrance forcing the alkene in the desired position. Morokuma reported that the rearrangement from **4ee** or **4ae** involves most of the activation energy, rather than the bond breaking, which points to an early transition state in which non-bonding interactions are most important. For wide bite angles one would expect a *trans* disposition of the diphosphine and as a result there is no *cis* relation between hydride and alkene. For bidentate ligands the movements involved seem unlikely and therefore we reject this transition state.

Gleich states that the geometry is less well defined and that just as well one can describe the transition state as a tbp structure in which the alkene assumes a position perpendicular to the trigonal plane in the transition state.

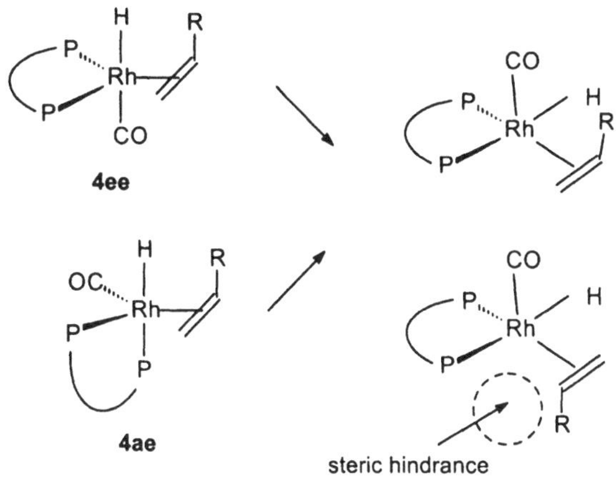

Figure 21. Square pyramidal transition state

The study of Gleich et al. involves a combination of quantum mechanics to calculate the geometry of the transition state and molecular mechanics to calculate the steric effects of the ligands. The tbp case is shown above in Figure 19. A wider bite angle will create more steric hindrance around the metal and a linear alkyl will result, but this could not be substantiated by molecular mechanics calculations using an estimated transition state [47c]. Thus according to the latter study dppe gives a calculated l:b ratio even higher than BISBI, as we hypothesized above for **4ae**, which might mean that there is another intermediate involved having a lower barrier and giving a low l:b ratio.

Table 9. Experimental versus calculated regioselectivities [61]

Ligand	l:b, exp.	l:b, calc.
BISBI	17.0	8.0
dppe	2.2	3.4
PPh_3	3.5	3.5

Gleich also assumes that the migratory insertion of alkene into rhodium hydride determines both the rate and the regioselectivity. In their calculations they start from a common alkene complex that leads to both linear and branched alkyl rhodium. The l:b ratios calculated for BISBI, dppe and PPh_3 agree quite well with those obtained from the experiment.

For PPh_3 the outcome seems surprising, because the bis-equatorial complex was used for the calculation for which one would expect a high l:b ratio (~12) according to experiments at high phosphine concentrations. For equatorial-apical coordination modes for both bidentates and PPh_3 they also calculate very higher l:b ratios, but indeed the transition state energies are much higher by 7 kcal/mol and therefore these pathways do not contribute to the product formation.

Energetics. While a lot of progress has been made both experimentally and theoretically, the kinetics and energetics of the hydroformylation sequence remain unclear, even if we only consider the "type-I" kinetics. Above we discussed the following scheme, now modified for propene:

$$\mathrm{RhHL_2(CO)_2}\ (\mathbf{2}) \rightleftharpoons \mathrm{RhHL_2(CO)}\ (\mathbf{3}) + \mathrm{CO} \qquad (1)$$

$$\mathrm{RhHL_2(CO)_2}\ (\mathbf{3}) + \mathrm{C_3H_6} \rightleftharpoons \mathrm{RhHL_2(CO)(C_3H_6)}\ (\mathbf{4}) \qquad (2)$$

$$\mathrm{RhHL_2(CO)(C_3H_6)}\ (\mathbf{4}) \rightarrow \mathrm{Rh(1\text{-}C_3H_7)L_2(CO)}\ (\mathbf{5l}) \qquad (3\text{-linear})$$

$$\mathrm{RhHL_2(CO)(C_3H_6)}\ (\mathbf{4}) \rightleftharpoons \mathrm{Rh(2\text{-}C_3H_7)L_2(CO)}\ (\mathbf{5b}) \qquad (3\text{-branched})$$

$$\mathrm{Rh(2\text{-}C_3H_7)L_2(CO)}\ (\mathbf{5b}) + \mathrm{CO} \rightarrow \mathrm{Rh(2\text{-}C_3H_7)L_2(CO)_2}\ (\mathbf{6}) \qquad (4)$$

In this scheme we have assumed that insertion to give the linear alkyl rhodium species is irreversible and rate-determining and rapid pre-equilibria exist for CO and propene association and dissociation. From deuteration studies we know that the conditions can be chosen in such a way that this is valid. At high temperatures and low pressures this is no longer true. The free-energy diagram for the kinetics showing irreversible insertion have been drawn in Figure 22, reactions (1), (2), and (3-linear).

For the formation of branched alkyls either from 1-alkenes or internal alkenes this scheme must be modified, because now often the insertion is reversible, dependent on conditions and concentrations. A potential scheme showing the competition between the backward reaction (3-branched) and the complexation of CO reaction (4) is shown in Figure 23. At low temperatures and sufficiently high pressures the formation of 2-alkyl species from 1-alkenes can also be irreversible. We have drawn the barriers for the backwards reaction (β-elimination) and forward reaction (CO complexation) at about the same height to indicate the competition between the two steps.

The 2-alkyl species shown in Figure 23 can also be formed from 2-alkenes and this species should have the same energy. We know that in most catalytic systems the hydroformylation of internal alkenes is one or two orders of magnitude slower. From the viewpoint of microscopic reversibility this leads to an interesting point. It would seem therefore that in most systems a relatively fast hydrogen exchange must take place for internal alkenes while no or little hydroformylation occurs.

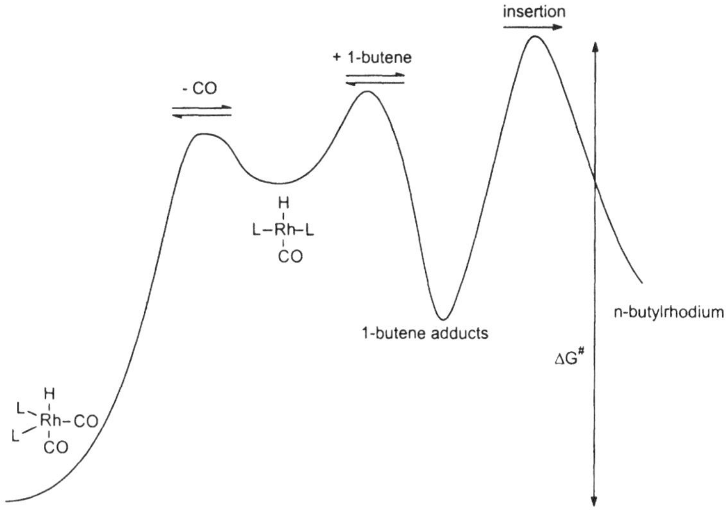

Figure 22. Energy diagram Type-I kinetics in which hydride migration is rate-determining for 1-alkenes

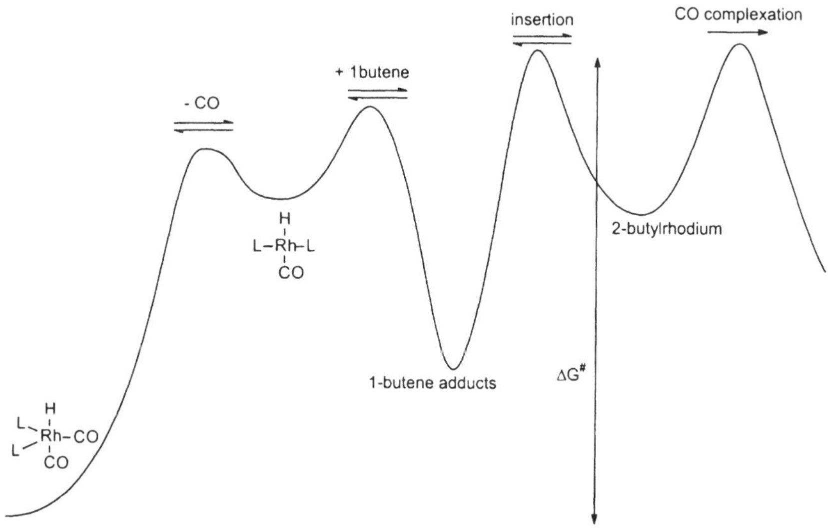

Figure 23. Reversible migratory insertion of alkene to 2-alkylrhodium

For bulky phosphite (Chapter 3) systems it is known that cyclooctene undergoes deuterium exchange at a rate an order of magnitude higher than that of hydroformylation [48]. The pathway for insertion of 1-butene to give 2-butylrhodium might well be very similar to the pathway leading to 1-butylrhodium, but likely the complexation of CO is more cumbersome in the branched alkyl species. Β-elimination may lead to 2-butene, which forms a less stable complex than 1-butene (not shown). Clearly, energy differences are very small and entropy and concentrations will have large influence on the actual kinetics.

The few data known on CO dissociation from the starting pentacoordinate rhodium hydrides show that the rate of these processes are only two orders of magnitude higher than those of the overall hydroformylation reaction. For this reason this barrier was drawn as it is in Figures 22 and 23. One might speculate that coordination of the alkene is rate-determining. A simplified scheme is shown in Figure 24.

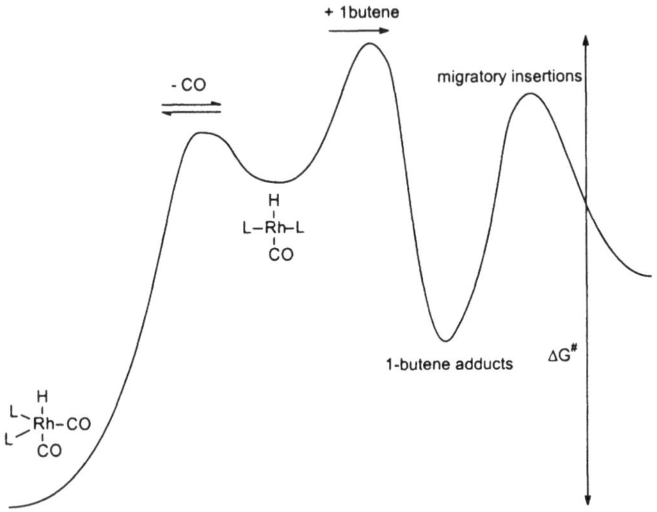

Figure 24. Type-I kinetics for rate-determining alkene complexation

The regioselectivity can be determined in two ways, by the formation of distinct stereoisomeric alkene complexes, assuming slow rotation, or by the migration reactions leading to linear and branched alkyl species. The rate-determining and selectivity determining steps do not coincide in the latter case. The kinetic equation does not distinguish between the mechanism of Figure 22 and that of Figure 24.

In spite of all recent findings we conclude that there is still a lot to be done before the kinetics, the mechanism, and ligand effects in hydroformylation will be fully understood.

References

1 Bath, S. S.; Vaska, L. *J. Am. Chem. Soc.* **1963**, *85*, 3500.
2 (a) Evans, D.; Osborn, J. A.; Wilkinson, G. *J. Chem. Soc. (A)*, **1968**, 3133. (b) Evans, D.; Yagupsky, G.; Wilkinson, G. *J. Chem. Soc. A* **1968**, 2660. (c) Brown, C. K.; Wilkinson, G. *J. Chem. Soc. A* **1970**, 2753.
3 Jardine, F. H. *Polyhedron*, **1982**, *1*, 569.
4 Jardine, F. H.; Osborn, J. A.; Wilkinson, G.; Young, J. F. *Chem. and Ind. (London)* **1965**, 560.
5 (a) Pruett, R. L. *Adv. Organomet. Chem.* **1979**, *17*, 1. (b) Cornils, B. in *New syntheses with Carbon Monoxide*, (Ed. Falbe, J.), Springer, Berlin, **1980**, Chapter 1. (c) Trzreciak, A. M.; Ziólkowski, J. J. *Coord. Chem. Rev.* **1999**, *190-2*, 883. (d) Pino, P. *J. Organomet. Chem.* **1980**, *200*, 223. (e) Falbe, J. (Ed.), *New Syntheses with Carbon Monoxide*, Springer-Verlag, New York, **1980**. (f) Leeuwen, P. W .N. M.; van Koten, G. *Catalysis, an Integrated Approach,* (Ed. van Santen, R. A.; van Leeuwen, P. W. N. M.; Moulijn, J. A.; Averill, B. A.) ElsevierScience Publishers, Amsterdam, 3rd Ed. **1999**, Chapter 6. (g) Frohning, C. D.; Kohlpaintner, C. W. in Cornils, B.; Herrmann, W. A. Eds. *Applied Homogeneous Catalysis with Organometallic Compounds,* VCH-Wiley: New York, **1996**; Vol. 1, pp 3-25. (h) Tolman, C. A.; Faller, J. W. in *Homogeneous Catalysis with Metal Phosphine Complexes,* Ed. Pignolet, L. H. Chapter 2, Plenum Press, New York, **1983**, pp 81-109.
6 (a) Hegedus, (b) Cotton, Wilkinson, (c) Masters, (d) Parshall G. W.; Ittel, S. D. *Homogeneous Catalysis*, 2nd Ed. **1992**, Wiley, New York.
7 (a) Heck, R. F.; Breslow, D. S. *J. Am. Chem. Soc.* **1961**, *83*, 4023. (b) Heck, R. F.; Breslow, D. S. *J. Am. Chem. Soc.* **1962**, *84*, 2499. (c) Heck, R. F. *Acc. Chem. Res.* **1969**, *2*, 10.
8 Brown, J. M., Kent, A.G. *J. Chem. Chem. Soc. Perkin Trans. II*, **1987**, 1597.
9 Dahlenburg, L.; Mirzaei, F.; Yardimcioglu, A. *Z. Naturforsch.***1982**, *37b*, 310.
10 Deutsch, P. P.; Eisenberg, R. *Organometallics*, **1990**, *9*, 709.
11 Chan, A. S. C.; Shieh, H. S.; Hill, J. R. *J. Chem. Soc., Chem. Commun.* **1983**, 688.
12 Castellanos-Paéz, A.; Castillón, S.; Claver, C.; van Leeuwen, P. W. N. M.; de Lange, W. G. J. *Organometallics*, **1998**, *17*, 2543.
13 (a) Oswald, A. A.; Hendrikse, D. E.; Kastrup, R.V.; Irikura, K.; Mozeleski, E. M.; Young, D. A. *Phosphorus and Sulfur,* **1987**, *30,* 237. (b) Oswald, A. A.; Merola, J. S.; Mozeleski, E. J.; Kastrup, R.V.; Reisch, J. C. *Adv. Chem. Series* **1981**, *104*, 503. (c) MacDougall, J .K.; Simpson, M. C.; Green, M. J.; Cole-Hamilton, D. J. *J. Chem. Soc. Dalton Trans.* **1996**, 1161. (d) Lawrenson, M. J. (to B.P.) UK Patent, 1,254,222 and 1,284,615; *Chem.Abstr.* **1972**, *76*, 33787 and **1972**, *77*, 125982.
14 Moser, W. R.; Papite, C. J.; Brannon, D. A.; Duwell, R. A.; Weininger, S. J. *J. Mol Cat.* **1987**, *41*, 271.
15 (a) Unruh, J. D.; Christenson, J. R. *J. Mol Cat.* **1982**, *14*, 19. (b) Unruh, J. D.; Wells, W. J. Belgian patent 840,906 1976. (c) Unruh, J. D.; Segmuller, B. E.; Chupa. G. R.; Pryor, K. E. (to Hoechst Celanese) U.S. Patent 5,567,856; *Chem. Abstr.* **1998**, *125,* 328099.
16 Buhling, A.; Kamer, P. C. J.; van Leeuwen, P. W. N. M. *J. Mol. Catal. A: Chemical*, **1995**, *98*, 69.
17 Janecko, H.; Trzeciak, A. M.; Ziolkowski, J. J. *J. Mol. Cat.* **1984**, *26*, 355.
18 van Rooy, A.; de Bruijn, J. N. H.; Roobeek, C. F.; Kamer, P. C. J.; van Leeuwen, P. W. N. M. *J. Organomet. Chem.* **1996**, *507*, 69.
19 Palo, D. R.; Erkey, C. *Organometallics*, **2000**, *19*, 81.

20 (a) Hayashi, T.; Tanaka, M.; Ogata, I. *J. Mol. Catal.,* **1979**, *6*, 1. (b) Hjortkjaer, J.; Toromanova-Petrova, P. *J. Mol. Cat.* **1989**, *50*, 203. (c) Neibecker, D.; Réau, R. *J. Mol. Catal.,* **1989**, *53*, 219.

21 Torrent, M.; Solà, M.; Frenking G. *Chem. Rev.* **2000**, *100*, 439.

22 Van Leeuwen, P. W. N. M.; Roobeek, C. F. *J. Organomet. Chem.***1983**, *258,* 343.

23 Diéguez, M.; Claver, C.; Masdeu-Bultó, A. M.; Ruiz, A.; van Leeuwen, P. W. N. M.; Schoemaker, G. C. *Organometallics,* **1999**, *18,* 2107

24 (a) Cavalieri d'Oro, P.; Raimondo, L.; Pagani, G.; Montrasi, G.; Gregorio, G.; Andreetta, A. *Chim. Ind. (Milan),* **1980**, *62*, 572. (b) Gregorio, G.; Montrasi, G.; Tampieri, M.; Cavalieri d'Oro, P.; Pagani, G.; Andreetta, A. *Chim. Ind. (Milan),* **1980**, *62*, 389.

25 (a) Strohmeier, W.; Michel, M. *Z. Phys. Chem.Neue Folge,* **1981**, *124*, 23 (in German). (b) Hjortkjaer, J. *J. Mol. Cat.* **1979**, *5*, 377.

26 (a) Musaev, D. G.; Morokuma, K. *Adv. Chem. Phys.* **1996**, *95*, 61. (b) Musaev, D. G.; Matsubara, T.; Mebel, A. M.; Koga, N.; Morokuma, K. *Pure Appl. Chem.* **1995**, *67*, 257.

27 (a) Nair, V. S.; Mathew, S. P.; Chaudhari, R. V. *J. Mol. Catal., A* **1999**, *143*, 99. (b) Divekar, S. S.; Deshpande, R. M.; Chaudhari, R. V. *Catal. Lett..* **1993**, *21*, 191. (c) Deshpande, R. M.; Chaudhari, R. V. *Ind. Eng. Chem. Res.* **1988**, *27*, 1996.

28 Palo, D. R.; Erkey, C. *Ind. Eng. Chem. Res.* **2000**, *38*, 3786.

29 (a) Heil, B.; Marko, L. *Chem. Ber.* **1968**, *101*, 2209. (b) Feng, J.; Garland, M. *Organometallics* **1999**, *18*, 417.

30 Gerritsen, L. A.; Klut, W.; Vreugdenhil, M. H.; Scholten, J. J. F. *J. Mol. Catal.* **1980**, *9*, 265.

31 Van der Veen, L. A.; Kamer, P. C. J.; van Leeuwen, P. W. N. M.; *Organometallics,* **2000**, *19*, ***.

32 Iwamoto, M.; Yuguchi, S. *J. Org. Chem.* **1966**, *31*, 4290.

33 (a) Dang, T. P.; Kagan, H. B. *J. Chem. Soc., Chem. Commun.* **1971**, 481. (b) Knowles, W. S.; Sabacky, M. J.; Vineyard, B. D.; Weinkauff, D.J. *J. Am. Chem. Soc..* **1975**, *97*, 2567.

34 (a) Sanger, A. R. *J. Mol. Catal.* **1977/8**, *3*, 221. (b) Sanger, A. R.; Schallig, L. R. *J. Mol. Catal.* **1977/8**, *3*, 101.

35 Castellanos-Páez, A.; Castillón, S.; Claver, C.; van Leeuwen P. W. N. M.; de Lange, W. G. J. *Organometallics*, **17**, *1998*, 2543.

36 Freixa, Z.; Pereira, M. M.; Pais, A. A. C. C.; Bayón, J. C. *J. Chem. Soc., Dalton Trans.* **1999**, 3245.

37 (a) Consiglio, G.; Botteghi, C.; Salomon, C.; Pino, P. *Angew. Chem.* **1973**, *85*, 665. (b) Pittman Jr., C. U.; Hirao, A. *J. Org. Chem.* **1978**, *43*, 640.

38 Kawabata, Y.;Hayashi, T.; and Ogata I. *J. Chem. Soc., Chem. Commun.*, **1979**, 462.

39 Van Leeuwen, P. W. N. M.; Roobeek, C. F. Brit. Patent Appl. 33,554, **1981** (to Shell Resarch); *Chem. Abstr.* **1982**, *96*, 6174. Van Leeuwen, P. W. N. M.; Roobeek, C. F. *J. Mol. Catal.*, **1985**, *31*, 345-53.

40 Hughes, O. R.; Unruh, J. D. *J. Mol. Catal.* **1981**, *12*, 71.

41 Hughes, O. R.; Young, D. *J. Am. Chem. Soc.* **1981**, *103*, 6636.

42 (a) Nettekoven, U.; Kamer, P. C. J.; Widhalm, M.; van Leeuwen, P. W. N. M. *Organometallics*, **2000**, *19*, ***. (b) The first reexamination is believed to be in error, since it places hydride in an equatorial position: Fung, D. C. M.;James, B. R. *Gazz. Chim. Ital.* **1992**, *122*, 329.

43 (a) Devon, T. J.; Phillips, G. W.; Puckette, T. A.; Stavinoha, J. L.; Vanderbilt, J. J. (to Texas Eastman) U.S. Patent 4,694,109, **1987**; *Chem. Abstr.* **1988**, *108*, 7890. (b) Devon, T. J.; Phillips, G. W.; Puckette, T. A.; Stavinoha, J. L.; Vanderbilt, J. J. (to Texas Eastman) U.S. Patent 5,332,846, **1994**; *Chem. Abstr.* **1994**, *121*, 280,879.

44 Casey, C. P.; Whiteker, G. T.; Melville, M. G.; Petrovich, L. M.; Gavney, J. A., Jr.; Powell, D. R. *J. Am. Chem. Soc.* **1992**, *114*, 5535.
45 Casey, C. P.; Whiteker, G. T. *Isr. J. Chem.* **1990**, *30*, 299.
46 Dierkes, P.; van Leeuwen, P. W. N. M. *J. Chem. Soc. Dalton Transactions*, **1999**, 1519.
47 (a) Yamamoto, K.; Momose, S.; Funahashi, M.; Miyazawa, M. *Synlett.* **1990**,711. (b) Casey, C. P.; Whiteker, G. T. *J. Org. Chem.* **1990**, *55*, 1394. (c) Casey, C. P.; Petrovich, L. M. *J. Am. Chem. Soc.*, **1995**, *117*, 6007.
48 Jongsma, T.; Challa, G.; van Leeuwen, P. W. N. M. *J. Organometal. Chem.* **1991**, *421*, 121.
49 van Rooy, A.; Orij, E. N.; Kamer, P. C. J.; van Leeuwen, P. W. N. M. *Organometallics*, **1995**, *14*, 34 and unpublished.
50 Casey, C. P.; Paulsen, E. L.; Beuttenmueller, E. W.; Proft, B. R.; Petrovich, L. M.; Matter, B. A.; Powell, D. R. *J. Am. Chem. Soc.* **1997**, *119*, 11817.
51 Casey, C. P.; Paulsen, E. L.; Beuttenmueller, E. W.; Proft, B. R.; Matter, B. A.; Powell, D. R. *J. Am. Chem. Soc.,* **1999**, *21*, 63.
52 Van der Made, A. W.; van Leeuwen, P. W. N. M., unpublished.
53 Kranenburg, M.; van der Burgt, Y. E. M.; Kamer, P. C. J.; van Leeuwen, P. W. N. M. *Organometallics*, **1995**, *14*, 3081.
54 van der Veen, L. A.; Keeven, P. H.; Schoemaker, G. C.; Reek, J. N. H.; Kamer, P. C. J.; van Leeuwen, P. W. N. M.; Lutz, M.; Spek A. L. *Organometallics*, **2000**, *19*, 872.
55 Van der Veen, L. A.; Boele, M. D. K.; Bregman, F. R.; Kamer, P. C. J.; van Leeuwen, P. W. N. M.; Goubitz, K.; Fraanje, J.; Schenk, H.; Bo, C. *J. Am. Chem. Soc.* **1998**, *120*, 11616.
56 Freeman, M. A.; Young, D. A. *Inorg. Chem.* **1986**, *25*, 1556.
57 Lazzaroni, R.; Raffaelli, A.; Settambolo, R.; Bertozzi, S.; Vitulli, G. *J. Mol. Catal.* **1989**, *50*, 1.
58 van der Veen, L. A.; Kamer, P. C. J.; van Leeuwen, P. W. N. M. *Organometallics*, **1999**, *18*, 4765.
59 van der Veen, L. A.; Kamer, P. C. J.; van Leeuwen, P. W. N. M. *Angew. Chem. Int. Ed. Engl.* **1999**, *38*, 336.
60 (a) Koga, N.; Qian Jin, S.; Morokuma, K. *J. Am. Chem. Soc.* **1988**, *110*, 3417. (b) Matsubara, T.; Koga, N.; Ding, Y.; Musaev, D. G.; Morokuma, K. *Organometallics,* **1997**, *16*, 1065.
61 Gleich, D.; Schmid, R.; Herrmann,W. A. *Organometallics,* **1998**, *17*, 4828.
62 Tatsumi, K.; Hoffmann, R.; Yamamoto, A.; Stille, J. K. *Bull. Chem. Soc. Jpn.* **1981**, *54*, 1857.

Chapter 5

Asymmetric hydroformylation

Carmen Claver† and Piet W.N.M. van Leeuwen‡

†Departament de Química Física i Inorgánica. Universitat Rovira i Virgili, Plaza Imperial Tarraco, 43005, Tarragona, Spain; ‡Institute of Molecular Chemistry, University of Amsterdam, Nieuwe Achtergracht 166, 1018 WV, Amsterdam, The Netherlands

5.1 Introduction

In this chapter we will concentrate on the asymmetric hydroformylation of vinyl aromatics, which are model substrates of interest to the pharmaceutical industry [1]. In 1993 and 1995 some reports were published which described the state of the art in hydroformylation with both rhodium and platinum systems [2, 3]. A report about carbonylation appeared in 1999 [4].

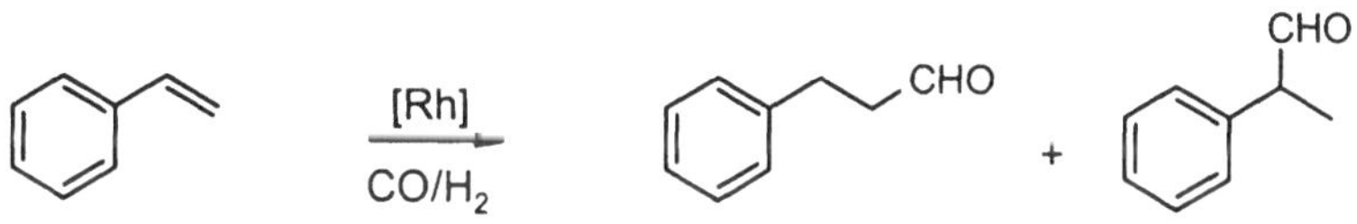

Figure 1. Hydroformylation of styrene

Highly enantioselective hydroformylation catalyzed by chiral metal complexes has been obtained with only a few catalytic systems. Many chiral phosphorus ligands have been used in Pt(II) and Rh(I) systems in the asymmetric hydroformylation of styrene. The first highly enantioselective examples of asymmetric hydroformylation of styrene were reported by Consiglio et al. in 1991 and used Pt-Sn systems [5, 6]. They achieved an ee of 86%. Platinum catalysts have several disadvantages: they have low reaction rates, they hydrogenate the substrate and their regioselectivity to the

P.W.N.M. van Leeuwen and C. Claver (eds.), Rhodium Catalyzed Hydroformylation, 107–144.

branched aldehyde is low. Pt-diphosphite/$SnCl_2$ systems also have low selectivity. When the appropriate diphosphite has been used, ee's have been as high as 90% [7].

Modified Rh systems show considerable activity in the hydroformylation of styrene to the branched aldehydes. Chiral diphosphine rhodium complexes that lead to high activity and high regioselectivity in branched aldehyde have been studied. These systems do not provide high enantioselectivity [2, 3]. Stanley reported that a tetraphosphine ligand can be used to form a bimetallic complex [8] and provide ee's up to 85% using vinyl esters.

The first reports on asymmetric hydroformylation using diphosphite ligands revealed no asymmetric induction [9, 10]. In 1992, Takaya et al. published the results of the asymmetric hydroformylation of vinyl acetate (ee=50%) with chiral diphosphites [11]. In 1992, an important breakthrough appeared in the patent literature when Babin and Whiteker at Union Carbide reported the asymmetric hydroformylation of various alkenes with ee's up to 90%, using bulky diphosphites derived from homochiral (2R,4R)-pentane-2, 4-diol, UC-PP* (see Figure 2). Van Leeuwen et al. studied the influence of the bridge length, bulky substituents and cooperativity of chiral centers on the performance of the catalyst [12, 13].

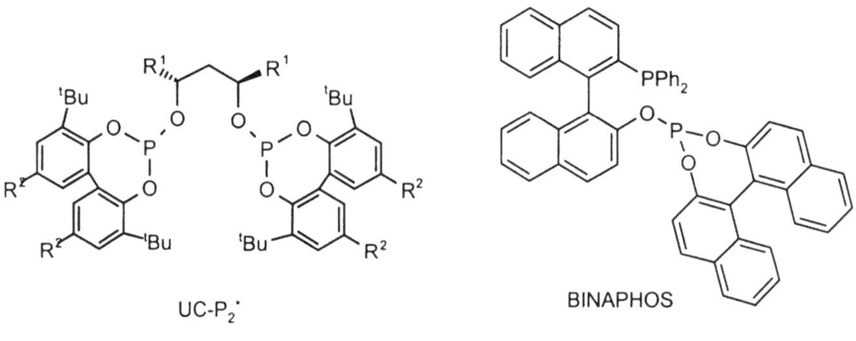

Figure 2. Chiral diphosphites and phosphine-phosphite ligands

The catalyst discovered by Takaya et al. using (R,S) BINAPHOS, a phosphine-phosphite ligand of C_1 symmetry [14] has provided ee's as high as 96% as well as total conversions and high regioselectivities. The origin of the stereodifferentiation in rhodium catalyzed hydroformylation has been discussed in theoretical reports [15, 16].

5.2 Rhodium systems with chiral diphosphite ligands

5.2.1 C_2 Symmetric chiral diphosphite ligands

Optically active diols are useful building blocks for the synthesis of chiral diphosphite ligands. Chiral diphosphites based on commercially available optically active 1,2 and 1,4-diols, 1,2:5,5-diisopropylidene-D-mannitol, L-α,α,α,α-tetramethyl-1,3-dioxalan-4,5-dimethanol and L-diethyl tartrate, **1-3,** were first used in the asymmetric hydroformylation of styrene [17].

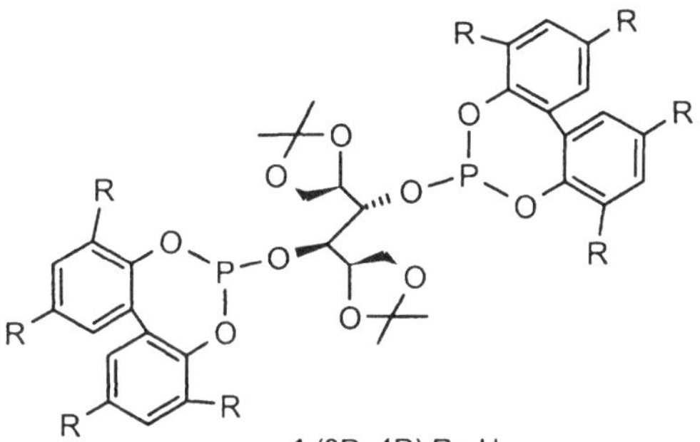

1 (3R, 4R) R= H

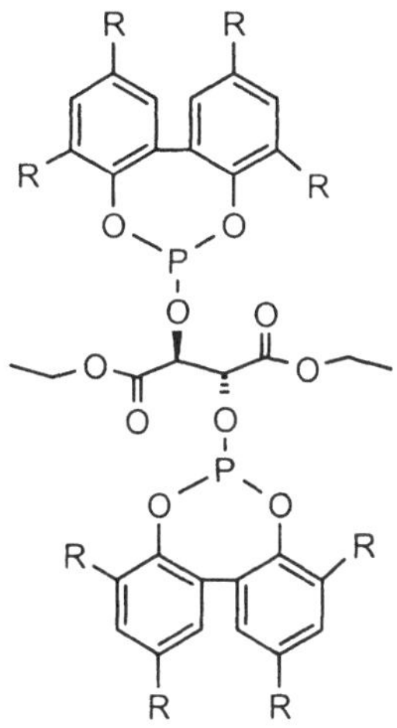

3a (2R, 3R) R= H
3b (2R, 3R) R=Bu ^{t}

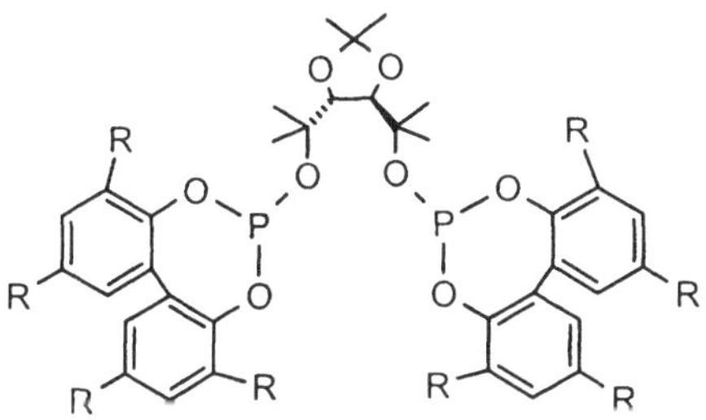

2 (2R, 3R) R= H

Figure 3. Chiral diphosphites used in rhodium asymmetric hydroformylation

Ee´s reached 20 % and regioselectivities to the branched aldehyde were high with the most bulky ligand under mild reaction conditions. The less bulky and more flexible diphosphites hardly produce enantioselectivity.

Chiral diphosphites, **4-7** with C_2 symmetry based on (2R,3R)-butane-2, 3-diol, (2R,4R)-pentane-2,4-diol and (2S,5S)-hexane-2,5-diol have been synthesized in order to study the effect of the length of the bridge (see Figure 4). In this series, *tert*-butyl and methoxy substituents were introduced at the *ortho* and *para* positions of the biaryl moieties to vary the properties of the ligands [13].

Diphosphite ligands containing biaryl moieties have low energy barrier for interconversion in atropisomers, which could lead to the formation of several diastereomers (see Figure 5). However, the rotation around the biaryl axis of the diphosphite ligands containing bulky substituents in the biaryl moieties is hindered. The advantage of using bisnaphthol derivatives is that interconversion around the bisnaphthol bond is energetically highly unfavorable, so stable diastereomeric diphosphites were obtained in a pure form. Because bulky substituents in the aromatic bisphenol positions significantly affect catalyst performance, derivatives based on 2,4-pentanediol with *ortho* substituents, having increasing steric bulk trimethyl-triethyl and tert-butyl dimethylsilyl have also been used, **8** and **9** [18].

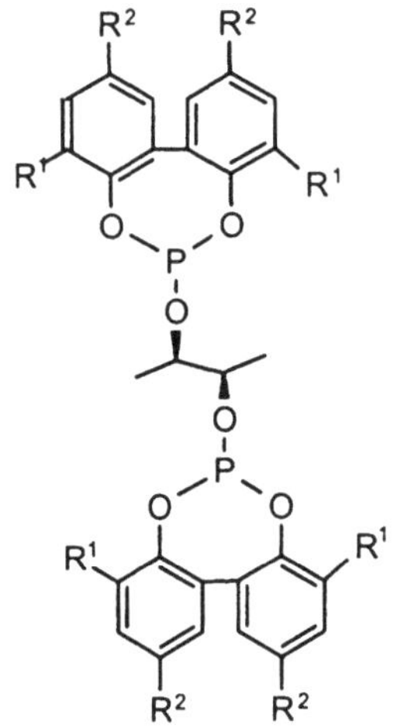

4a (2R, 3R) $R^1 = R^2 = Bu^t$
4b (2R, 3R) $R^1 = Bu^t$, $R^2 = OMe$

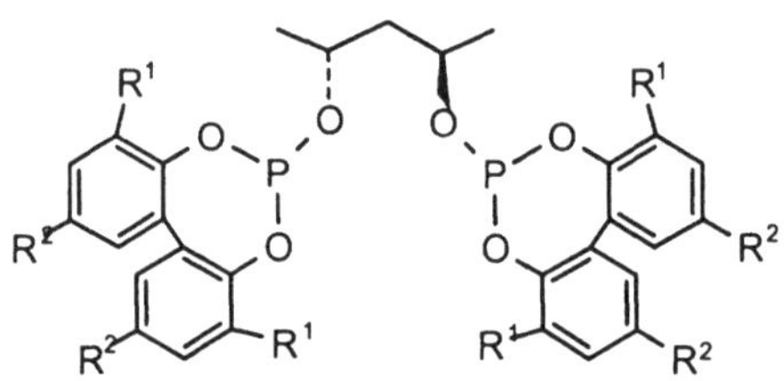

5a (2R, 4R) $R^1 = R^2 = H$
5b (2R, 4R) $R^1 = R^2 = Bu^t$
5c (2R, 4R) $R^1 = Bu^t$, $R^2 = OMe$

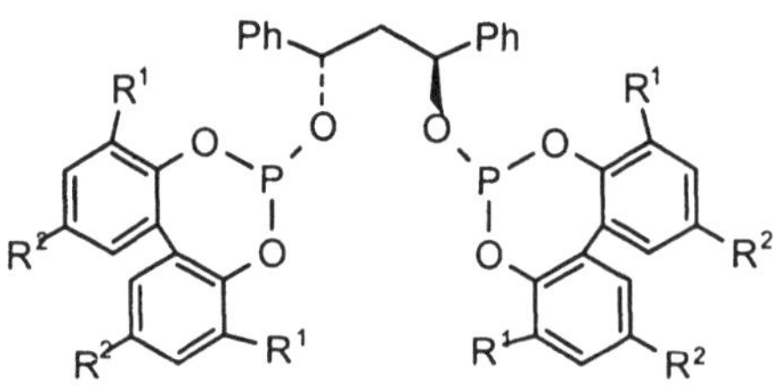

6 (1S, 3S) $R^1 = Bu^t$, $R^2 = OMe$

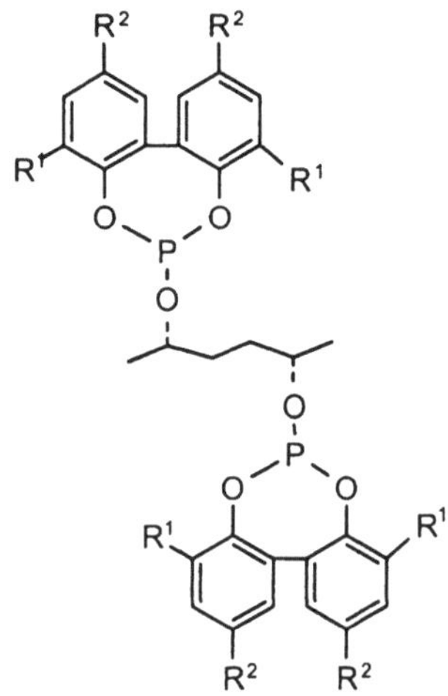

7a (2S, 5S) $R^1 = R^2 = Bu^t$
7b (2S, 5S) $R^1 = Bu^t$, $R^2 = OMe$

Figure 4. Variation of the length of the bridge in diphosphite ligands

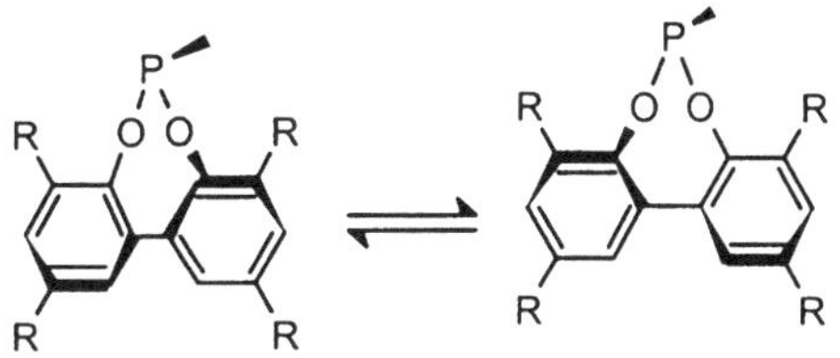

Figure 5. Atropisomerism in bisphenol moieties

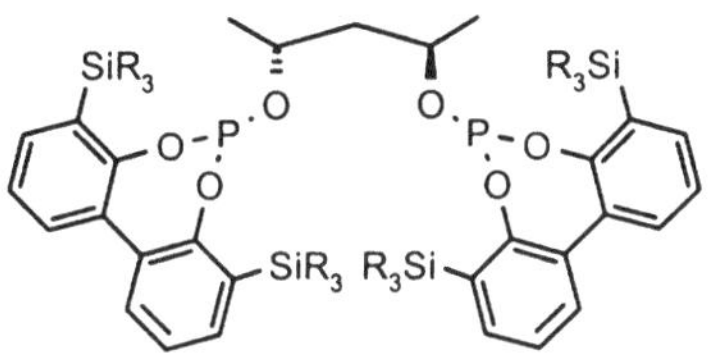

8a (R = Me)
8b (R = Et)
8c (R = *tert*-Bu Me$_2$)

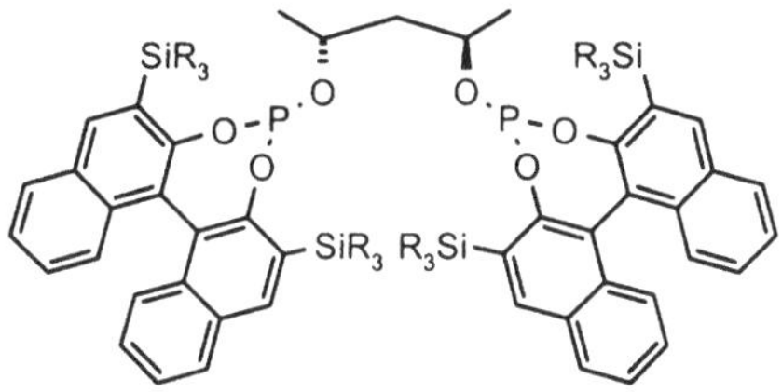

9a (R = Me)
9b (R = Et)
9c (R = *tert*-Bu Me$_2$)

Figure 6. Bisphenol and bisnaphthol pentane-2,4-diol diphosphites

The trimethylsilyl group is the substituent that provides the best enantioselectivity, and is used to prepare the optically pure bisnaphthol substituents in diphosphites **9** [18]. One of the interesting aspects of this study is that it analyses the solution structures of the rhodium diphosphite system with NMR and IR spectroscopy. The [RhH(CO)$_2$(diphosphite)] complexes form a mixture of isomers containing an equatorial-equatorial (**ee**) or apical-equatorial (**ae**) coordination of the diphosphite in a trigonal bipyramidal structure. Owing to the C$_2$ symmetry, the phosphorus atoms are equivalent in the free ligand. However, the trigonal bipyramidal hydridorhodium dicarbonyl [RhH(CO)$_2$(diphosphite)] complex, the putative resting state of the hydroformylation catalyst, has C$_1$ symmetry [13, 18, 19].

5.2.2 Catalyst preparation and hydroformylation results

The chiral diphosphites **4-9** have been used in the rhodium catalyzed asymmetric hydroformylation of styrene. The catalysts are prepared *in situ* by adding the diphosphite L to [Rh(acac)(CO)$_2$] (acac = acetylacetonate) as a catalyst precursor. Under typical hydroformylation conditions the active catalyst precursor [RhH(diphosphite)(CO)$_2$] is formed (See 5.2.3). Two important features have been found for the preparation of efficient catalyst using these diphosphites: (1) An excess of diphosphite is used to exclude the formation of $[RhH(CO)_4]$ which is a highly active achiral hydroformylation catalyst. (2) The $[RhH(diphosphite)(CO)_2]$ can form from $[Rh(acac)(CO)_2]$ and the diphosphite in an incubation time of about 15h under hydroformylation conditions. Enantioselectivities were high under relatively mild reaction conditions (Tables 1 and 2). The regioselectivity for branched aldehyde (2-phenylpropanal) always exceeded 90%. When the reaction temperature was decreased from 40 to 25 °C rates were lower but the enantiomeric excess increased considerably (Table 1).

Table 1. Hydroformylation of styrene with chiral rhodium diphosphite catalysts

Diphosphite	T °C	Turnover frequency[b]	Conversion[c] %	% 2-phenyl-propanal	% ee (Absolute Configuration)
4a	40	66	74	93	19 (S)
4b	40	177[d]	99	92	25 (S)
4b	25	31	40	93	34 (S)
5a	40	117	81	80	11 (R)
5b	40	113	89	96	50 (S)
5c	40	207[d]	98	94	67 (S)
7a	40	19	26	93	1 (R)
7b	40	19	26	92	7 (R)
6	40	165	99	90	47 (S)
6	25	40	45	95	62 (S)

All reactions were carried out with 13.3 mmol of styrene.CO/H_2 = 10 bar. Styrene:catalyst molar ratio is 421, P:Rh molar ratio is 2.5. [b] In mol styrene (mol Rh^{-1})h^{-1} determined after 2 h of reaction by GC. [c] Of styrene after 5 h. [d] After 1 h of reaction.

The influence of the structure of various diphosphites on the activity and regioselectivity in the rhodium catalytic hydroformylation has been studied by modifying the diphosphites previously reported by Union Carbide. In the hydroformylation of 1-octene, a bulky diphosphite coordinating in a bis-equatorial fashion is required if the regioselectivity is to be high in linear aldehyde. A kinetic study was made to determine several parameters affecting the reaction rates and regioselectivity in the hydroformylation of 1-octene and styrene [20].

The catalytic systems formed with the diphosphites based on (2R,4R)-pentane-2,4-diol in which the bisphenol moiety is substituted with methoxy groups at the *para* position, **5c** and **6**, give the highest activity. The catalyst based on (2S,5S)-hexane-2,5-diols show lower catalytic activity than those based on (2R,4R)-pentane-2,4-diol (Table 1).

The highest regioselectivity was obtained with trimethylsilyl and *ter*-butyl substituents at the *ortho* position. The reaction rates decreases as the steric bulk of the *ortho* substituents $C(CH_3)_3 < Si(CH_3)_3 < Si(t\text{-}Bu)(CH_3)_2 < Si(CH_2CH_3)_3$ increases. The bulky groups $Si(t\text{-}Bu)(CH_3)_2$ and $Si(CH_2CH_3)_3$

don't improve the enantiomeric excess. The steric bulk in the *ortho* positions is optimal with the trimethylsilyl substituent, **9a** [18].

An interesting connection is found between the enantiomeric excess and the structure of the diphosphite. Enantiomeric excesses are highest for the backbones of the diphosphites based on (2R,4R)-pentane-2,4-diol, **5** and **6,** which form eight-membered rings in the catalyst. Ee's are low for the diphosphites based on (2S,5S)-hexane-2,5-diol **7** and moderate for the backbones of the diphosphites based on (2R,3R)-butane-2,3-diol, **4** (Table 1). In most cases when the ligands are based on (R,R) diols the (S)-aldehyde is formed predominantly. If the configuration is inverted at the chiral carbon atom C-2 and C-5 in the (2S,5S)-hexane-diol (R)-aldehyde is predominantly formed [13,18].

5.2.3 Characterisation of [RhH(L)(CO)$_2$] intermediates. Solution structures of hydroformylation catalysts

The solution structures of the trigonal bipyramidal hydridorhodium complexes containing diphosphite ligands, [HRh(L-L)(CO)$_2$], which are the resting states in the hydroformylation reaction, have been analyzed in detail [13, 18-21] (see also Chapter 3).

^{31}P and ^{1}H NMR spectroscopy at variable temperature, together with IR studies, allow to establish a relation between the results from catalysis, the coordination mode of the diphosphites to the rhodium, and the stability of the hydride rhodium diphosphite intermediates.

Hydride rhodium diphosphites have been prepared and analyzed "in situ" by adding one equivalent of diphosphite to the catalyst precursor [Rh(acac)(CO)$_2$] under standard hydroformylation reaction conditions. Relatively long reaction times are required for a complete conversion into [RhH(L-L)(CO)$_2$], which explains the incubation times in the hydroformylation experiments. The complexes showed broadened doublets at room temperature in the ^{31}P-{^{1}H} NMR spectra, caused by a rhodium to phosphorus coupling and exchange phenomena. In the case of ligands **4a** and **4b**, which form a seven-membered ring with the metal, the $^{1}J_{Rh-P}$ coupling constants are relatively small, about 210 Hz. Furthermore, both ^{31}P and ^{1}H NMR spectra showed a large $^{2}J_{P-H}$ coupling constant, about 100 Hz, which suggests a time-averaged *cis, trans* relationship between the phosphorus and the hydrogen atoms bonded to the rhodium (see Figure 7). Large $^{2}J_{P-H}$ coupling constants are characteristic of phosphorus and hydrogen atoms bonded *trans* to rhodium [19, 22, 23, 24]. The $^{2}J_{P-H}$ coupling constants for a *cis* relationship between the phosphorus and the hydrogen atom bonded to the rhodium are small (< 3Hz).

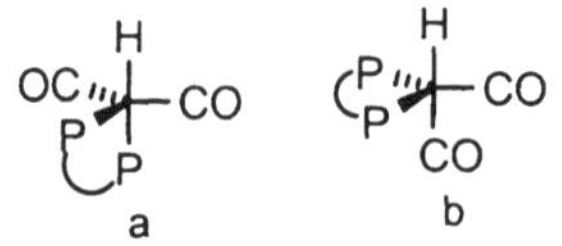

Figure 7. Ee and ea rhodium hydride dicarbonyl diphosphite

The mode of coordination depends on the length of the bridge [13, 18, 19]. For ligands **5-7** derived from 2,4-pentanediol and 2,5-hexanediol, which form an eight- or nine-membered ring with the metal, the $^{1}J_{Rh\text{-}P}$ coupling constants are in the range 231-237 Hz. Stable hydride rhodium diphosphite dicarbonyl complexes are formed. Typically, coupling constants of rhodium to an equatorial phosphorus are in the range of 220-246Hz [13,19,25]. The $^{2}J_{P\text{-}H}$ coupling constants are < 3Hz, which is also indicative of a *cis* disposition between the phosphorus and the hydrogen atom bonded to the rhodium (Figure 7b). The two phosphorus atoms are diastereotopic in the **ee** complexes but at room temperature they are in fast exchange. Evidence for the rapid exchange process was obtained from variable temperature NMR experiments between 293 and 203 K. At temperatures between 293 and 253 line broadening occurred. On further cooling to 223K an ABX double doublet appeared, indicative of two chemically inequivalent phosphorus atoms P_1 and P_2 in the C_1-symmetric complex.

All complexes of these diphosphite ligands showed fluxional behavior on the NMR time scale, although the exchange parameters depended on the mode of coordination (**ea** or **ee**). Ligand exchange in trigonal bipyramidal HML_4 complexes has been explained by so-called Berry-type rotations [26], in one step two apical ligands exchange positions with two equatorial ligands. However, a Berry mechanism is unlikely to be operative in RhH(L-L)$(CO)_2$ complexes containing bidentate diphosphite, since phosphorus exchange requires two successive Berry-type interconversions via a high-energy intermediate which has a hydride ligand equatorially coordinated to rhodium. There are no known examples of trigonal bipyramidal HML_4 complexes, which have an equatorial coordinated hydride ligand. X-ray and infrared studies on such complexes always show the hydride in an apical position. It is also unlikely that phosphorus-rhodium-phosphorus bite angles can vary freely between 90 and 180° without considerable energy strain.

The low-energy turn-stile rearrangement postulated by Meakin *et al.* [22, 23] for monophosphites has been proposed to explain the fluxional behavior of these complexes containing flexible diphosphite. It is believed that the phosphorus atoms can easily exchange without changing the phosphorus-rhodium-phosphorus bite angle appreciably.

Figure 8 shows a simultaneous bending motion of the hydride and the carbon monoxide ligands in the hydride rhodium phosphite dicarbonyl complexes containing equatorial-apical coordinating phosphite, (A) and the motion of the hydride and phosphite functions for bis-equatorially coordinating diphosphite, (B) respectively. For diphosphites complexes, containing a bridge between the two phosphorus atoms, the last hydride movement may not be facile, as the hydride must pass the bridge.

The energy barrier for the fluxional processes is lower when the steric bulkiness of the *ortho* substituent of the ligand decreased. The lowest energy barriers correspond to equatorial-axial phosphorus exchange in complexes containing seven-membered chelate rings.

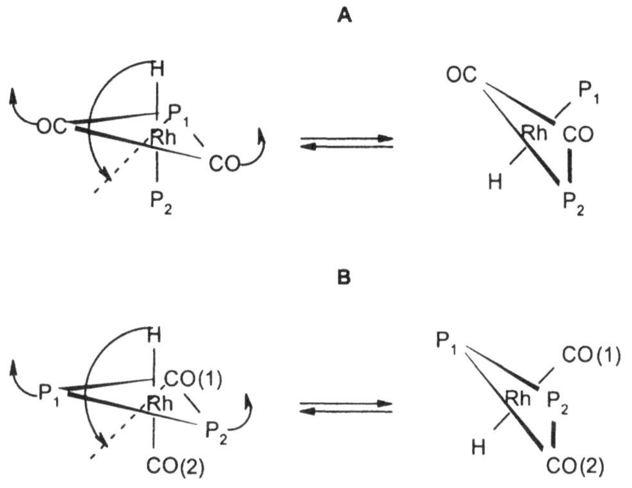

Figure 8. (A) Equatorial-axial phosphorus exchange. (B) Equatorial-equatorial phosphorus exchange

This seems logical since this process involves the movement of two carbonyl ligands, which should be easier than the movement of the two phosphite moieties that is required for equatorial-equatorial exchange. Furthermore, the energy barrier for phosphorus-phosphorus exchange is lower for the nine-membered bis-equatorial chelate rings than for the eight-membered ones, as the former are more flexible owing to the reduced steric congestion.

5.2.4 Structure versus stability and enantioselectivity

Both the structure and stability of the hydridorhodium dicarbonyl diphosphite catalysts play a role in asymmetric induction. As described above, short-bridged diphosphites forming seven-member rings coordinate in an equatorial-apical fashion, while larger bidentate ligands forming eight- or nine-member chelate rings show preferentially bis-equatorial coordination. As well as differences in coordination of the diphosphites to the rhodium, there are also differences in the stability of [RhH(diphosphite)$(CO)_2$] complexes. The complexes formed with ligands coordinating in a seven or nine membered ring are rather unstable. The IR spectra of the characteristic orange solution in toluene often show the presence of $[Rh_4(CO)_{12}]$ and carbonyl-bridged dirhodium complexes containing diphosphite. Signals also appear in the ^{31}P NMR spectrum, which may be ascribed to the hydrolysis of the diphosphite ligand (see Chapter 9). The complexes containing ligands that form an eight-membered chelate ring give yellow solutions in toluene and the IR and NMR spectra show that decomposition is considerably less than in the complexes containing ligands forming seven or nine membered rings. IR absorptions of rhodium-carbonyl cluster are practically absent. Enantiomeric excesses are highest with the relatively stable [RhH(diphosphite)$(CO)_2$] complexes in which the phosphorus donor atoms coordinate both equatorially to the rhodium, forming an eight-membered ring. Low enantiomeric excesses are obtained with the rather unstable [RhH(diphosphite)$(CO)_2$] complexes which contain a nine-membered ring having bis-equatorial phosphites (20%), or those containing seven-membered rings and axial-equatorial phosphite coordination [13].

It should be noted that coordination in an equatorial-axial manner leads to two intermediate species b' and c', having effectively opposite absolute configurations (see Figure 9), thus explaining the low enantioselectivities [13, 16]. As we will see later when discussing BINAPHOS, this is not a general rule for enantioselectivity.

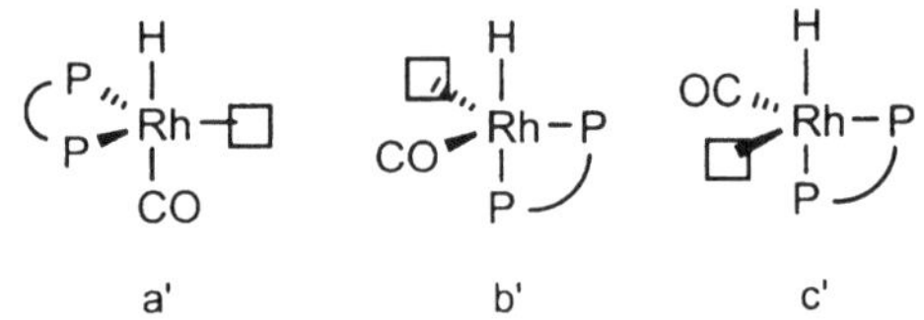

Figure 9. The intermediates for equatorial-equatorial and equatorial-apical coordination

5.2.5 Chiral cooperativity and effect of terminal substituents in diastereomeric diphosphite ligands

Diastereomeric diphosphites based on homochiral pentane-2,4-diol ligand backbones substituted with bulky 2,2'-bisphenols and bulky, 2,2'-bisnaphtol with different configurations **8-13** have been prepared to study the effect of chiral cooperativity on the asymmetric hydroformylation of styrene (see

Figures 4 and 9). Stable diastereomeric diphosphites were obtained in a pure form using the bisnaphthol derivatives [18].

The absolute configuration of the diastereomers is given as (S,2R,4R,S) (S,2S,4S,S), (R,2R,4R,R), (R,2R,4R,S)**10-13** (see Figure 10). The indicators S and R refer to the absolute configuration of the chiral axis, while the indicators, 2R, 4R and 2S, 4S refer to the absolute configuration of the carbon atoms C-2 and C-4 in the pentane-2,4-diol backbone. The structures with the configurations (S,2S,4S,S) and (R,2R,4R,R) are mirror images.

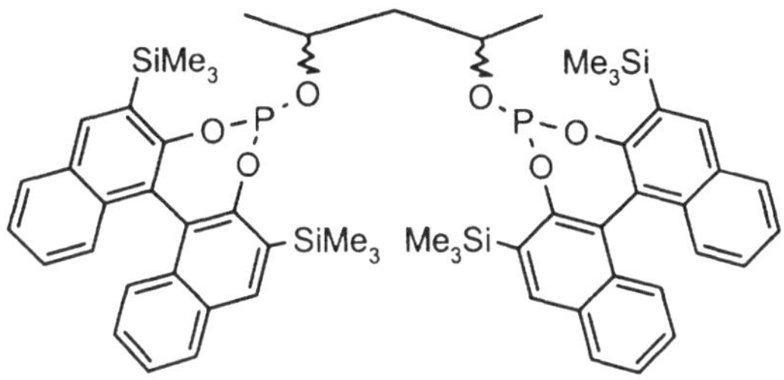

10: (S, 2R, 4R, S)
11: (S, 2S, 4S, S)
12: (R, 2R, 4R, R)
13: (R, 2R, 4R, S)

Figure 10. Diastereomeric diphosphite pentane-2,4-diol derivatives

The fact that the diphosphite ligands behave differently in the formation of the $RhH(diphosphite)(CO)_2$ complexes demonstrates that both the absolute configuration of the 2,4-pentanediol ligand backbone and the chiral bisnaphthol substituents determine the stability and catalytic of the rhodium complexes. The ligands with configuration (S,2S,4S,S) **11** and (R,2R,4R,R) **12**, lead to undefined mixtures of complexes and ligand decomposition. However the ligands (S,2R,4R,S) **8**, **10**, and (R,2R,4R,S) **13** form well defined, stable complexes. When there was an excess of the statistical mixture of diastereomers **9**, only the complexes derived from **10** and **13** formed. When both the pentane backbone and the bisnaphthol substituents have all S (**11**) or all R (**12**) configurations the ee is low. In situ NMR studies show that the $[RhH(CO)_2(L\text{-}L)]$ complexes of these ligands cannot be synthesized selectively.

The low enantioselectivity and relatively high reaction rates observed have been attributed to the coexistence of highly active rhodium species in which the ligands coordinate as monodentates. In contrast, the ligands with different configurations at the pentane bridge backbone and such bisnaphthol substituents, as ligands **10** and **13**, both form stable rhodium complexes, but only diastereomer **10** gives a high ee (Table 2). These results have been attributed to an effect known as chiral cooperativity between the central backbone and the biaryl substituents.

Curiously, the results of rigid **10** are the same as for **8a**, the bisphenol derivative in which the ligand can adopt freely several diasteromeric conformations. These results may indicate that the bisphenol derivative leads predominantly to the (S,2R,4R,S) conformation in the $[RhH(CO)_2(diphosphite)]$ complex, as can indeed also be inferred from a comparison of the low temperature ^{31}P NMR spectra.

Table 2. Hydroformylation of styrene with $RhH(diphosphite)(CO)_2$

Diphosphite	T °C	Turnover frequency[b]	Conversion[c] %	% 2-phenyl-propanal	% ee (Absolute Configuration)
8a	40	45	21	89	67 (S)
8a	25	9	26[d]	93	87 (S)
9a	25	14	20	91	47 (S)
10	50	133	43	83	58 (S)
10	25	17	38[d]	88	69 (S)
10	15	11	12[d]	92	86 (S)
11	40	259	89	91	18 (R)
11	25	45	21	94	40 (R)
12	25	28	18	95	38 (S)
13	25	4	2	91	23 (S)

Styrene:catalyst molar ratio is 1000. CO/H_2 = 20 bar. P:Rh molar ratio is 2.2. [b] TOF in mol styrene.(mol Rh^{-1})h^{-1} determined after 1 h by GC. [c] Of styrene after 5 h. [d] After 24 h of reaction.

The ee is highest, 86%, when the diastereomer **10** is used and it is in the same range as for **8**, which has rotational freedom around the biaryl axis. When the diphosphite ligands are bisnaphthol derivatives **9a** and **10, the** three diastereomers of **9a** resulted in low asymmetric inductions in comparison with the single isomer **10**.

The formation of the $[RhH(CO)_2(diphosphite)]$ complexes using one equivalent of ligand **9a** results in the formation of a statistical mixture of the diastereoisomeric complexes **10**, **12** and **13**. The turnover frequency in the hydroformylation reaction corresponds to the average of this statistical mixture [18]. Also, the observed enantiomeric excess corresponds to the average value, when the different reaction rates of the diastereomeric complexes are considered. ^{31}P NMR analysis of the rhodium complexes derived from the ligand **9a** showed that the most stable complex is formed by ligand **10**, the one that also induces the highest enantioselectivity. Using an excess of **9a**, a mixture of rhodium hydride complexes derived only from **10** and some **13** is formed. This explains the increased ee when an excess of **9a** is used.

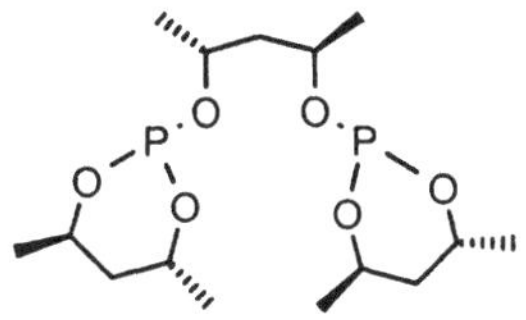

14a (2R, 4R)-bis(4R,6R)

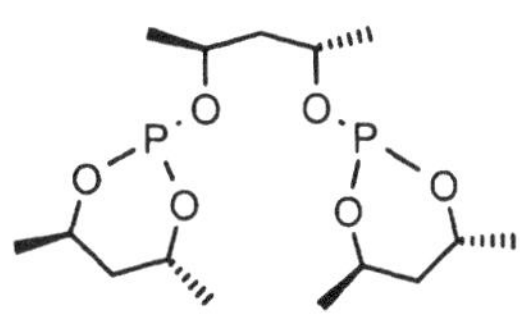

14b (2S, 4S)-bis(4R,6R)

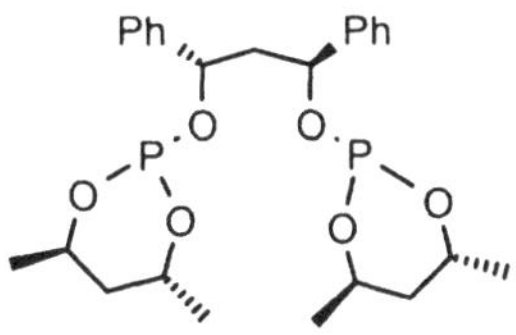

15a (1S, 3S)-bis(4R,6R)

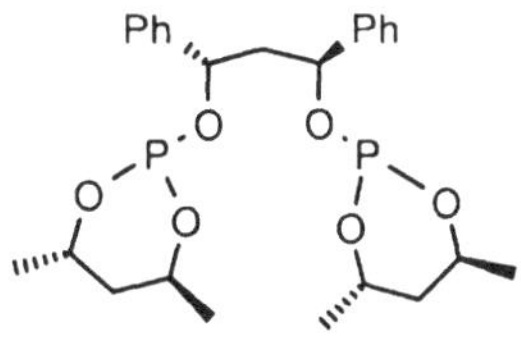

15b (1S, 3S)-bis(4S,6S)

Figure 11. Diphosphite ligands containing chiral chelate backbones and chiral dioxaphosphorinane moieties

Kollar et al. prepared diastereomers of diphosphites which were synthesized by combining different modifications in the bridge and the phosphorus substituents [27, 28]. These ligands lead to active rhodium systems, and are considerably affected by chiral cooperativity. Some of these diastereomers prepared from (2S,4S) 2,4-pentenediol and (1S,3S) diphenyl propane 1,3-diol form rhodium complexes in which the ligand is coordinated in an eight-membered ring, **14, 15**, but the enantioselectivity obtained in the experiments described are no higher than 24% [27].

Diastereomers derived from atropisomeric backbones, such as the bisnaphthol derivatives, form nine-membered rings **16-19** (see Figure 12). The diphosphites **17** and **19** are sterically less demanding than the diphosphites **16** and **18**. The diastereomers differ only in the axial configuration in the backbone; the configurations in the terminal groups are the same. The results obtained in asymmetric hydroformylation of styrene suggest that catalysts containing **16** and **18** have a higher preference for the formation of the branched alkylrhodium intermediate than those containing **17** and **19,** which is probably due to their different steric constraints. The enantioselectivities were always low or moderate, never exceeding 24% [28, 29]. The systems having the same absolute configuration in the chelate backbone and the terminal groups, R-bis(4R,6R)**17b,** and S-bis(S)**19a**, provide slightly higher enantiomeric excesses than the corresponding S-bis(4R,6R), and R-bis(S).

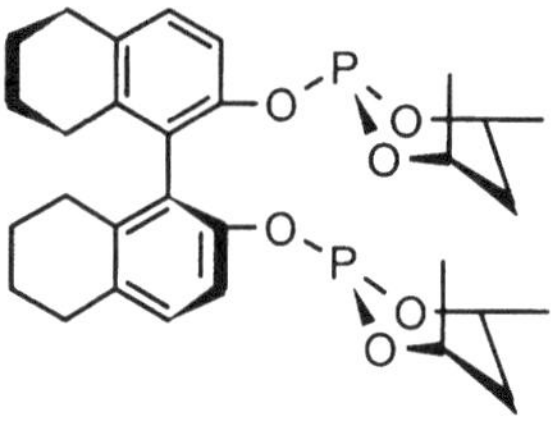

16a S-bis(4R, 6R)
16b R-bis(4R, 6R)

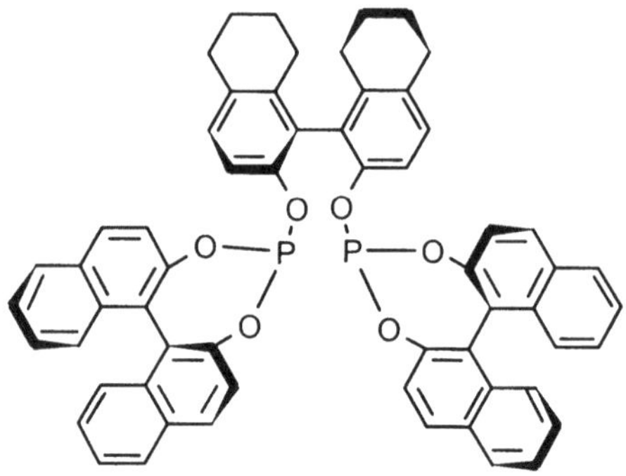

18a S-bis(S)
18b R-bis(S)

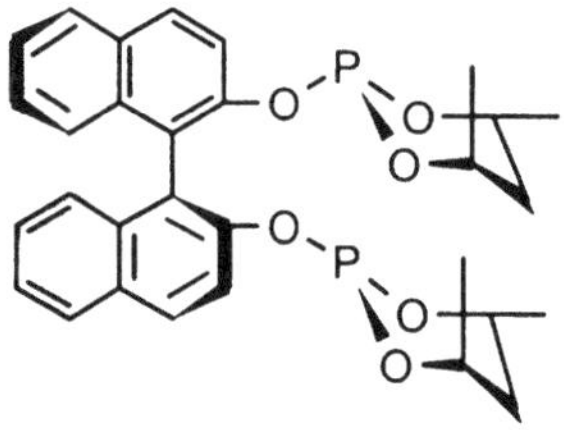

17a S-bis(4R, 6R)
17b R-bis(4R, 6R)

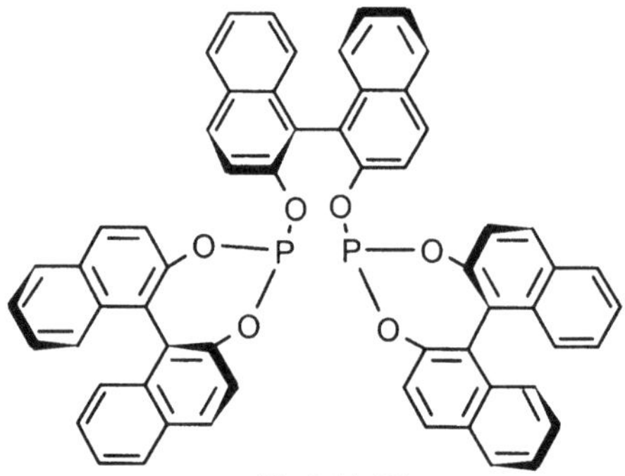

19a S-bis(S)
19b R-bis(S)

Figure 12. Chiral diphosphites forming nine membered ring coordinated to the metal

Systematically varying the chirality at both the chelate backbone and the terminal groups affected the enantioselectivity of the catalyst. The axial chirality on the bridge determines the product configuration with a cooperative effect from the terminal groups. When both the bisnaphthol backbone and the terminal dioxaphosphepine units have the same chiral denominator, ee's are lower than when they have different configurations, as can be seen from the systems discussed above [18, 28].

Chiral diphosphite ligands containing a spiro backbone in the bridge and forming a complex with an eight-membered ring have also been used as catalysts in the asymmetric hydroformylation of styrene and derivatives (Figure 13) [30]. At 10 bar of CO/H_2 and 40 °C regioselectivity to 2-phenylpropanal is 97% and enantioselectivity 69% using the (1S,5S,6R)-(*cis,trans*)-diol phosphite shown.

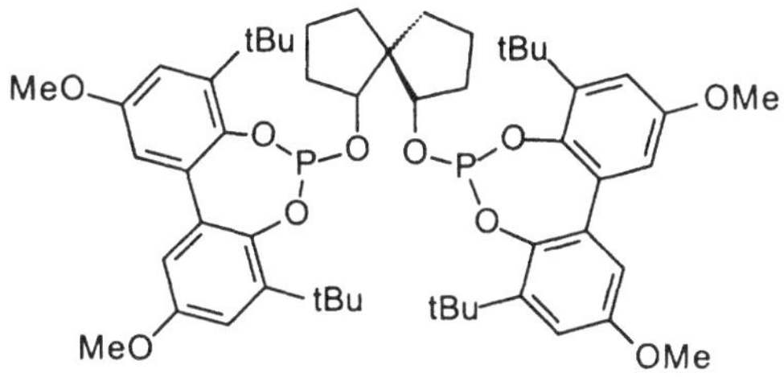

Figure 13. Chiral diphosphites derived from spiro[4,4]nonane-1,6-diol

Decreasing the temperature from 40 to 25 °C led to a slight increase of enantiomeric excess at the cost of a much lower conversion. Diphosphites with the same spiro backbone, and 1,1'-bisnaphthol substituents of different chirality, afforded the opposite prevailing enantiomer in the product. This shows that the terminal bisnaphthol groups dictate enantioselection. The catalytic efficacy of these diphosphite systems is mainly controlled by the large bulky substituents. No cooperative effect between central and axial chirality was observed in this case [30].

5.2.6 C_1 Sugar backbone derivatives. Diphosphinite and diphosphite ligands

Sugar derivatives have been used as ligands in homogeneous catalysis and are among the most easily accessible chiral ligands. They have provided high enantiomeric excesses for several asymmetric processes. Their versatility and the presence of several stereogenic centers allow the syntheses of several series of ligands [31]. C_2 chiral ligands are extensively used in asymmetric homogeneous catalysis and can be considered the best choice, because their symmetry diminishes the number of diastereomeric transition states, but C_1 chiral ligands have proved to be of value in enantioselective catalysis as well. C_1 symmetric ligands can cause stereochemical restrictions because of such electronic effects as the *trans*-influence, differentiation in the metal environmental and enantioface discrimination towards the coordination of the prochiral substrate. C_1 ligands also cause differences in diastereomeric transition states [15, 16]. In the rhodium-catalyzed hydroformylation of styrene and related substrates, the first carbohydrate ligands were tetrahalose derivative diphosphines. When these systems were used under ambient conditions regioselectivities were high for branched aldehydes, but zero ee's were obtained [32]. Later, Rajan–Babu applied in asymmetric hydroformylation diphosphinite ligands, which had previously been used in asymmetric hydrocyanation [33].

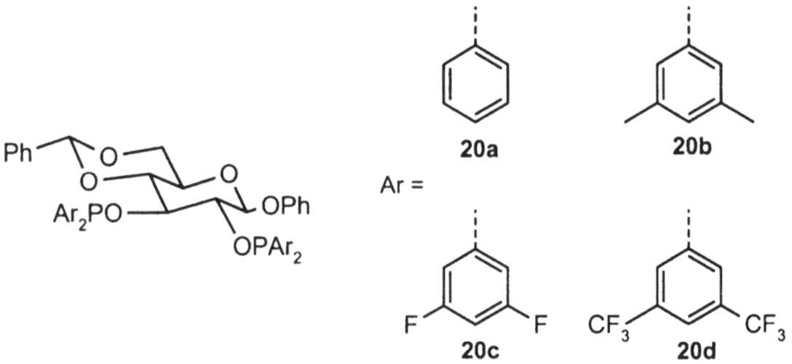

Figure 14. Diphosphinite carbohydrate derivatives

Vicinal carbohydrate diphosphinites **20** provide high regioselectivities for the branched aldehyde, but the enantioselectivities are moderate. In the case of 6-methoxy-2-vinylnaphthalene, however, the ee increased from 10% to 38% when electron deficient aryl groups containing two fluorine atoms or two trifluoromethyl substituents are introduced on the aryl groups (**20c-d**). The enantioselectivity further improves to 51 and 72% when apolar solvents like hexane and triethylsilane respectively are used. These ligands are quite sensitive to the alkene structure and lead to large differences in enantioselectivity for similar substrates. The highest ee for 2-vinylnaphthalene does not exceed 30% and asymmetric induction was even lower for other alkenes such as styrene and vinyl acetate [34].

Diphosphites derived from 1,2-O-isopropylidene-α-xylofuranose and the corresponding ribose derivatives, differently modified are effective C_1 chiral ligands in the rhodium hydroformylation of styrene. As described for C_2 diphosphites in the section 5.2.2, the diphosphite ligands based on 1,3-diols give rise to hydridorhodium diphosphite hydroformylation catalysts, which are relatively stable in comparison to those based on 1,2 and 1,4-diols, and enantioselectivities are higher. 1,2-O-isopropylidene-α-xylofuranose, which has a three-carbon atom bridge between the hydroxy groups, is an appropriate starting material for a series of useful diphosphites [35].

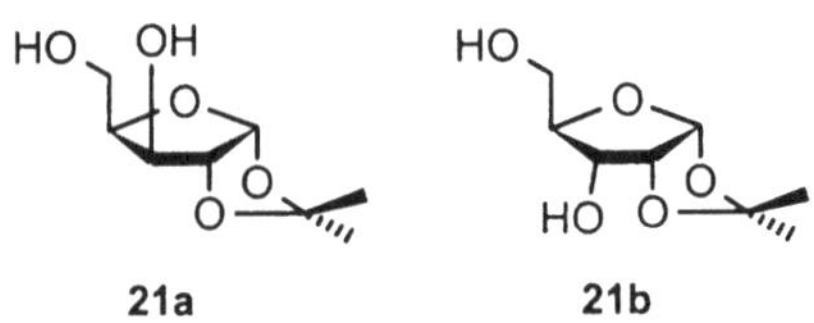

Figure 15. 1,2-O-isopropylidene-D-xylofuranose **21a** and ribofuranose **21b**

Because the primary hydroxy at C-5 is more reactive towards phosphorochloridites than the secondary hydroxy group at C-3, different substituents can be introduced into the molecule. Using bisphenol phosphorochloridites having various steric bulk at the *ortho* and *para* positions, several xylose and ribose derivative ligands were synthesized and used in rhodium asymmetric hydroformylation [35, 36].

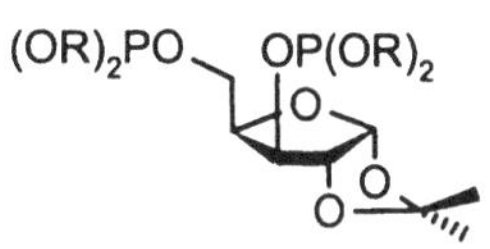

22a $R_1 = R_2 = H$
22b $R_1 = Bu^t$; $R_2 = OMe$
22c $R_1 = R_2 = Bu^t$

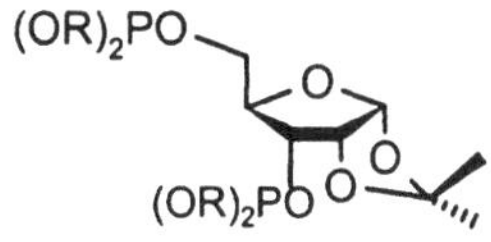

23a $R_1 = R_2 = H$
23b $R_1 = Bu^t$; $R_2 = OMe$
23c $R_1 = R_2 = Bu^t$

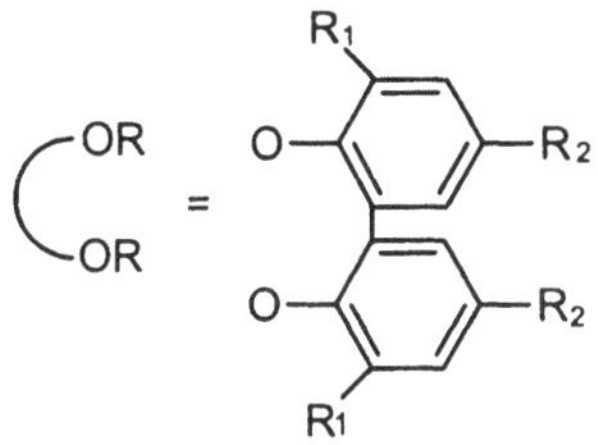

Figure 16. Xylose (**22**) and ribose (**23**) diphosphite derivatives

Regioselectivities, as high as 97% to the branched aldehydes were found under relatively mild conditions, T= 25-40 °C, 9-45 bar of CO/H_2 pressure. As expected, the enantioselectivity depends on the substitution of the ligand. The highest enantioselectivity was found for the ligands having bulky *tert*-butyl substituents at the *ortho* position of the bisphenol moieties. Enantioselectivities up to 64% have been obtained with these substituted ligands [35]. Ligands based on 1,2-O-isopropylidene-α-xylofuranose gave (S)-aldehydes, while the related ligands with a ribofuranose backbone lead to the (R)-aldehydes with similar rates and regioselectivities and enantioselectivities of the same magnitude for each enantiomer. The inversion of the asymmetric has been rationalized by the proposal that the absolute configuration at the stereogenic carbon atom C-3 produces two diastereomeric rhodium complexes that are effectively mirror images [36]. The solution structures for bisphenol derivatives are consistent with the presence of two diastereomers for the $[RhH(diphosphite)(CO)_2]$ intermediates, diphosphites of which coordinate bisequatorially to the rhodium in a fast exchange on the NMR time scale [36].

Selke et al. prepared sugar diphosphites derived from glucopyranose. They had different cyclic substituents on the phosphorus atoms and they were tested in the rhodium hydroformylation of vinyl acetate, allyl acetate and *p*-methoxystyrene. The highest ee obtained for allyl acetate is 32% and for p-methoxystyrene 20%. The conclusions are in accord with the general features reported by van Leeuwen for diphosphites derived from vicinal diols [17, 37].

5.3 Phosphine-phosphite rhodium catalysts

5.3.1 Introduction

In 1992, Takaya et al. reported on the use of a chiral diphosphite derived from bisnaphthol in the asymmetric hydroformylation of vinyl acetate [11], but the enantioselectivity achieved was only 50%. They noted that diphosphites led to more stable hydroformylation catalysts than diphosphines. This observation prompted Takaya, Nozaki et al. to synthesize the chiral phosphine-phosphite ligands (R,S)- and (R,R)-BINAPHOS (**24**, figure 17), which were expected to combine the high enantioselectivity obtained with diphosphines such as BINAP in asymmetric hydrogenation, with the apparently efficient coordination of the phosphite moiety [38]. Indeed, the Rh(I) complex of C_1-symmetric (R,S)-BINAPHOS provided much higher enantioselectivities than either C_2 symmetric diphosphine ligands or diphosphite ligands, *viz.* more than 90% ee for a wide variety of both functionalized and internal alkenes [38, 39,40, 41].

Ever since the successful results obtained for (R,S)-BINAPHOS, the preparation of other unsymmetrical ligands to be used in the asymmetric hydroformylation has become a recurrent theme in the literature [3, 4, 36, 42]. The following section will discuss those ligands, which are structurally related to BINAPHOS, while other phosphine-phosphite systems will be discussed separately in section 5.3.5.

5.3.2 Rhodium complexes with BINAPHOS and related ligands.

The chirality and the steric properties of BINAPHOS were varied systematically by synthesizing of **24-27** shown in Figures 17 and 18 [14]. The flexible ligand (R)-**26** derived from bisphenol provides a single species, for which the NMR data indicate that there is a fast exchange between the two diastereomers (R,S) and (R,R) via a rotation around the bisphenol axis. (S,R)-**28a** and (R,R)-**28b** were obtained as a mixture of two diastereomers, owing to the rotation around the biphenyl axis. According to the NMR spectra a 55:45 mixture (S,R vs. R,R) is obtained that cannot be separated by chromatography. This suggests that the exchange is too fast at room temperature on the laboratory time scale, but slow on the NMR time scale.

The first R/S indicator refers to the binaphthyl (biphenyl) bridge and the second to the bisnaphthol (bisphenol) moiety. Thus for **26**, which lacks the bisphenol chirality, we write in the table (R,--), and for **28**, which contains a non-chiral biphenyl bridge, we write (--,R).

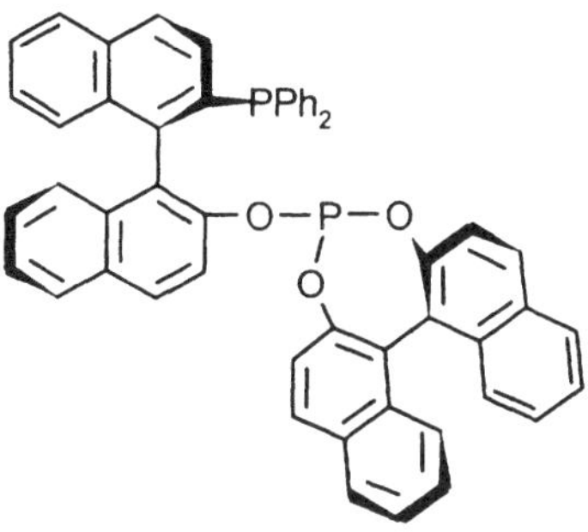

24 (R,S)-BINAPHOS

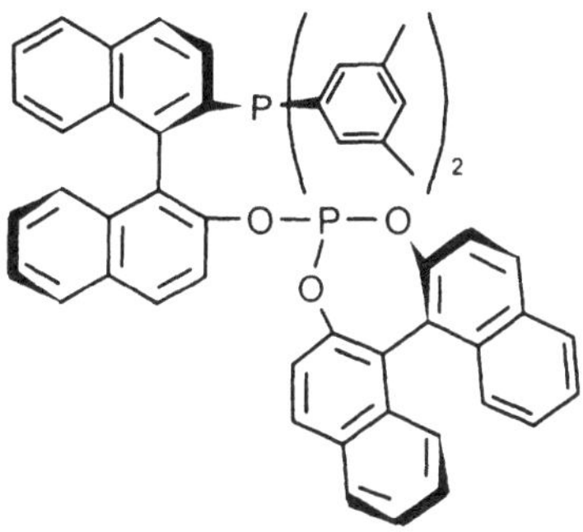

25 (R,S)

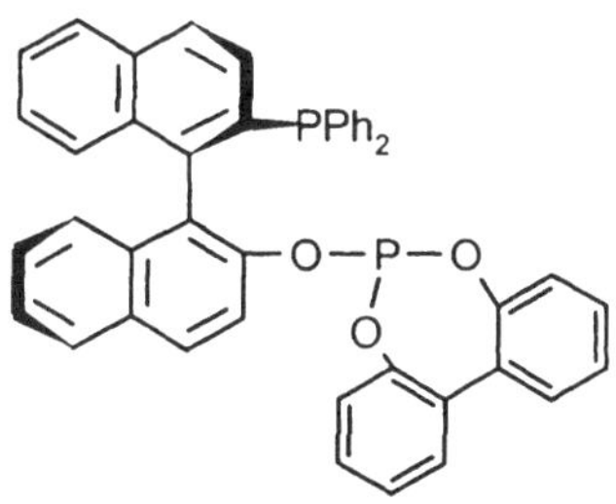

26 (R)

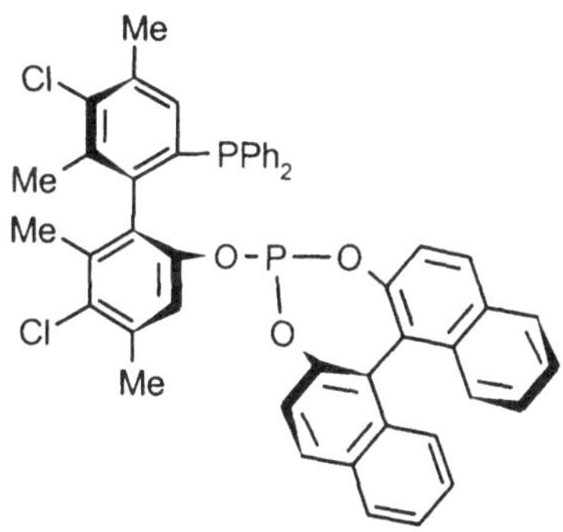

27a (S,R)
27b (R,R)

Figure 17. (R,S) BINAPHOS and related ligands

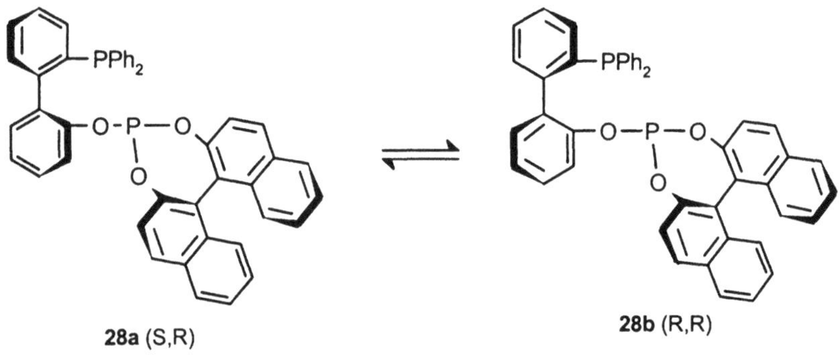

Figure 18. (S,R) and (R,R) equilibrium

The [Rh(acac)(phosphine-phosphite] complexes were used as precursors for the asymmetric hydroformylation of styrene after adding 3 equivalents of free ligand. The results are shown in Table 3. The ee of the product decreases with increasing temperature. The ratio of CO to H_2 does not affect the regio- and enantioselectivity.

For all ligands the absolute configuration of the product is the same as that of the absolute configuration of the binaphthyl (biphenyl) bridge. Thus, it seems that the binaphthyl bridge directs the position of the diphenylphosphino fragment. Remarkably, the best combination is formed when the absolute configurations for the binaphthyl and bisnaphthol moieties are opposites (i.e. S,R or R,S diastereomers). Enantioselectivities were much lower with (R,R)-ligands. R,R diastereomer **24** still gives the R-aldehyde but the ee is low (26%).

Table 3. Asymmetric hydroformylation of styrene using ligands **24-28**

Ligand	T °C	CO/H_2 bar/bar	Time h	Conversion (%)	b/l	% ee (Absolute configuration)
24 (S,R)	60	50/50	43	>99	88/12	94 (S)
24 (R,S)	80	50/50	16	>99	86/14	89 (R)
24 (R,S)	60	10/90	40	>99	88/12	92 (R)
24 (R,R)	60	50/50	38	>99	86/14	25 (R)
25 (R,S)	60	50/50	37	>99	90/10	85 (R)
26 (R,--)	60	50/50	43	>99	91/9	83 (R)
27 (S,R)	60	50/50	42	>99	90/10	94 (S)
27 (R,R)	60	50/50	40	95	92/8	16 (R)
28 (--,R)	60	50/50	40	98	89/11	69 (S)

Styrene 20 mmols in benzene. [Rh] = 0.010mmol. L = 0.040 mmol. The ee's were determined by GLC analysis of the corresponding 2-arylpropionic acids derived by Jones oxidation of the products

The results obtained for ligands **26** and **28,** which contain only one fixed stereocenter, are interesting and very informative about the system. Ligand **26**-(R,--), in which only the binaphthyl bridge has a predetermined absolute configuration R, leads to an ee of 83% (R-aldehyde), which is quite close to the value of 94% for (R,S)-BINAPHOS. This suggests that in the formation of the complex the binaphthyl bridge controls the conformation of the bisphenol moiety so that the final conformation is (R,S). Likewise, **28**-(--,R) gives 69% ee of the S-aldehyde, which suggests that the complex formed assumes an (S,R) conformation, since now the R-bisnaphthol induces the formation of an S-diphenyl bridge. The control by the binaphthyl bridge in **26** is somewhat more efficient than that of the bisnaphthol moiety in **28**.

In the free ligand **28** we saw a slight preference for the S,R diastereomer, but in the complexes there is a strong preference for the S,R or R,S conformation. The NMR spectra of the complexes of the two non-rigid ligands **26** and **28** led to this conclusion but other data also support this observation. In the free ligands with R,S configurations the through-space coupling constants between the two phosphorus nuclei are always larger than those in R,R ligands. This shows that in R,S ligands the ligand has a natural tendency to bring the donor atoms into close proximity. Secondly, in competition experiments between R,R and R,S ligands the latter give the most stable complexes and are prevalently formed in the mixtures. In diphosphites, the bisphenol configurations were also observed to be controlled by the backbone and the most stable configurations lead to the highest ee's in hydroformylation catalysis [4, 14, 18, 38].

5.3.3 [RhH(CO)$_2$(BINAPHOS)] complexes; models for enantioselectivity

Thus, the sense of the enantioselection is predominantly controlled by the phosphine moiety and opposite configurations of the binaphthyl and bisphenol moieties give the highest ee. Complexes [RhH(CO)$_2$(L-L)] (L-L= **24**-(R,S), **24**-(R,R), **26**-(R,--), **27a**-(S,R), **27b**-(R,R) have been characterized by ^{31}P and ^{1}H NMR spectroscopy. In all cases a single species was observed in which the phosphine occupies an equatorial position and the phosphite an apical position [43]. The NMR data indicate that the hydride is oriented *cis* to phosphine and *trans* to phosphite. The $^2J_{P\text{-}H}$ coupling constant (phosphite – hydride) is ~160 Hz which corresponds to an apical hydride *trans* to phosphite in a trigonal bipyramidal rhodium complex [14, 24], except for **24**-(R,R) for which it is smaller. All data point to a high preference for a coordination mode with equatorial phosphine and apical phosphite. This is unexpected, since for a d^8 electron configuration one would expect the strongest acceptor ligand to be in the equatorial plane. As yet there is no explanation for this phenomenon.

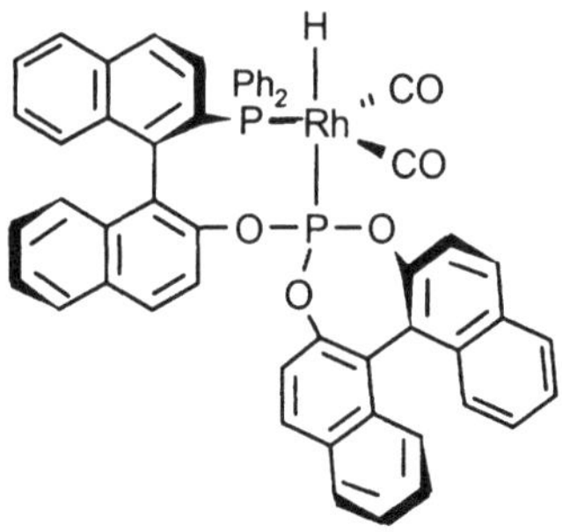

Figure 19. $RhH(CO)_2$(BINAPHOS) complex

The spectroscopic data of phosphine-phosphite complexes show slight differences between the (R,S) and (S,R) complexes on the one hand and the (R,R) and (S,S) complexes on the other. In the latter compounds the data are less clear-cut, which could be due to structural deviations of the monohydride complexes from an ideal trigonal bipyramid or to equilibria between isomers. The two **ea** isomers cannot be distinguished by IR spectroscopy as their absorptions tend to coincide [49].

^{31}P NMR spectroscopy shows that the complex [$RhH(CO)_2$(R,S-**24**)] is a single species at 60 ° C and 25 ° C in toluene-d_8 under 1 bar of CO. No apical-equatorial interchange was observed at any temperature. The unique dissymmetric environment in a single catalytically active species created by the phosphine-phosphite seems to be an important factor in the high enantioselectivity obtained.

Mechanistic studies have been made of the asymmetric hydroformylation of the BINAPHOS system [4, 44]. Deuterioformylation experiments (Chapter 2) have been used to study the alkene insertion step [45]. Studies using [$RhH(CO)_2$(R,S-BINAPHOS)] indicate that the alkene-insertion is irreversible for styrene at CO/H_2 total pressures between 20 and 100 bar, but the insertion is partially irreversible at a total pressure of 1 bar [44].

The models which explain the enantiofacial selection in the Rh-BINAPHOS systems assume that the alkene insertion in the Rh-H bond is irreversible. The old model proposed by Consiglio and Pino [46] has been used by Takaya [14]. The model involves partitioning the space around the metal center into four quadrants and for each of these the steric encumbrance is estimated. This simple model differentiates between the enantiofaces of the alkene and it has often been successfully applied. For $RhH(CO)_2$(R,S-BINAPHOS) the aldehyde enantiomer expected from this model was indeed obtained [15].

While the simplicity of this model is attractive, the many studies on asymmetric hydrogenation have shown that such models can be deceptive and that more advanced studies are needed. Gleich et al used a semi-quantitative theoretical model that combined quantum mechanics and molecular mechanics. Their findings matched previously reported experimental data; the aldehyde found experimentally had the lowest barrier of formation and the (R,R)-BINAPHOS complex should give much lower ee's than the (R,S)-BINAPHOS complex [14, 15].

5.3.4 Separation studies for the BINAPHOS system

The importance of recovering of the catalyst used in hydroformylation is described in chapters 7 and 8. Chiral phosphine-phosphite Rh(I) complexes have been immobilized on polystyrene matrices so that they can be used in the hydroformylation of alkenes. After introducing a vinyl group into the phosphine-phosphite **24**, the resulting ligand can be copolymerized with styrene derivatives to provide a highly cross-linked polymer matrix. The high levels of catalytic activity and selectivity obtained in the homogeneous hydroformylation were reproduced in the immobilized system [47].

The BINAPHOS system has also been modified by using the perfluoroalkyl-substituted derivative of (R,S)-BINAPHOS (see Chapter 10) in the rhodium catalyzed hydroformylation of vinyl arenes in liquid or supercritical carbon dioxide as solvent This system leads to similar catalytic activity and the same level of enantiocontrol as the rhodium-(R,S)-BINAPHOS system in organic solvents [48].

5.3.5 Chiral Phosphine-Phosphite Ligands containing a stereocenter in the backbone

Van Leeuwen et al designed a series of phosphine-phosphite ligands that contain a phosphine with a stereogenic phosphorus atom, a biaryl phosphite and a backbone with a stereocenter generated from chiral epoxides. The moieties combine to form diastereomers **29** and **30** (Figure 20). Systematically varying the ligands shows how the stereocenters affect the enantioselectivity. The substituents on the chiral phosphine are in close vicinity to the metal. The stereocenter in the backbone is close to the phosphite moiety.

The behavior of the substituted bisphenol **29** and bisnaphthol **30** derivatives show what influence the biaryl moiety has. Complexes $[RhH(CO)_2(\text{phosphine-phosphite})]$ have been used as catalysts in the asymmetric hydroformylation of styrene.

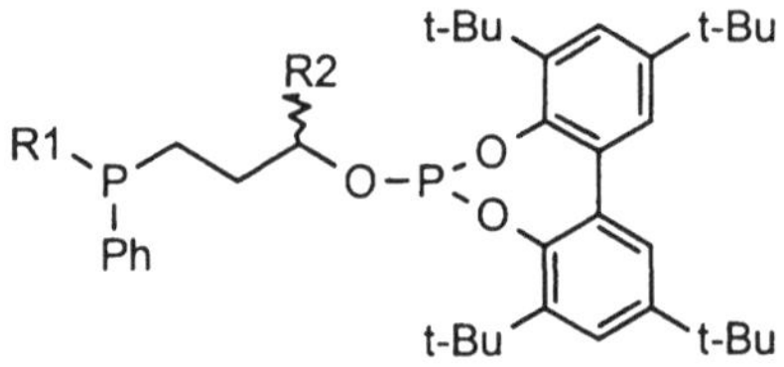

29a (S_p,S_c) R_1=1-Naphthyl, R_2=Ph
29b (S_p,S_c) R_1=2-Anisyl, R_2=Ph
29c (S_p,R_c) R_1=1-Naphthyl, R_2=Ph
29d (--,S_c) R_1=Ph, R_2=Ph

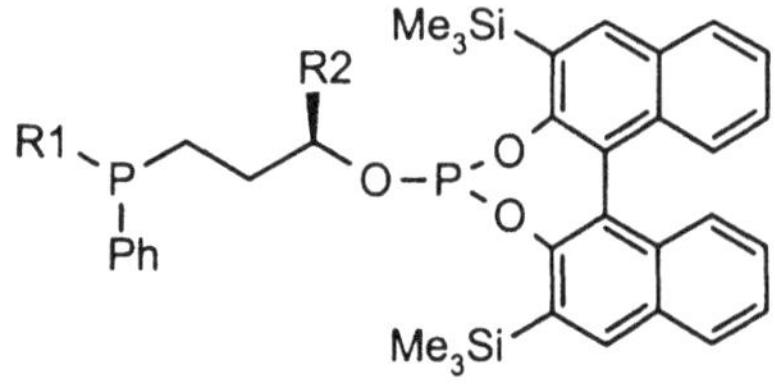

30a (S_p,S_c,R) R_1=1-Naphthyl, R_2=Ph
30b (S_p,S_c,S) R_1=1-Naphthyl, R_2=Ph

Figure 20. Examples of phosphine-phosphite ligands

Table 4. Asymmetric hydroformylation of styrene catalyzed by Rh/phosphine-phosphite ligands

Ligand	Conversion % (h)	b/l [a]	TOF [b]	ee % [c]
29a S_PS_C	55 (24)	11	45	63 (S)
29b S_PS_C	24 (25)	20	18	57 (S)
29c S_PR_C	7 (6)	-	24	33 (R)
29d --S_C	69 (25)	2	56	18 (S)
30a S_PS_CR	95 (18)	9	111	40 (S)
30b S_PS_CS	26 (19)	8	25	4 (S)

Styrene:Rh= 2000. Ligand to rhodium ratio is 4. T °C = 50. [a] Branched to linear aldehydes ratio.[b] TOF in mol styrene (mol Rh)$^{-1}$h^{-1} determined at the end of the reaction time by GC. [c] Enantiomeric excess. Absolute configuration is given in parentheses.

The results in Table 4 show regioselectivities for the branched aldehyde larger than 92% and ee's as high as 63% at 20 bar and 50 °C. A ligand to rhodium ratio of ~4 is required to obtain enantiomeric excesses. The results of the hydroformylation reaction using bisphenol and bisnaphthol derivatives show the different effects of the substituents in the backbone, the substituents at the phosphine, and the phosphite moieties. A phosphine with a stereogenic phosphorus atom must be present if enantioselectivity is to be high. The 1-naphthyl group of ligand **29a** and the *o*-anisyl group of ligand **29b** both have a considerable effect on the enantioselectivity.

The configuration of the stereocenter in the backbone controls the absolute configuration of the major enantiomer, although the enantiomeric excess for **29d** containing the phenyl group in the bridge is small. The configuration of the bisnaphthol moiety in the phosphite has also an important influence on the ee. The diastereomer containing the (S)-bisnaphthol moiety **30b** provides only 4% ee, while ligand **30a** with a (R)-bisnaphthol moiety results in an ee of 40%. The results obtained with the ligands **29a**, **30b**, **30a**, show that ligand **29a** leads predominantly to the same conformation as ligand **30a**. The conformation of the bisphenol moiety at the phosphite is controlled by the substituent at the stereocenter in the backbone, which also controls the configuration of the product.

The coordination mode of the ligands in the [RhH(CO)$_2$(phosphine-phosphite] complexes has been studied using IR and NMR spectroscopy under 20 bar of CO/H$_2$. The HP-IR spectrum shows the presence of a single species and HP-NMR spectroscopy shows that the phosphine and the phosphite occupy an apical and equatorial position respectively, as would have been expected from the donor and acceptor properties. This is in contrast with the coordination mode found for the [RhH(CO)$_2$(phosphine-phosphite)] complex of BINAPHOS in which the phosphine occupies an equatorial position and the phosphite an apical position [49].

The results of hydroformylation show that both the configuration of the phosphite moiety, coordinated at an equatorial position, and the chiral phosphine at the apical position are important for obtaining high enantioselectivity. The experiments done using a ligand containing a stereogenic bisnaphthol group shows a cooperative effect of the bisnaphthol moiety and the group attached to the stereocenter in the backbone [49]. In the case of the BINAPHOS catalyst, the equatorially coordinating phosphine (or rather the binaphthyl bridge) has a slightly larger influence on the ee than the apically coordinating phosphite, although the results show that a strict separation is impossible.

5.4 Diphosphine rhodium catalysts

5.4.1 Introduction

Diphosphines are among the most efficient chiral ligands in homogeneous catalysis. Their characteristics and how they are used as ligands in the hydroformylation reaction are described in chapter 4. Many chiral diphosphines have been used as ligands in rhodium systems for asymmetric hydroformylation. Despite all the effort made to apply diphosphines in asymmetric hydroformylation, the enantioselectivities of rhodium-diphosphine systems are not as high those of rhodium-diphosphite or rhodium phosphine-phosphite systems.

5.4.2 C_1 diphosphines as chiral ligands

C_1 diphosphines have been much less extensively studied than C_2 chiral diphosphines. As an example of a C_1 diphosphine we mention BPPM, **31**, which had previously shown a high activity and enantioselectivity in Pt/Sn based hydroformylation catalysts giving ee's up to 96%, albeit with a low chemo- and regioselectivity, characteristic of the Pt/Sn systems. In the catalytic systems based on rhodium, BPPM was used in the asymmetric hydroformylation of several substrates such as N-vinylphthalimide, tricarbonylchromium complexes and arylalkenes, but the ee's were always low at 10-14% [2-4].

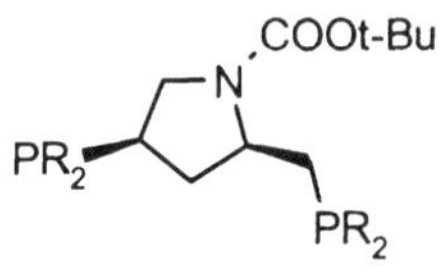

31a BPPM R = Ph
31b BPPM-DBP R_2= DBP

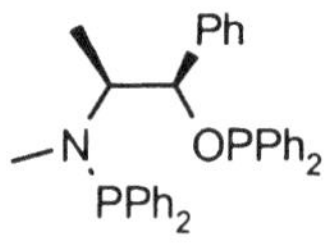

32 EPHOS

Figure 21. C_1 chiral diphosphines

Mortreux et al. prepared diphosphine **32**, EPHOS, derived from (S,R)-ephedrine. This phosphinite-phosphorus amide was applied in the hydroformylation of styrene in the presence of a rhodium chloride precursor as well as an electrochemically generated rhodium catalyst. The enantioselectivity increased to 31% (96% regioselectivity) when electrochemical reduction was applied. This is the first non-phosphine system in which the structures of the resting state of the catalyst were investigated under 1 bar of CO/H_2. ^1H and ^{31}P NMR data show that there is a major hydrido rhodium complex $[RhH(CO)_2(diphosphine)]$ where the ligand coordinates in an **ea** mode, in equilibrium with the $[Rh(\mu\text{-}CO)(CO)(S,R\text{-}EPHOS)]_2$ dinuclear complex [50]. A bidentate ligand such as EPHOS, having two very distinct absorptions for the amido phosphorus and the phosphinite phosphorus nuclei gives NMR spectra that are readily interpreted. The hydride resonance shows a large coupling constant (116 Hz) with the amidite phosphorus atom in *trans* position and smaller coupling constants with rhodium (10.1 Hz) and the phosphinite phosphorus nucleus (9.9 Hz). The phosphorus spectra are in accord with this.

5.4.3 C_2 diphosphines as chiral ligands

Many chiral C_2 diphosphines have been used as ligands in rhodium systems for asymmetric hydroformylation. It can be observed that the chiral diphosphines providing the highest enantioselectivities in asymmetric hydrogenation lead to low chiral induction in hydroformylation. The C_2-symmetric diphosphines, which were developed for square planar coordination in hydrogenation catalysts, will coordinate in an apical-equatorial manner. The intermediate four-coordinate species can undergo an approach from the alkene at either side of the complex. As suggested for phosphites [10], these two sides of the complex may lead to opposite, enantiospecific binding modes of the alkene. The scheme presented in 5.2.4 for diphosphites illustrates this.

DIOP and related ligands. The first reports on asymmetric hydroformylation using DIOP as an asymmetric diphosphine date back to 1973 [51]. Throughout the seventies and eighties platinum and rhodium systems with DIOP were extensively studied. Several modifications have been reported, as for instance the introduction of different substituents at the phosphorus atoms, the application to different substrates, the study of the reaction conditions. DIOP and several DIOP type ligands were used in the asymmetric hydroformylation of unsaturated nitrogen compounds, aliphatic alkenes, N-acyl-1-aminoacrylic acid derivatives, unsaturated carboxylic acids and esters (see Figure 23). In the hydroformylation of alkyl α -N-acylaminoacrylates using rhodium complexes of DIOP, **33**, and a related ligand DIOCOL, **34**, the reaction takes place with complete selectivity for the branched aldehyde and an ee of 60 %.

It has been said that the regiospecific addition of carbon monoxide to the most substituted carbon is due to the polarization of the double bond and to the acylamide group's chelation control [52]. Although the enantioselectivity is not exceptional, it is higher than normal in the rhodium hydroformylation of vinylidenic alkenes.

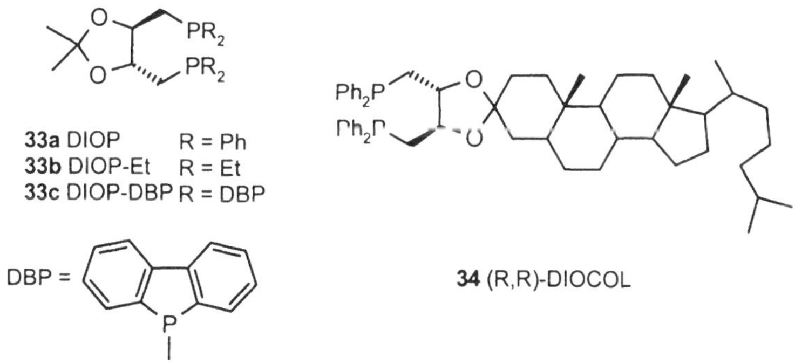

Figure 22. DIOP derivatives and DIOCOL diphosphines

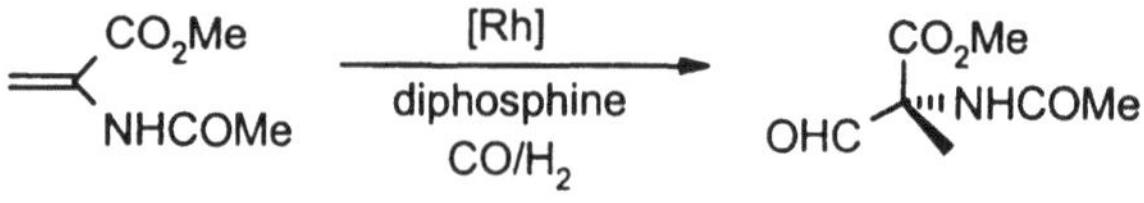

Figure 23. Hydroformylation of acrylic acid derivatives

Many C_2 diphosphines have been used in the asymmetric hydroformylation reaction. The enantioselectivities, however, are always low or moderate, generally no higher than 30% [2-4].

Diphosphines with small bite angles. BDPP and related diphosphines.
In 1996 a rhodium system was described which provided ee's as high as 60% and which had a regioselectivity for 2-phenylpropanal of 95% despite having a diphosphine as simple as BDPP, **35** (see Figure 24). The ligand to rhodium ratio had to be 2 and the temperature as low as 65 °C to obtain this enantioselectivity [53]. Other diphosphines such as DIOP, **33**, BINAP, or CHIRAPHOS, **36**, used under the same conditions provide lower ee's (12-25%) [53] (see Table 5). Diphosphines such as DIOP or BINAP form seven membered rings when coordinating to rhodium. On the basis of previous findings described in chapter 4, one might consider the formation of both bis-equatorial and equatorial-apical intermediates, especially for the more flexible DIOP. As outlined in sections 5.2 and 5.3 the occurrence of a variety of species does not seem to be a good starting point for obtaining high ee's. On the other hand, CHIRAPHOS or BDPP with smaller bite angles between 80-90° [54] probably form equatorial-apical bipyramidal intermediates only, which limits the number of possible structures. In section 5.4.4 it will be shown that BDPP indeed forms only **ae** species [55]. It was also argued in 5.2 that **ae** coordination leads to four-coordinate intermediates, the two faces of which may be effectively mirror images leading to opposite enantiomeric products.

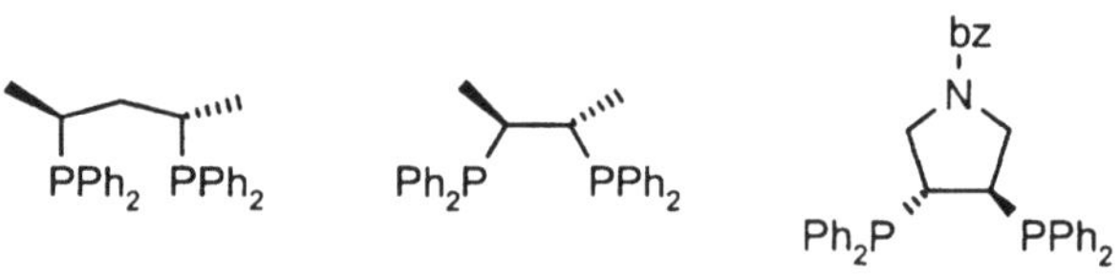

35 BDPP **36** CHIRAPHOS **37** DEGUPHOS

Figure 24. BDPP, CHIRAPHOS and DEGUPHOS

For the rhodium catalytic systems containing BDPP or CHIRAPHOS the regioselectivities depend in the same way on pressure and temperature. The amount of branched aldehyde increases when the pressure is raised and/or when the temperature decreases. The same behavior had previously been observed in the hydroformylation of styrene with the unmodified rhodium catalyst (chapter 2) [45, 56].

Table 5. Asymmetric hydroformylation of styrene with C_2 chiral diphosphines

Ligand	P (bar)	T °C	L/Rh[a]	Conversion [b] %	b/l	ee % [c]
33 (S,S)	10	65	2	94	61/39	12 (S)
35 (S,S)	8	65[d]	2	33	95/5	43 (S)
35 (S,S)	8	80	1	76	74/26	7 (S)
35 (S,S)	8	80	2	76	79/21	51 (S)
35 (S,S)	30	80	2	45	94/6	58 (S)
36 (R,R)	8	80	1	98	85/15	15 (R)
36 (R,R)	8	80	2	99	97/3	25 (R)
36 (R,R)	30	80	2	100	95/5	21 (R)
37 (R,R)	8	65[d]	2	97	94/6	>2 (S)

10 mmol of styrene and 0.0125 mmol of $[Rh_2(\mu\text{-}(OMe)_2(cod)_2]$ in toluene [a] Ligand to rhodium ratio. [c] Enantiomeric excess by GC. Absolute configuration is given in parentheses. [d] Time is 22 h.

The enantioselectivities for the system Rh/CHIRAPHOS are in the range of 15-28% ee and there was not significant change in the enantioselectivity when the ligand to rhodium molar ratio was increased. However, the ee's of the catalytic system with BDPP the ee depends on the ligand/rhodium molar ratio, and can vary from 10% at a diphosphine-rhodium ratio of 1 until 58% for a diphosphine/rhodium ratio of 2 [53, 56].

A related bis(diphenylphosphino)pyrrolidine ligand, DEGUPHOS **37**, has been used under the same conditions as those for BDPP and CHIRAPHOS. The enantioselectivity however was nearly null. An NMR study showed that there were five-coordinated rhodium species, in which the ligand is coordinated in both bridging and chelating modes. The bridging coordination mode also involves monodentate species and is considered to be the cause of the lack of enantioselectivity [57].

Several diaryl phosphines have been sulphonated to form a rhodium catalytic precursor for the hydroformylation of styrene in an aqueous solution. The enantioselectivities reported for (S,S)-BDPPTS are in the range of 14-17% only [58].

Diphosphines having wide bite angles. BISBI and related diphosphines

Herrmann et al. reported how chiral diphosphines with wider natural bite angles were applied in the asymmetric hydroformylation of styrene. The diphosphines related to the BISBI ligand, **38-40** coordinate in a bis-equatorial mode to the metal. NAPHOS **39** and a 2,2',6,6'-tetrasubstitued derivative of BISBI named BIPHOLOPHOS **40** have been prepared and sulphonated to be used in mono- and biphasic asymmetric hydroformylation.

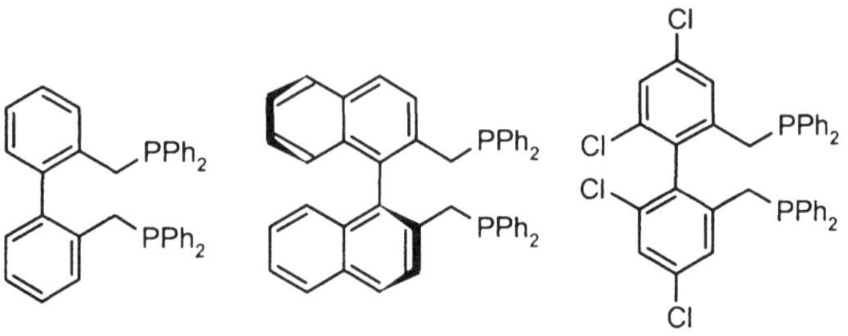

38 BISBI **39** NAPHOS **40** BIPHOLOPHOS

Figure 25. BISBI and related ligands

The rhodium system formed using (-)BIPHOLOPHOS provides an ee of 15% at 40 °C, whereas only racemic product was detected at 80 °C. The results have been best at 100 bar and 40 °C for this ligand: the ee was 34% with a regioselectivity in branched aldehyde of 96%. The binaphthyl backbone in NAPHOS **39** seems to produce a better enantiodiscrimination than the substituted biphenyl backbone [59, 60]. The NMR characterization of the pentacoordinate [RhH(NAPHOS)$(CO)_2$] complex show **ee** coordination of these diphosphines [59].

As a continuation of an earlier study on biphasic hydroformylation [61] Herrmann et al. prepared chiral ligands which were highly soluble in water by direct sulfonation of the enantiopure diphosphines NAPHOS and BIPHOLOPHOS. They then applied them in the aqueous/organic biphasic hydroformylation of styrene [60]. The enantioselectivities obtained in the two-phase hydroformylation of styrene are in general lower than those obtained by the corresponding homogeneous catalysts in organic solvents. The highest ee achieved in a two-phase hydroformylation of styrene is 18%, using the sulphonated NAPHOS ligand [59]. This trend has been observed in many other catalytic asymmetric reactions; application of water-soluble catalysts normally decreases the optical induction.

5.4.4 The Rh/BDPP system. HPNMR and HPIR studies under hydroformylation conditions.

Characterization of the intermediates. The solution structure of the species formed during the reaction of [Rh(μ-OMe)(cod)]$_2$ with (-)-BDPP under hydroformylation conditions has been determined at 20 bar CO/H_2 at room temperature [55].

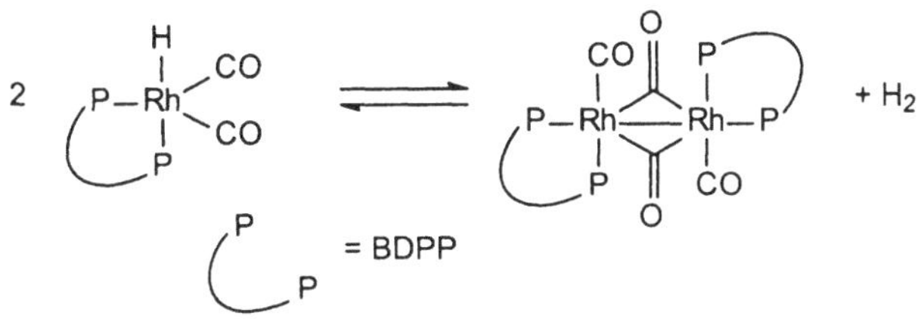

Figure 26. Mononuclear and dinuclear species in equilibrium

The ^{31}P NMR spectrum reveals the presence of two species in equilibrium, the [RhH(BDPP)$(CO)_2$] hydride complex and the dimer [Rh(BDPP)(μ-CO)(CO)$]_2$. The mononuclear [RhH(BDPP)$(CO)_2$] complex shows only one doublet at 29.8 ppm ($^1J_{Rh-P}$)=112 Hz and the ^1H NMR spectrum revealed a double triplet in the hydride region $^1J_{Rh-H}$= 11, $^2J_{H-P}$= 57 Hz, indicative of a coupling of the hydride with the rhodium nucleus and two degenerated phosphorus nuclei. Small *cis* phosphorus-hydride coupling constants are reported for complexes with **ee** coordinating diphosphine ligands [62] and relatively large phosphorus-hydride coupling constants are characteristic of complexes containing *trans* P-Rh-H arrangements [22, 24]. These studies revealed that the diphosphine BDPP is equatorially-axially coordinated, in accord with the calculated bite angle of 91° [54], and that a fast exchange of the two phosphine groups occurs. The "turn-stile" mechanism is the same as that described for diphosphites (5.2) [21].

The monomer–dimer equilibrium. Since the work of Wilkinson it has been known that complexes of the type [RhH$(CO)_2$(L-L)] and [Rh(L-L)(μ-CO)(CO)$]_2$ are in equilibrium with one another. The equilibrium constant depends on the ligand used and the concentrations of rhodium and hydrogen. Many authors have studied on the formation of these dinuclear species, but no quantitative data had been reported before the study involving BDPP [50, 62]. In the Rh-BDPP system, both species can be observed and the equilibrium constant is measured by ^{31}P NMR. The data indicate that at room temperature, relatively high catalyst concentrations, and partial H_2 pressures of below 5 bar, rhodium will be present predominantly as the dimeric complex. More monomer is formed when the H_2 pressure is raised, and hence, the rate of hydroformylation increases with the pressure, regardless the mechanism of the rhodium hydride catalyst. The equilibrium, therefore, is of great importance to the overall kinetics observed during hydroformylation.

In addition the ^{31}P NMR data show the presence of other species containing more than two phosphorus atoms, in which the diphosphine acts as monodentate. Hughes et al. [63] have described related complexes (see chapter 4).

In situ HPIR asymmetric hydroformylation. Under actual catalytic conditions, high pressures, low concentrations, at 65°C and 10 bar, the IR spectra were measured. They revealed strong carbonyl absorptions at 1988 and 1944 cm^{-1}, characteristic of the mononuclear hydride dicarbonyl complex, and only weak carbonyl absorptions at 1745 and 1724 cm^{-1}, due to a small quantity of the dinuclear complex. The substrate was then added and the hydroformylation reaction was monitored. The absorptions corresponding to $RhH(BDPP)(CO)_2$ are retained throughout the reaction indicating that the hydride is the resting state under these "standard" hydroformylation conditions (see Chapter 4), at least for BDPP [55].

5.5 Mechanistic considerations

5.5.1 Regioselectivity

Regioselectivity is a key issue in hydroformylation. While for simple 1-alkenes the emphasis is on the formation of linear aldehydes (see previous chapters), in asymmetric hydroformylation the branched, chiral aldehyde is the desired one. Many functional alkenes have a high propensity for the formation of the branched product. In this chapter we have concentrated on styrene, but before discussing the regioselectivity, it is instructive to have a look at other alkenes (see Figure 27).

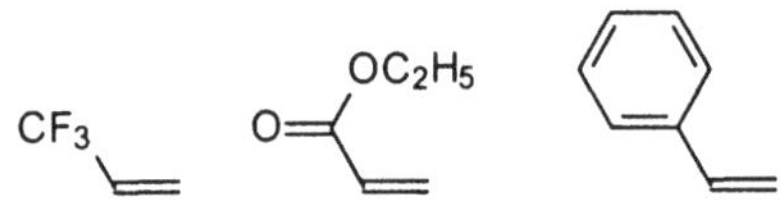

Figure 27. Functionalized alkenes

3,3,3-Trifluoropropene was studied by Ojima [64] and rhodium catalysts gave a highly branched product at very high rates. If we assume that alkene insertion into the rhodium-hydride bond is irreversible, the regioselectivity is determined in this step. Theoretical studies point to an early transition state for this reaction [15]. What would determine this early transition state? Both frontier orbital combinations would lead to a secondary rhodium alkyl bond. The electron withdrawing CF_3-group would also stabilize the secondary alkyl anion and a late transition state would lead to the same alkyl species. Thus the branched product would form provided that no equilibration occurs between the linear and the branched species. If there were to be equilibration, the result would change drastically, because the primary alkyl group would migrate much faster than the electronically stabilized secondary alkyl group [65]. These are the electronic considerations. For steric reasons one would expect a preference for the linear alkyl.

Ethyl acrylate, the hydroformylation of which has been studied many times, is an interesting case. Tanaka et al. [66] found that the ratio of l:b could be varied from 100:0 (80 °C and 1 bar) to 1:99 (40 °C and 30 bar) by using a rhodium catalyst and triphenyl phosphite as the ligand. Electronic arguments lead to the same result as for trifluoropropene because the frontier orbitals on ethyl acrylate are the same. Thus, high pressures lead to the expected result.

Migration of the stabilized ester enolate, however, is expected to be much slower than that of the linear alkyl, irrespective of whether it is η^3 or η^1 bonded. At low pressure the catalyst "rests" in the enolate state of the cycle for the branched intermediate and the formation of the enolate can be reversed. The product-forming route is now via the linear alkyl, which undergoes a relatively rapid migration to form the acyl species (see Figure 28). The role of CO may be either to form a species containing more CO molecules or to trap the acyl species after migration. In the absence of more data this remains speculation.

"Steric control" in the figure is an oversimplification, because this route is also taken when migration of the ester enolate is too slow, irrespective of steric control as outlined above.

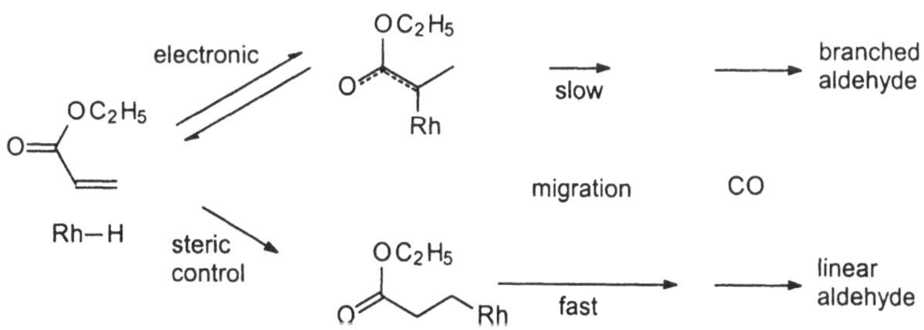

Figure 28. Hydroformylation of ethyl acrylate

In a broad range of conditions the rate laws for the rate of formation of the linear and the branched product are different, which explains the enormous influence that these have on the product distribution. At low CO pressures the enolate-rhodium species may be the resting state of the catalyst. Species of this type often undergo hydrogenation instead of hydroformylation.

For styrene the reactivity pattern is less obvious. The assumption that there is an early transition state does not lead to a simple conclusion, because the coefficients of both the LUMO and the HOMO of styrene are higher on the terminal carbon atom. If the interaction of the HOMO on the Rh-H located on the hydrogen with the LUMO of styrene were to dominate the branched alkyl would be formed. Clearly the resonance stabilized branched alkyl is more stable than the linear phenylethyl species. Thus, electronically the branched species is preferred as most authors have argued (see also Chapter 2). In addition the branched alkyl species may be stabilized by η^3 coordination, but the homo-allylic linear species is also stabilized by alkene-rhodium interaction.

Insertion of styrene does not always lead to branched alkyl species. Insertion of styrene into a cationic palladium hydride species may give purely linear alkyl species, but perhaps not for steric reasons [67]. An early transition state for this process may involve the interaction of the LUMO for PdH^+, which has the highest coefficient on palladium, with the HOMO of styrene, which has the highest coefficient on the terminal carbon atom.

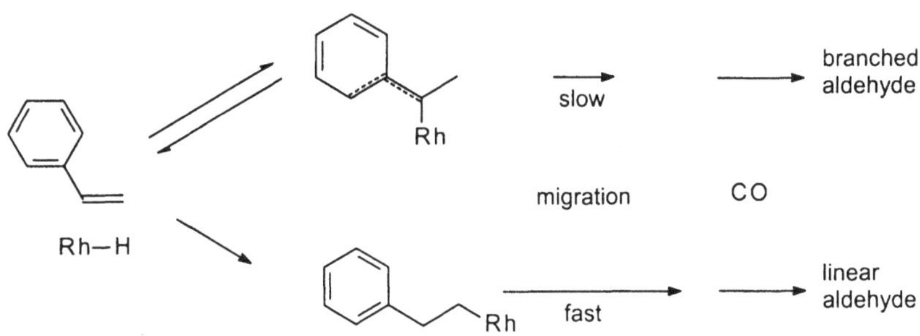

Figure 29. Hydroformylation of styrene

The influence of the conditions on the product distribution is the same as for ethyl acrylate (see Figure 29). High temperatures and low pressures give linear product (up to l:b=3:1), and low temperatures and high pressures lead to branched products (l:b=1:99). For carbonyl based catalysts, phosphite based catalysts, and phosphine based catalysts this has been well documented (Chapters 2, 3, and 4). The reversibility of the branched alkyl formation is also accounted for.

5.5.2 Enantioselectivity and conclusions

Two types of ligands have been successfully applied in asymmetric hydroformylation using rhodium catalysts: diphosphite ligands and phosphine-phosphite ligands. For both ligands, complexes with the formula $RhH(L\text{-}L)(CO)_2$ have been characterized as the resting state of the catalyst. The formation of only one diastereomeric complex and alkene adduct is thought to be crucial if enantioselectivity is to be high.

In the case of the diphosphites, compounds containing seven-, eight-, and nine-membered rings have been studied. The bis-equatorial coordination mode that involves an eight-membered ring provides the highest enantioselectivities. Chiral cooperativity between the central backbone and the aryl substituents as well as the influence of the terminal groups has been observed for several systems.

In $RhH(BINAPHOS)(CO)_2$ the ligand coordinates in an equatorial-apical fashion. Both apical-equatorial isomers of the phosphine-phosphite bidentates have been identified. For BINAPHOS and related systems, the unique dissymmetric environment created by phosphine-phosphite seems to be determinant for enantioselectivities to be high. As for diphosphites, phosphine-phosphite systems require a well-tuned cooperation between the chiral centers in the ligand.

High enantioselectivities have not still been found for rhodium chiral diphosphine systems in asymmetric hydroformylation.

As pointed out in Chapters 3 and 4 the intermediate alkene adducts of the diphosphite catalysts probably contain bis-equatorial bidentate ligands and the intermediates of BINAPHOS contain an equatorial–apical phosphine-phosphite ligand. Thus the structures involved in the migratory insertion are different for the two systems.

The models used for rationalizing the enantioselectivities have been discussed in section 5.3.3. The simple four-quadrant model was cited [46], which gave good results for many substrates, but often failed for styrene. Gleich and Herrmann [15,16] have applied computational methods to hydroformylation using a combination of MM and QM. The predictive power of this approach will certainly increase in the near future.

References

1 (a) Sheldon, R. A. *Chemistry and Industry*, **1990**, 212. (b) Botteghi, C.; Paganelli, S;. Schionato, A; Marchetti, M. *Chirality*, **1991**, *3*, 355.

2 Consiglio, G. *Catalytic Asymmetric Syntheses*, I. Ojima editor VCH Publishers, New York, **1993**.

3 (a) Agbossou, F.; Carpentier, J. F.; Mortreux, A. *Chem Rev.* **1995**, *95*, 2485. (b) Gladiali, S.; Bayón, J .C.; Claver, C. *Tetrahedron Asymmetry,* **1995**, *7*, 1453.

4 Nozaki, K. *"Comprehensive Asymmetric Catalysis"*; Jacobsen, E. N.; Pfaltz, A., Yamamoto, H. Ed.; Springer Heidelberg, **1999**; Vol.I , pp 381-409.

5 Stille, J. K.; Su H.; Brechot, P.; Parrinello, G.; Hegedus, L. S. *Organometallics,* **1991**, *10,* 1183.

6 a) Consiglio, G., Nefkens, S. C. A, *Tetrahedron Asymmetry*, **1990**, *1* 417. b) Consiglio, G.; Nefkens, S. C. A, Borer, A. *Organometallics*, **1991**, *10*, 2046.

7 Cserépi-Szücs, S. Bakos, J. *Chem. Commun.* **1997**, 635.

8 Stanley, G. G. *Catalysis of Organic Reactions*, ed. Scaros, M. G.: Prunier, M. L.: Marcel Dekker Inc. New York, **1995**, p. 363.
9 Clark jun. C. E.*: Diss Abstr.Int. B*, **1973**, *36*, 2220.
10 Wink, J. D.; Kwok ,T. J. ; Yee A. *Inorg. Chem.* **1990**, *29*, 5007
11 Sakai, N.; Nozaki, K.; Mashima, K.; Takaya, H.*Tetrahedron . Asymmetry*, **1992**, *3*, 583.
12 Babin, J. E.; Whiteker, G. T. *Pat*, WO 93/03830, 1992.
13 Buisman G. J. H., Vos, E. J.; Kamer, P. C. J.; van Leeuwen, P. W. N. M. *J. Chem. Soc. Dalton Trans*, **1995**, 409.
14 (a) Sakai, N.; Mano, S.; Nozaki, K.; Takaya, H. *J. Am. Chem.Soc.* **1993**, *115*, 7033. (b) Nozaki K.; Sakai, N.; Nanno, T.; Higashijima, T.; Mano, S.; Horiuchi, T.; Takaya, H. *J. Am. Chem. Soc.* **1997**, *119*, 4413.
15 Gleich, D.; Schmid, R.; Herrmann, W. A.; *Organometallics*, **1998**, *17*, 2141
16 Gleich, D.; Herrmann, W. A.; *Organometallics*, **1999**, *18*, 4354
17 Buisman, G. J. H.; Kamer, P. C. J.; van Leeuwen, P. W. N. M.; *Tetrahedron Asymmetry*, **1993**, *4,* 1625.
18 Buisman, G. J. H.; van der Veen, L. A.; Klootwijk,: de Lange, W. G. J.; Kamer, P. C. J.; van Leeuwen, P. W. N. M.; Vogt D. *Organometallics*, **1997**, 16, 2929.
19 van Leeuwen, P. W. N. M.; Buisman , G. J. H.; van Rooy, A.; Kamer, P. C. J., *Recl. Trav. Chim. Pays-Bas*, **1994**, *113*, 61.
20 van Rooy, A.; Kamer, P. C. J.; van Leeuwen, P. W. N. M.; Goubitz, K. L Fraanje, J.; Veldman, N.; Spek, A. L.; *Organometallics,* **1996**, *15*, 83
21 Buisman, G. J. H.; van der Veen, L. A.; Kamer, P. C. J.; van Leeuwen, P. W. N. M.; *Organometallics*, **1997**, *16*, 5681.
22 Meakin, P.; Muetterties, E. L.; Jesson, J. P. *J. Am. Chem. Soc.* **1972**, *94*, 5271.
23 Meakin, P.; Jesson, J. P.; Tebbe, F. N.; Muetterties, E. L. *J. Am. Chem. Soc.* **1971,** *93*, 1797.
24 Hyde, E. M.; Swain, J. R.; Verkade, J. G.; Meakin, P.; *J. Chem. Soc. Dalton Trans*, **1976**, 1169.
25 van Rooy, A.; Kamer, P. C. J.; van Leeuwen, P. W. N. M.; Veldman, N.; Spek, A. L. *J. Organometal. Chem.* **1995**, *494*, C15.
26 (a) Berry, R. S.*; J. Chem. Phys.*; **1960**, *32*, 933. (b) Muetterties, E. L. *Acc.Chem. Res.*,**1970**, *3*, 266.
27 Cserepi-Szucs, S.; Tóth, I.; Párkány, L. Bakos, J.; *Tetrahedron: Asymmetry*, **1998**, *9,* 313.
28 Cserepi-Szucs, S.; Huttner, G.; Zsolnai, L. Bakos, J. J. *Organomet. Chem.*, **1999**, *586*, 70.
29 Cserepi-Szucs, S.; Huttner, G.; Zsolnai, L.; A. Szölösy, A.; Hegedüs, C.; Bakos J. *Inorganica Chimica Acta*, **1999**, *296*, 222.
30 Jiang ,Y.;. Xue, S.; Li, Z.; Deng, J.; Mi, A.; Chan, A. S. C. *Tetrahedron: Asymmetry*, **1998**, *9,* 3185.
31 (a) Reetz, M. T.; Neugebauer, T. *Angew. Chem. Int. Ed.* **1999**, *38*, 179. (b) RajanBabu, T. V.; Radetich, B.; You, K.; Ayers, T. A.; Casalnuovo, A.L.; Calabrese, J. C.; *J. Org. Chem.* **1999**, *64*, 3429 and references cited therein.
32 Brown, J. M.; Cook, S. F., *Tetrahedron*, **1986**, *42*, 5105.
33 RajanBabu, T. V.; Casalnuovo, A. L. *Pure and Appl. Chem.* **1994**, *66* 1535.
34 RajanBabu, T. V.; Ayers T.A. *Tetrahedron Letters* **1994**, *35*, 4295.
35 Buisman, J. H. G.; Martin, M. E.; Vos, E. J.; Kamer, P. C. J.; van Leeuwen, P. W. N. M *Tetrahedron Asymmetry*, **1995**, *6*, 719.

36 Pàmies, O.; Net, G.; Ruiz, A.; Claver, C. *Tetrahedron Asymmetry*, **2000,** *11*, 1097.

37 Kadyrov R.; Heller, D.; Selke, R. *Tetrahedron Asymmetry*. **1998**, *9*, 329.

38 Sakai, N.; Mano, S.; Nozaki, K.; Takaya, H.; *J. Am. Chem. Soc.*; **1993**, *115*, 7033.

39 Sakai, N.; Nozaki, K.; Takaya, H. *J. Chem Soc.Chem. Comm.* **1994**, 395.

40 Higashijima, T.; Sakai, N.; Nozaki, K.; Takaya, H. *Tetrahedron Letters,* **1994**, *35*, 2023.

41 a) Horiuchi, T.; Ohta, T.; Nozaki, K.; Takaya, H. *Chem Comm.* **1996**, 155. (b) Horiuchi T.; Ohta, T.; Shirakawa, E.; Nozaki, K.; Takaya, H. *J. Org. Chem.*; **1997**, *62*, 4285.

42 (a)Naieli ,S.; Carpentier, J-F.; Agbossou, F.; Mortreux, A.; Nowogrocki ,G.; Wignacout J. P. *Organometallics,* **1995**, *14*, 401 (b) Kless, A.; Holz, J.; Heller, D.; Kadyrov, R.; Selke, R.; Fischer, C.; Börner, A.; *Tetrahedron Asymmetry,***1996**, *7*, 33. (c) Basoli, C.; Botteghi, C.; Cabras, M. A.; Chelucci, G.; Marchetti, M. *J. Organomet. Chem.* **1995**, *488,* C20. (d).; Chelucci, G.; Cabras, M. A.; Botteghi, C., Basoli, C., Marchetti, M. *Tetrahedron Asymmetry,***1996,** *7*, 33.

43 Vaska, L. J. *J. Am. Chem.Soc.* **1996**, *88*, 4100.

44 Horiuchi, T.; Ohta, T.; Shirakawa, E.; Nozaki, K., Takaya, H. *Organometallics*, **1997**, *16*, 2981.

45 (a) Lazzaroni, R.; Raffaelli, A.; Settambolo, R.; Bertozzi, S.; Vitulli ,G.; *J. Mol. Cat.* **1989**, *50*, 1(b)) Lazzaroni, R; Ucello Barretta, G.; Bennetti, M.; *Organometallics*, **1989**, *8*, 2323 (c) Raffaelli A.; Pucci, S.; Settambolo, R.; Ucello Barretta, G.; Lazzaroni, R.; *Organometallics*, **1991**, *10*, 3892, (d) Lazzaroni, R.; Settambolo, R.; Ucello Barretta, G.; *Organometallics*, **1995**, *14*, 4644 (e) Lazzaroni, R.; Ucello Barretta G., Scamuzzi S.; Settambolo, R.; Caiazzo, A.; *Organometallics*, **1996**, *15*, 4657.

46 Consiglio, G.; Pino, P.; *Top Curr. Chem.*, **1982**, *105*, 77.

47 Nozaki, K.; Itoi, Y.; Sibahara, F.; Shirakawa , E.; Ohta , T.; Takaya, H.; Hiyama, T. *J. Am. Chem. Soc.* **1998,** *120*, 4051.

48 Francio, G.; Leitner, W. *Chem. Commun.* **1999**, 1663.

49 Deerenberg, S.; Kamer, P. C. J.; van Leeuwen, P. W. N. M. *Organometallics*, **2000**, *19*, 2065.

50 Pottier, Y.; Mortreux, A.; Petit, F.; *J. Organometal. Chem.* **1989**, *370*, 333.

51 Salomon, C.; Consiglio, G.; Botteghi, C.; Pino P. *Chimia,* **1973**, *27*, 215.

52 Gladiali, S.; Pinna, L. *Tetrahedron: Asymmetry*, **1991**, *2*, 263.

53 Masdeu-Bultó, A. M.; Orejon, A.; Castellanos, A.; Castillon, S.; Claver, C. *Tetrahedron Asymmetry*, **1996**, *7*, 1829.

54 Dierkes, P.; van Leeuwen, P. W. N. M. *J. Chem. Soc. Dalton Trans.* **1999**, 1519.

55 Castellanos-Páez, A.; Castillón, S.; Claver, C.; van Leeuwen, P. W. N. M.; de Lange, W. G. J. *Organometallics*, **1998**, *17*, 2523.

56 Diéguez, M.; Pereira, M. M.; Masdeu-Bultó, A. M.; Claver, C.; Bayón, J. C. *J. Mol. Cat.* **1999**, *143*, 111.

57 Freixa, Z.; Pereira, M. M.; Pais, A. A. C. C.; Bayon, J. C. *J. Chem. Soc. Dalton Trans.* **1999**, 3245.

58 Miquel-Serrano, M. D., Masdeu-Bultó, A. M.; Claver, C.; Sinou, D. *J. Mol. Cat.* **1999**, *143*, 49.

59 Eckl, R. W.; Primereir, T.; Herrmann, W. A. *J. Organometal. Chem.* **1997**, *532*, 243.

60 Rampf, F. A.; Splieger, M.; Herrmann, W. A. *J. Organometal. Chem.* **1999**, *582*, 204

61 Cornils, B.;. Herrmann, W. A (eds) *Applied Homogeneous Catalysis with Organometallic Complexes*, VCH, Weinheim, **1996**. Chap 4.1.

62 Casey, C. P.; Whiteker, G. T.; Melville, M. G.; Petrovich, L. M.; Gavney, J. A.;Jr.; Powell, D. R. *J. Am. Chem. Soc.***1992**, *114*, 5535.

63 Hughes, O. R.; Young, D. A. *J. Am. Chem. Soc.* **1981**, *103*, 6636.
64 Ojima, I.; Kato, K.; Okabe, M.; Fuchikami, T. *J. Am. Chem. Soc.* **1987**, *109*, 7714.
65 Cotton, J. D.; Dunstan, P. R. *Inorg. Chim. Acta,* **1984**, *88*, 223.
66 Yamashita, H.; Roan, B. L.; Sakakura, T.; Tanaka, M. *J. Mol. Catal.* **1993**, *81*, 255.
67 Pisano, C.; Mezzetti, A.; Consiglio, G. *Organometallics*, **1992**, *11*, 20.

Chapter 6

Hydroformylation in Organic Synthesis

Sergio Castillón† and Elena Fernández‡

†Departament de Química Analítica i Química Orgànica. ‡Departament de Química Física i Inorgànica. Universitat Rovira i Virgili. Plaça Imperial Tàrraco1, 43005 Tarragona, Spain.

6.1 Introduction

The discovery of new catalytic systems that enabled the regio- and steroselectivity of hydroformylation to be controlled has meant that this reaction has emerged as a flexible and important tool in organic synthesis. Some interesting reviews on asymmetric hydroformylation (see chapter 5 in this book) [1a-h], hydroformylation tandem reactions, [2,3] other general aspects of the reaction and how it can be applied in the synthesis of fine chemicals [4a-g] give an account of the impressive progress of hydroformylation over the last decade. Rhodium-catalyzed hydroformylation takes place under mild conditions and it is compatible with the most common functional groups present in an olefinic substrate [4a], and it is, therefore, a synthetically useful tool for the preparation of organic compounds [2].

This chapter discusses some general aspects of the hydroformylation of alkenes in organic synthesis. It focuses mainly on regio- and stereoselective processes, and analyzes the influence of the substrate and the catalysts. Practical methods which provide high yields and selectivities, and short-cuts compared to classical organic routes will be described. Particular attention will be paid to recent advances that have helped to enlarge the synthetic application of this reaction. Section 6.7 deals with the hydroformylation of alkynes, and such key aspects as hydroformylation in water-gas shift conditions and silylformylation, particularly efficient catalytic systems and the application of hydroformylation in organic synthesis.

P.W.N.M. van Leeuwen and C. Claver (eds.), Rhodium Catalyzed Hydroformylation, 145–187.

6.2 Hydroformylation of unfunctionalized alkenes

Control of selectivity, chemo-, regio-, and stereoselectivity, is the most important problem in the hydroformylation reaction. As far as chemoselectivity is concerned such competitive reactions as isomerization, double bond hydrogenation and aldehyde hydrogenation occur under hydroformylation conditions.

In general, rhodium catalysts do not provide hydrogenation products, and factors favoring isomerization are well established (see discussion below for factors favoring β-elimination).

It is generally accepted that in rhodium-catalyzed hydroformylation the formation of the most stable linear and branched alkyl rhodium complexes is assumed to be the step that determines the regioselectivity when there are no β-elimination processes (Figure 1), (see chapter 2). But the reverse process, the dissociation of the metal-alkyl via β-hydride elimination, plays an important role in the regioselectivity of the reaction, because it helps to equilibrate the alkyl species and can produce isomerization. In this case the regioselectivity is determined by the relative migratory ability of both metal-alkyl intermediates.

β-Elimination is favored by the temperature [5,6] and disfavored by the pressure. However, temperatures cannot be low in many cases because the activity of catalyst is low. This is the case of highly hindered olefins, because the coordination of the olefin to the alkene is the rate limiting step and high temperatures are required which result in secondary isomerization processes.

The regioselectivity is determined by the steric and electronic properties of the alkenes, both the metal and the ligand of the catalytic system and the reaction conditions, mainly temperature, pressure and the ratio of carbon monoxide to hydrogen.

The substituents in the double bond strongly determine the regioselectivity, and in general the formyl group is introduced into the less substituted alkene carbon. Thus, regioselectivity is complete in disubstituted terminal alkenes **2**, (Figure 2) to give the linear aldehyde. The monosubstituted alkenes **1** give mainly lineal aldehydes, but different percentages of the branched aldehydes are always obtained. The disubstituted alkenes such as **3** give mixtures, except when are symmetrically substituted. The main problem of alkene **4** is its low reactivity, which decreases with the substitution. As a result more drastic reaction conditions are required and side reactions such as isomerization or hydrogenation also usually take place [7,8].

Alkenes **1** and **2** are the synthetically most useful. Alkene **3** can also be useful in substrates containing structural elements which can control the regio- and, eventually, the stereoselectivity (i.e. heteroatoms which can provide chelation).

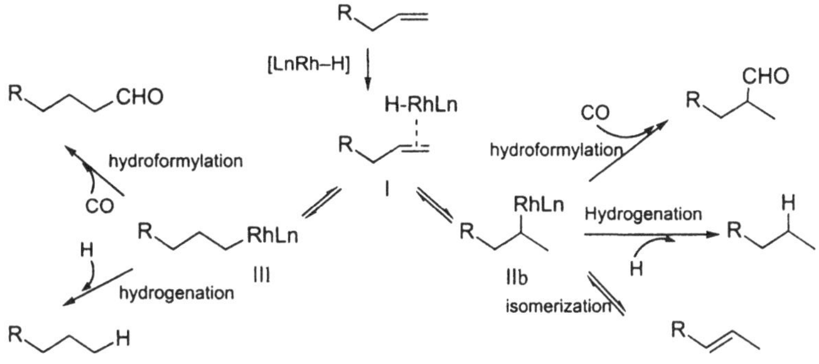

Figure 1. Competitive processes in the alkene hydroformylation

Hydroformylation of terminal alkenes. The hydroformylation of terminal alkenes has been widely investigated in an attempt to find systems that can control the regioselectivity. Although linear/branched mixtures are usually obtained, an excess of monophosphorus ligand helps to obtain preferentially the linear isomer. Recently, new ligands have been reported which overcome the regioselectivity problems.

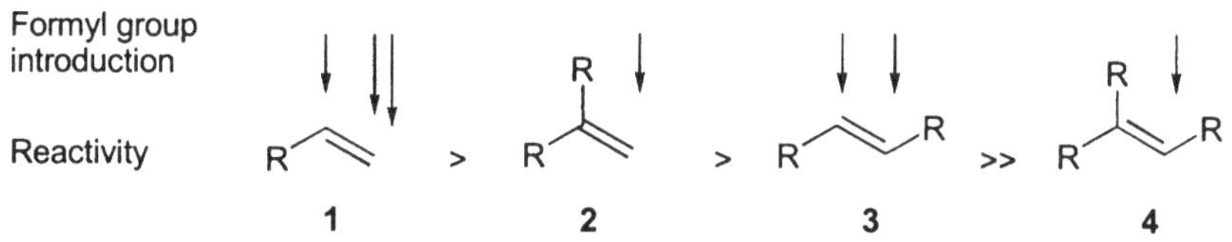

Figure 2. Selectivity and reactivity in the hydroformylation of differently substituted alkenes

Thus, monosubstituted terminal olefins are selectively hydroformylated to give the lineal aldehyde by using diphosphite ligands such BIPHEPHOS (**5**) (also known as UCC ligand (Figure 3) [9,10,11], or diphosphines such as BISBI (**6**)[12] or XANTPHOS (**7**)[13]. By combining these ligands with simple rhodium complexes such as [Rh(acac)$(CO)_2$] or [Rh(dipivaloyl-methanoate)$(CO)_2$], regioselectivities can be as high as 100%.

Since differently substituted alkenes are hydroformylated at different rates, mono- or disubstituted alkenes can be selectively hydroformylated in the presence of di- or trisubstituted ones, respectively [4b].

As has been explained above, isomerization is a competitive process that can be diminished, or suppressed, by choosing low reaction temperatures and an appropriate catalyst. However, isomerization can be advantageously used to hydroformylate internal olefins to give the linear aldehyde.

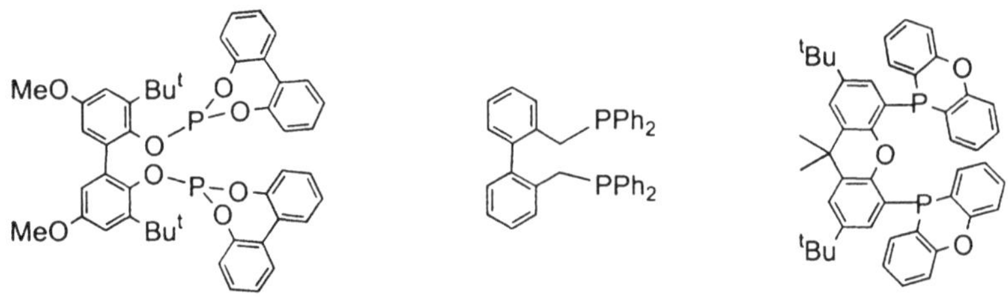

5 BiPHEPHOS **6** BISBI **7**

Figure 3. Phosphite and phosphine ligands for regioselective hydroformylation

Thus, by using XANTPHOS analogues [14] and appropriate reaction conditions such as high temperature and low pressure, 2- and 4-octene are hydroformylated to give a remarkable 90% of the lineal aldehyde (Table 1) (Figure 4) (see chapter 4). Under these conditions, the isomerization takes place very quickly and all the possible alkenes are present in equilibrium in the reaction medium, and the terminal alkene is hydroformylated faster than the internal ones. Moreover, the catalytic system proved to be remarkably regioselective towards the linear aldehyde.

8 **9** **10** **11**

Figure 4. Isomerization-hydroformylation of internal alkenes

Table 1. Isomerization-hydroformylation of alkenes

Substrate	Catalyst	P(bar)	T °C	Conv %	l/b %
2-octene	A[a]	2	120	22	9.2
4-octene	A[a,b]	2	120	67	4.4
2-butene	B[c]	48	120	5	1.5

[a]A:[Rh(CO)$_2$)dipivaloylmethanoate)]/**7**, [Rh]= 1mM, ratio [Rh]/P/substrate= 1:10:673, CO/H_2=1, toluene, 1h. [b] Same conditions than a, 17h. [c]B: Rh(OAc)$_3$/phosphine-phosphite ligand/$Ru_3(CO)_{12}$, CO/H_2=1, toluene, 17h.

Linear aldehydes are mainly obtained from 2-butene by using a dual catalytic system in which one metal catalyses the isomerization reaction and another metal the consecutive hydroformylation [15]. Although the results are still far from being synthetically useful, this work opens up new perspectives for this process (Figure 4).

As will be seen below, when coordinating atoms are present in the substrate chelation is an important factor in regioselectivity control. Linear or branched aldehydes can be obtained depending on the size of the chelate ring (see section 6.5).

Hydroformylation of 1,1-disubstituted alkenes. As has been mentioned above, 1,1-disubstituted alkenes are regioselectively hydroformylated to give 3,3-disubstituted propanal derivatives [16,17]. They are valuable intermediates in the preparation of various drugs [18,19].

Functional groups such as silyl ether, ester, acetal and halide, are compatible with hydroformylation conditions. Thus, ω-functionalized 3-substituted alkanals are synthesized from 2-alkyl-1-alkenes, and this methodology has been applied to the synthesis of muscone (**14**) (Figure 5). [20].

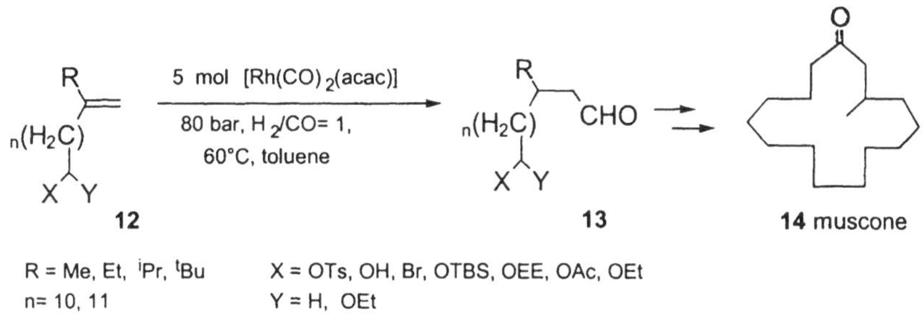

Figure 5. Selective hydroformylation of ω-alkenes. Synthesis of muscone

Acrolein acetals **15** bearing an alkyl or aryl substituent in the 2-position are regiospecifically hydroformylated to the corresponding succinaldehyde monoacetals **16** with rhodium catalysts (Figure 6) [21].

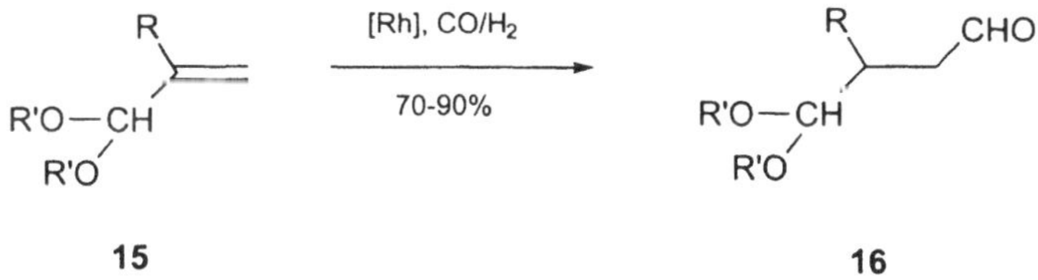

Figure 6. Hydroformylation of acrolein dimethylacetal. Synthesis of partially protected 1,4-dialdehydes

6.3 Hydroformylation of functionalized alkenes

The formyl group generated by rhodium (I) catalyzed hydroformylation is generally incorporated at the carbon atom contiguous to the heteroatom. This is the case of 1-monosubstituted and 1,2-disubstituted alkenes.

However, in 1,1-disubstituted alkenes the steric factors predominate and the formyl group usually goes to the less substituted alkene carbon (Figure 7) although there are many exceptions.

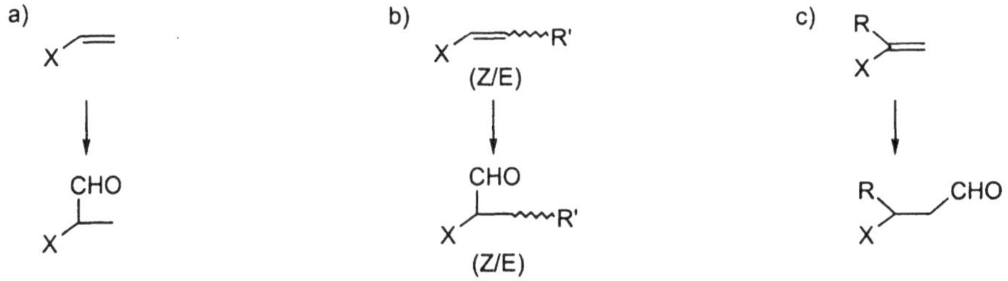

Figure 7. General trends in the hydroformylation of heteroatom substituted alkenes

Consequently, functionalized 1-alkenes such as vinyl esters, vinyl phthalimides, vinyl fluorides and vinyl ethers can lead to α-acyloxy-aldehydes **17** [22], α-phthalimido-aldehydes **18** [23], α-halo-aldehydes **20** [24], and α-alkoxy-aldehydes **21**, [5, 25], respectively, (Figure 8). Chelation usually helps to increase the regioselectivity, although the reactivity decreases significantly, as is the case of vinyl acetate [26].

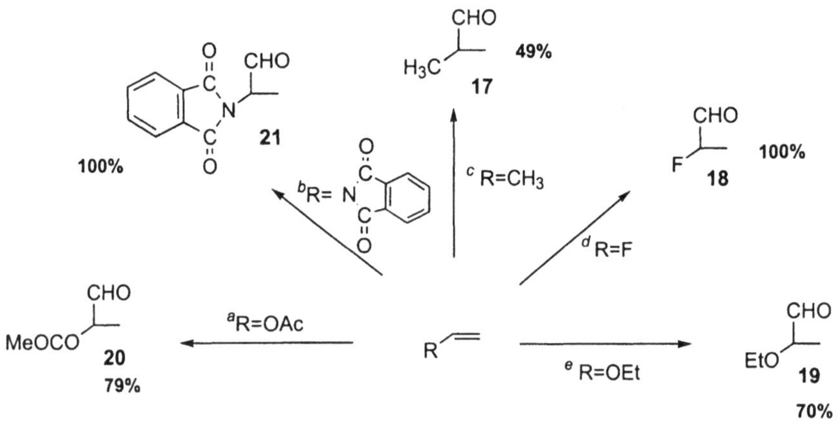

[a] $[Rh_4(CO)_{12}$, 217 bar, CO/H_2= 1, 60°C. [b] $[RhHCO(PPh_3)_3]$, 100 bar CO/H_2= 1, 100°C. [c] $Rh_6(CO)_{16}$, 120 bar, CO/H_2= 1, 70°C. [d] $Rh_4(CO)_{12}$, 110 bar, CO/H_2=1, 80°C. [e] $Rh_4(CO)_{12}$, 100 bar , CO/H_2 (1:1), 20°C.

Figure 8. Hydroformylation of substituted alkenes

Usually unsaturated functionalized cyclic alkenes such as dihydrofurans [27] and N-acyl-2-pyrrolines [23] are also very reactive towards rhodium-catalyzed hydroformylation. They principally give the expected α-formyl product, (Figure 9). Dihydropyrans [27] behave similarly but are much less reactive.

As has been seen above, ligands have a considerable influence on the control of regioselectivity because they stabilize a preferred M-alkyl intermediate or accelerate or avoid the β-elimination process. An example of this is the hydroformylation of 2,3- and 2,5-dihydrofuran [27]. Thus, the hydroformylation of both dihydrofurans **22** and **23**, using $P(O\text{-}o\text{-}^tBuC_6H_4)_3$ gives practically the same ratio of products **24:25** (Figure 9) (entries 2 and 4 Table 2).

The evolution of the reaction shows that **23** is isomerized into **22** before the hydroformylation reaction starts, because the M-β–alkyl intermediate is β-eliminated faster than CO is inserted. However, if the reaction is started from **23** and the ligands have small cone angles, such as $P(OMe)_3$, practically the only aldehyde detected is the 3-formyl derivative (entry 3). If it is started from **22**, the 3-formyl derivative is the main aldehyde detected (entry 1). This suggests that under these conditions there is no β-elimination process or that it is very slow.

These results show that the regioselectivity of the process can be controlled if the ligand and reaction conditions are selected. Thus, PPh_3 can be used to quantitatively convert **23** into **25**. The results are similar to when $P(OMe)_3$ is used in a high P/Rh ratio, at moderate temperatures and high CO pressures but the catalyst is more active using PPh_3. When $P(O\text{-}o\text{-}^tBuC_6H_4)_3$ is used at high temperatures and low CO pressure, however, the main aldehyde **24** is obtained.

However, 3,4-dihydro-2H-pyran and 5,6-dihydro-2H-pyran required more drastic conditions to be hydroformylated and when only $P(O\text{-}o\text{-}^tBuC_6H_4)_3$ was used as the ligand conversions were as high as 80% and the selectivity in 2-/3-formyl derivatives was 68/32 [27].

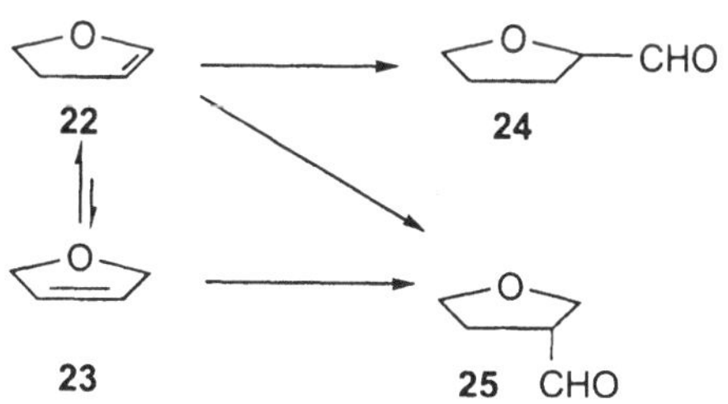

Figure 9. Hydroformylation of dihydrofurans

Reactive unsaturated halides, such as allyl chloride, destroy the catalytic activity of rhodium catalysts, probably because very stable inactive complexes are formed [7].

Table 2. Hydroformylation of dihydrofuran **24** and **25** using different phosphorus ligands.[a]

Entry	Substrate	Ligand	Conversion(%)	Ratio **24/25**
1	**22**	$P(OMe)_3$	13	29:71
2	**22**	$P(O\text{-}o\text{-}^tBuC_6H_4)_3$	99	76:24
3	**23**	$P(OMe)_3$	43	1:99
4	**23**	$P(O\text{-}o\text{-}^tBuC_6H_4)_3$	97	65:35

[a] 0.25 mol/% of $[Rh_2(\mu\text{-}CH_2)_3NMe_2)_2(cod)_2]$, Ligand/[Rh]= 10, CO/H_2= 1, 5 bar, 80°C, 1,2-dichloroethane.

The hydroformylation of vinyl arenes **26** has been widely studied and the branched aldehyde **28** is the main product of the reaction (Figure 10). When rhodium catalysts are used and pressures are high, regioselectivities are usually between 80 and 98%. The major regioisomer is the inverse of the one for alkyl monosubstituted olefins. The formation of η^3-complex **27** explains the regioselectivity and this has already been discussed in the literature (Figure 10) [24, 28, 29]. The regioselectivity can vary according to the substituents in the ring [24b, 30].

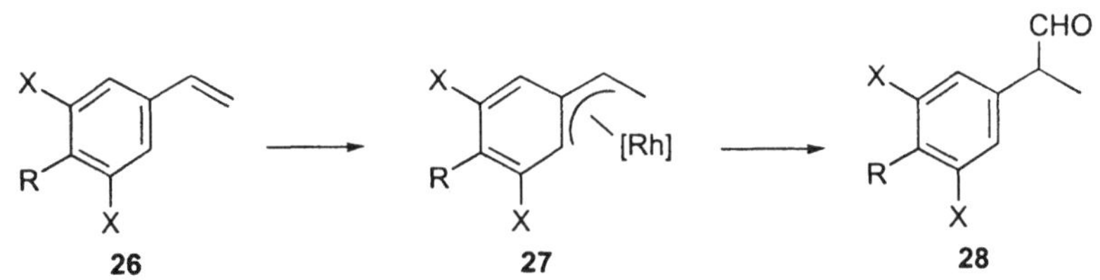

Figure 10. Hydroformylation of vinylarenes

In the rhodium catalyzed hydroformylation of 1,1-diphenyl alkenes the linear aldehyde is preferred. However, the branched metal-alkyl is formed faster, although the CO insertion is quicker in the linear metal-alkyl intermediate. On the contrary, in 1-phenyl-1-(2-pyridyl)-alkenes such as **30** the branched aldehyde **31** is obtained instead of the linear one **29** [31,32].

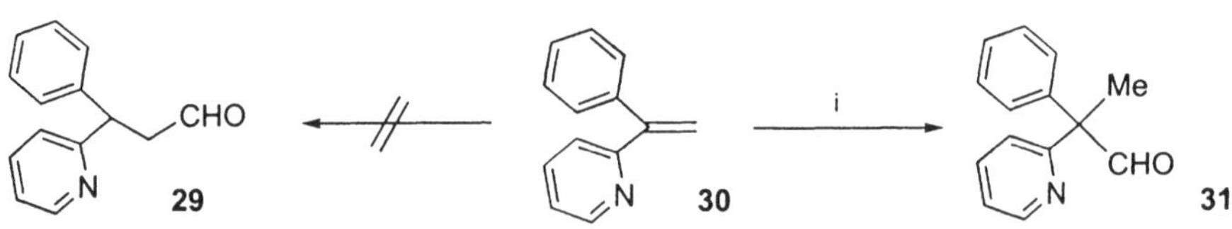

i) $[RhH(CO)(PPh_3)_3]$, substrate/[Rh]= 80, CO/H_2= 1, 100 bar, 80°C, benzene, 99%.

Figure 11. Hydroformylation of 1-phenyl-1-(2-pyridyl)-ethene

The chemoselectivity of 3- and 4-vinylpyridines when $Rh_4(CO)_{12}$ is used as the catalyst is very different (Figure 12). While the 3-vinylpyridine **32** is fully hydroformylated to give the branched aldehyde **33** (b:l =96:4), the 4-vinylpyridine **34** affords only the hydrogenation product **35**. This is because of the different nucleophilicity of the alkyl carbon atom in the rhodium intermediate, which in the case of 4-vinylpyridine is not able to give the migratory insertion [33]. However, when $Rh_4(CO)_{12}$ is modified with phosphorus ligands, the branched aldehyde is obtained [34]. In these conditions the carbon bonded to the metal becomes sufficiently nucleophilic to give the migratory insertion of CO, hence evolving to the acyl intermediate. Vinyl pyrrole also give the branched aldehyde, but temperature has a greater influence on 2-vinylpyrrole and the nitrogen needs to be protected if conversions are to be good [35]. Thus, 1-tosyl-2-vinyl-pyrrole and 1-tosyl-3-vinyl-pyrrole are hydroformylated in good yields and with a branched/linear ratio >94:6.

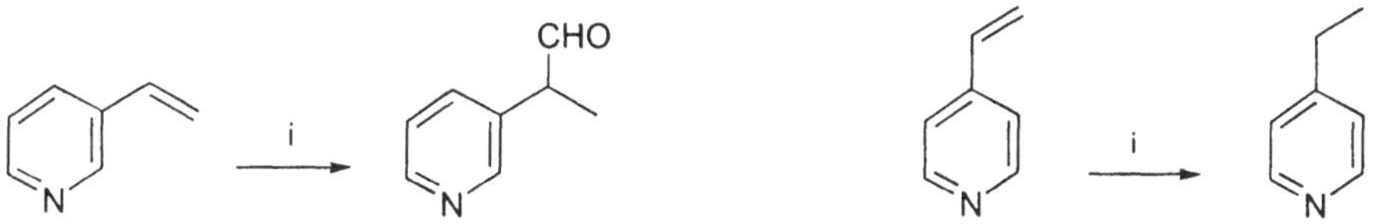

32 33 34 35

i) $[Rh_4(CO)_{12}]$, substrate/[Rh]= 300, CO/ H_2= 1, 100 bar, 60-110 °C, benzene.

Figure 12. Hydroformylation of vinylpyridines

Electron-poor alkenes are less reactive than common alkenes and hydroformylation using rhodium catalysts gives mainly the branched aldehyde (Figure 13). Interestingly, cobalt catalyst lead to the linear aldehyde, while for the electron-rich alkenes (Figure 8) both catalysts provide the same regioselectivity. Compared to simple olefins, the regioselectivity in the branched aldehyde is higher in the rhodium hydroformylation of electron acceptor-substituted olefins such as trifluoropropene **36** (X=CF_3 [24]), dimethoxyacrolein **37** (X=$(AcO)_2CH$ [23]) vinylsulphones **38** ($PhSO_2$ [36]), acrylonitrile **39** (X=CN [37]). This reflects also the higher stability of branched alkyl-metal intermediates (Figure 1). The regioselectivity shown by cobalt catalysts can be explained because branched alkyl-cobalt intermediates tend to isomerize more than the analogous rhodium catalysts.

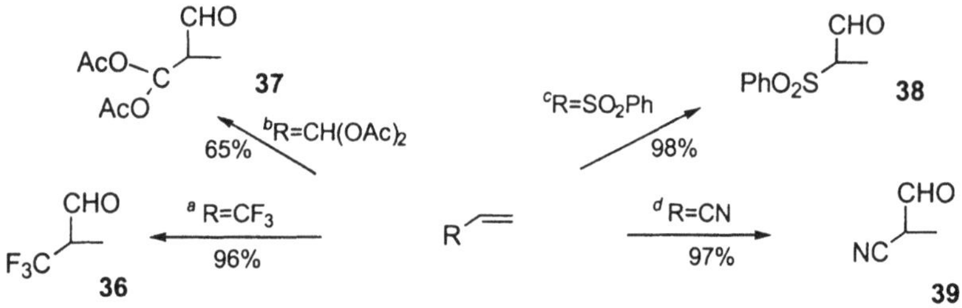

[a] $[Rh_6(CO)_{16}]$, CO/H_2=1, 110 bar, 80°C. [b][Rh], 200 bar CO/H_2=1, 100°C, [c]$[Rh^+(\eta^6C_6H_5B^-Ph_3)(cod)]$ /dppb, CO/H_2= 1, 40 br, 75°C.

Figure 13. Hydroformylation of electron-deficient alkenes

The chemo- and regioselectivity of rhodium catalyzed hydroformylation of these substrates depends heavily on the phosphorus ligand and on the reaction conditions. Thus, if $P(OPh)_3$ is used in hydroformylation of ethyl acrylate at low temperatures and high pressures, the branched aldehyde is almost exclusively obtained (Table 3). Under these conditions, no isomerization takes place and the regioselectivity reflects that the branched alkyl-metal intermediate is preferred. If the temperature is increased and pressure decreased, the regioselectivity is completely reversed and only the linear aldehyde is obtained. Significant amounts of hydrogenation product are also observed. Under these conditions, the isomerization is faster than hydroformylation, and the linear and branched alkyl-metal intermediates are equilibrated. In this context, the linear intermediate migrates faster than the branched intermediate and therefore generates the linear aldehyde. In the presence of electron-donating phosphines such as $P(O^iPr)_3$ hydrogenation takes place and gives ethyl propanoate, probably because the rhodium(III) intermediate stabilizes.

Table 3. Influence of phosphorus ligands and reaction conditions on the hydroformylation of ethyl acrylate.[a]

Ligand	T (°C)	Pressure (bar)	Conversion (%)	Branched (%)	Linea(%)	Hydrog. (%)
$P(OPh)_3$	40	30	100	96	1	1
$P(OPh)_3$	80	1	70	0	27.7	38
$P(O^iPr)_3$	40	1	96.4	7.9	0.7	87.8

[a] 0.005 mol of $[Rh(acac)(CO)_2]$ per mol of substrate. L/Rh= 5, CO/H_2=1, 17 h.

Since most hydroformylation experiments are performed at high pressures, the reported examples of hydroformylation of ethyl acrylate mainly gives the branched aldehyde. However, if sterically hindered and electron withdrawing phosphite ligands are used with high ligand/Rh ratios and high concentrations of catalysts at high pressures and low temperatures, the linear aldehyde can be preferentially obtained [38].

When the alkene is 1,1-disubstituted, the linear aldehyde is obtained (Figure 14) [28]. The behavior of amide derivatives, however, depends on the amino group substitution. Thus, hydroformylation of the 2-methyl-acrylamide **42** (R=H) gives aldehyde **46** and by subsequent cyclization with the formation of unsaturated lactams **47**, [39, 40] (Figure 14). Alternatively when R=Et, lactone **45** is formed by consecutive hydroformylation, reduction and cyclization (see section 6.6 for other consecutive processes).

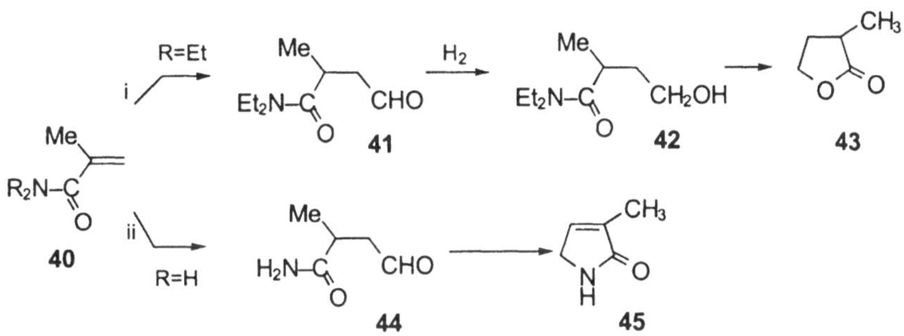

i) [$Rh_4(CO)_{12}$], substrate/[Rh]= 300, CO/H_2= 1, 300 bar, 130°C, 26h, toluene. ii)[$Rh_4(CO)_{12}$], CO/H_2= 1, 80bar, 120°C, 72h, toluene.

Figure 14. Hydroformylation of vinylamides

1,1-Disubstituted functionalized alkenes can give the branched derivative when they are sufficiently activated and the conditions are appropriate, (see for instance figure 22 in chapter 5).

6.4 Substrate directed stereoselectivity

This section deals with substrate-controlled stereoselective hydroformylation, since asymmetric hydroformylation is covered in chapter 5. The stereoselectivity of the hydroformylation reaction is the result of the *cis* addition of the proton and the formyl group to the less hindered face of the double bond [41]. The presence of heteroatoms in the substrate causes chelation, so the stereoselectivity can be controlled, (see section 6.5).

There are various ways of generating stereocenters by hydroformylation. In monosubstituted terminal alkenes, a stereocenter is generated when the branched aldehyde is obtained (Figure 15, a). This is the case when R is a phenyl group or a heteroatom. The regioselectivity in 1,1-disubstituted alkenes gives the linear aldehyde but a stererocenter is generated in the branched aldehyde when R^1 is different from R^2 (Figure 15, b). Two possible regioisomers can be achieved in the hydroformylation of 1,2-disubstituted alkenes so the formation of a stereocenter is only interesting in symmetric alkenes or when exist elements controlling the regioselectivity (Figure 15, c). And in trisubstituted alkenes, two stereocenters are generated as can be seen in Figure 15, d.

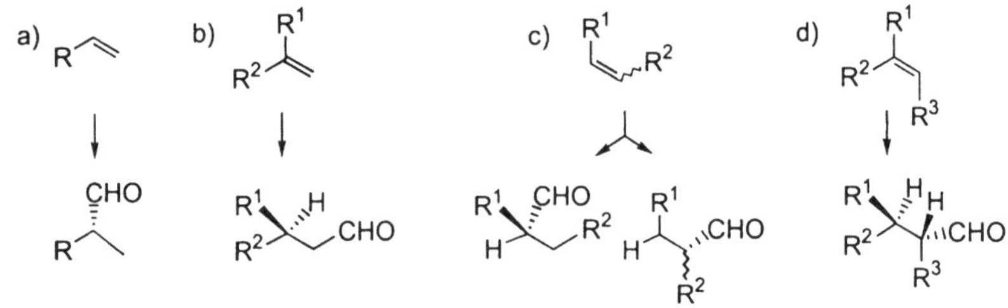

Figure 15. Stereocentres generated in the hydroformylation reaction

Rhodium catalysts can be used to hydroformylate differently substituted endocyclic alkenes such as α-pinene (**46**) from the less hindered face of the double bond to give 10-formylpinane (**48**) [42, 43]. Two stereocenters are created. Interestingly, a cobalt catalyst gives aldehye **49** as a consequence of a skeletal rearrangement followed by hydroformylation. The whole process takes place with no loss in optical activity. However, the related exocyclic alkene β-pinene (**47**) is stereoselectively hydroformylated to give *cis* **50** or *trans* **51** derivatives depending on the catalyst used [44] (Figure 16).

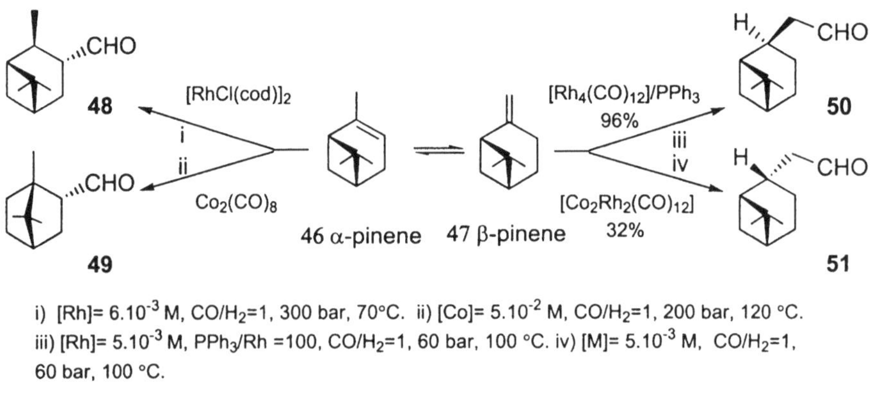

i) [Rh]= 6.10^{-3} M, CO/H_2=1, 300 bar, 70°C. ii) [Co]= 5.10^{-2} M, CO/H_2=1, 200 bar, 120 °C. iii) [Rh]= 5.10^{-3} M, PPh_3/Rh =100, CO/H_2=1, 60 bar, 100 °C. iv) [M]= 5.10^{-3} M, CO/H_2=1, 60 bar, 100 °C.

Figure 16. Stereoselective hydroformylation of α- and β-pinene

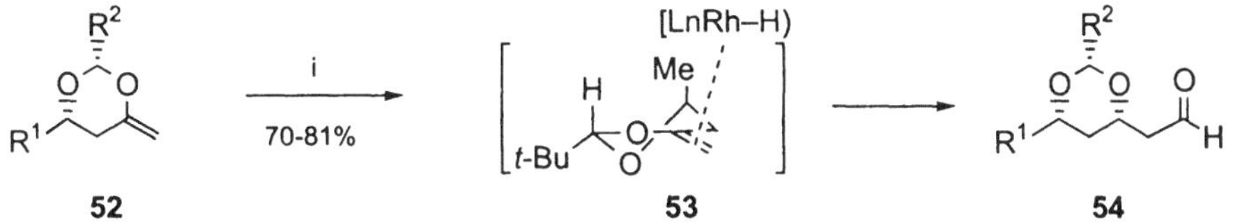

i) [Rh(acac)(CO)2] / PPh3 (ratio 1:4), H2/CO=1, 60 bar, 85°C, 16h, THF.

Figure 17. Stereoselective hydroformylation of exocyclic alkenes

3,5-Dihydroxy aldehydes **54** are also stereoselectively prepared by hydroformylation of enol ether **52** (R^1= H, Me, ${}^iPr$, $BnO(CH_2)_2$, R^2= ${}^tBu$, Me). This suppose an alternative procedure to the aldol reaction, (Figure 17) [45, 46].

The reaction is highly stereoselective and only the *syn* isomer **54** is observed. The process is kinetically controlled, and the olefin insertion is apparently the rate-determining step.

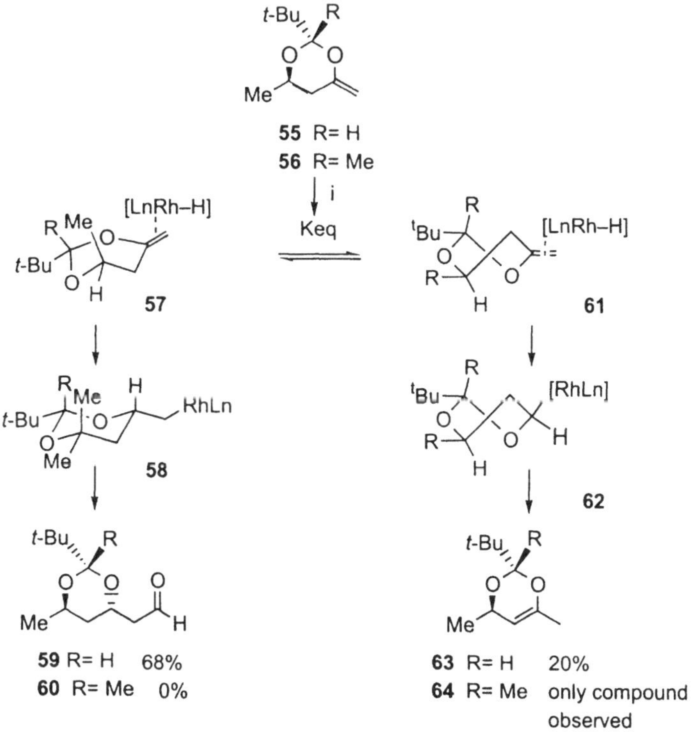

i) [Rh(acac)(CO)2] / PCy3 (from **4**), PPh3 (from **5**), ratio Rh/P= 1:4, H2/CO=1, 60 bar, 120°C,THF

Figure 18. Conformational control of the hydroformylation reaction

In these exocyclic alkenes the chemo- and regioselectivity of the process depends heavily on the conformation [46]. Thus, compound **55**, which in chair conformation has an axial methyl close to the coordinating point, led to the anti isomer **59** after hydroformylation, together with significant amounts of the isomerization compound **63**. Elevated catalyst loading and drastic conditions were required, and the results were best when PCy_3 was the auxiliary ligand.

Both hydroformylation and isomerization were found to occur from the same diastereoface of enol ether **55**. The fact that isomerization must take place through a tertiary rhodium-alkyl which has a 1,3-diaxial interaction in chair conformation seems to indicate that the intermediate has a boat conformation.

Compound **56,** the chair conformation of which is destabilized because of a 1,3-diaxial interaction, gives exclusively the isomerization product **64**, and no aldehydes are detected. To explain these results it is suggested that compounds with a preference for a chair-like conformation show good regioselectivity for the primary rhodium-alkyl, and substrates with a preference for a twist-boat conformation (substrate **56**) show good regioselectivity for the tertiary rhodium-alkyl leading to the isomerization products (compound **64**). Further, substrates without an overwhelming conformational bias (substrate **55**) react with lower regioselectivity.

The hydroformylation of glucal derivatives is a potential method for synthesizing 2-deoxy-C-glycosides. When the cobalt catalyst $[Co_2(CO)_8]$ was used at high pressures and temperatures, glucal derivatives were hydroformylated to give a mixture α/β of aldehydes **68** (Figure 19) [47, 48].

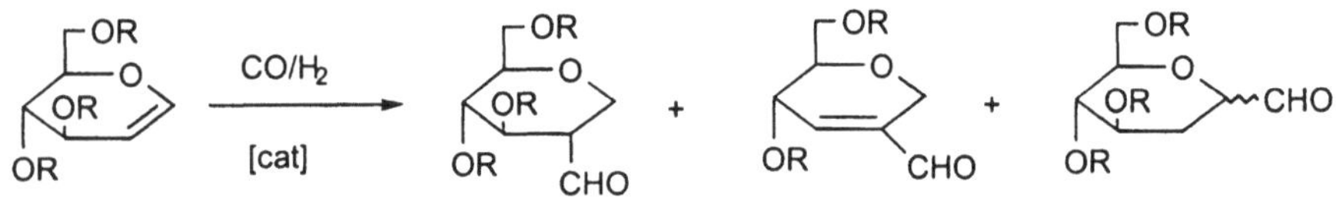

Figure 19. Hydroformylation of glucal derivatives

On the other hand, rhodium catalysts were only able to hydroformylate glycals in the presence of $P(O\text{-}o\text{-}^tBuC_6H_4)_3$. Independently of protecting groups, the main product was the α-2-formyl derivative **66** (sugar numbering) [49]. When the protecting group is acetate significant amounts of the elimination product **67** are also obtained, but when it is an ether-type group the elimination product is not formed and increasing amounts of compound **68** are obtained (Table 4).

Table 4. Hydroformylation of glucal derivatives **65**.

R	Catalyst	P(bar)	T°(C)	Conv. (%)[a]	**66**[b]	**67**[b]	**68**[b]
Ac	$Co_2(CO)_8$	300	200	>90			
Ac	[Rh]/P[c]	70	100	91	58	19	--
Bn	[Rh]/P[c]	50	100	99	68		19

[a] Percentage of transformed product. [b] Percentage of products identified related to the products of the reaction detected by GC. [c] [Rh]/P= $[Rh_2(\mu\text{-}OMe)_2(cod)_2]/P(O\text{-}o\text{-}{}^tBuC_6H_4)_3$

Although glucal derivatives can exist in a conformational equilibrium, the hydroformylation reaction is stereoselective and prefers the attack from the α face. The stereoselectivity is much better when the formyl group is in C2 and not in C1. Apparently, the stereoselectivity in the former case is controlled by the substituent at C3, which adopts a pseudoaxial arrangement (vinylogous effect).

The regioselectivity obtained in the hydroformylation of **65** is the inverse of the regioselectivity for the hydroformylation of 3,4-dihydro-2H-pyran with the same catalytic system [27], or for the hydroformylation of **65** with cobalt catalyst (Table 4).

The regioselectivity in dihydro-pyran and glucals may be different when the same catalytic system is used because C2-metal intermediates form more quickly than C1-metal intermediates and because isomerization requires a conformational arrangement in the molecule (Figure 20). This arrangement is more difficult in substituted rings, such as glucals, than in non substituted rings such as dihydropyrans.

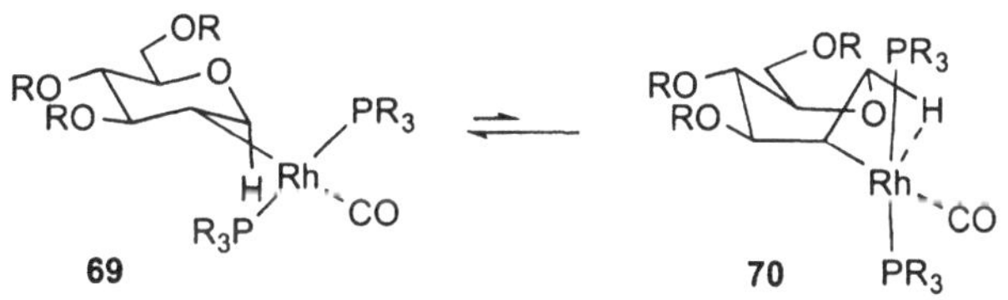

Figure 20. Conformational equilibrium for β–elimination in Rh-alkylglucals

Double bonds not directly bonded to cycles can also be stereoselectively hydroformylated, but there must be an efficient sterical discrimination of both faces of the double bond must. This is the case of 1-methylvinyl-C-β-glucoside **71** which is hydroformylated to give a 2-substituted aldehyde **72** in excellent yield and diastereoselectivity of 99% (Figure 21) [50]. The bulky substituent at position 2 of sugar blocks the conformation shown in **71** and forces the rhodium to be coordinated from the back face of the double bond.

Similarly, the hydroformylation of 2-phenyl-4-(prop-2-enyl)-1,3-dioxane **73** affords an all-anti-stereotriad **75** which becomes a valuable intermediate in the total synthesis of the macrolide antibiotic bafilomycin A [51] (Figure 22). The conformation for intermediate **74** was calculated to be the most stable and the equatorial methyl group appears to cause the high stereocontrol observed. As can be deduced from these representative examples six-member rings seem to provide good stereoselectivities.

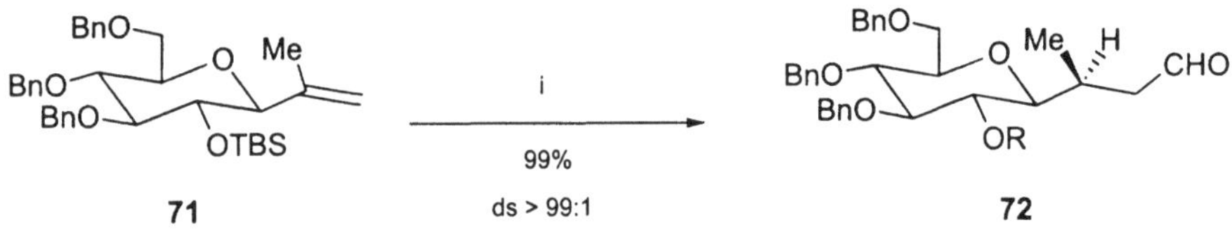

i) [Rh(CO)$_2$acac], toluene, H_2/CO=1:1, 80bar, 80°C, 48h.

Figure 21. Stereoselective hydroformylation of vinyl-C-glycosides

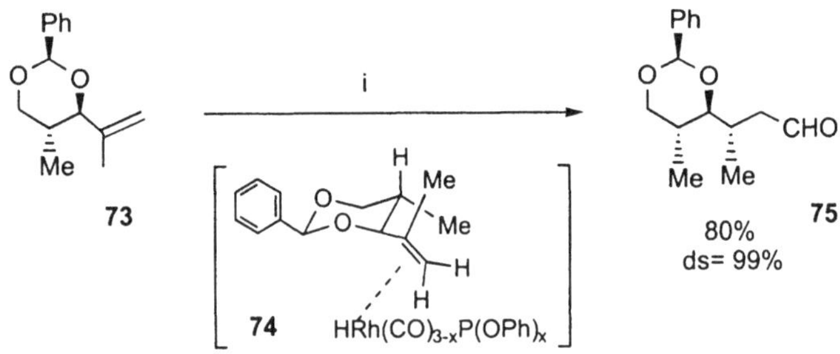

i) [Rh(CO)$_2$(acac)]/4P(OPh)$_3$, toluene, H_2/CO=1, 20 bar, 70 °C.

Figure 22. Stereoselective hydroformylation of exocyclic alkenes

6.5 Control of the regio- and stereoselectivity by heteroatom-directed hydroformylation

Remote substituted groups in the substrate, which are able to chelate to the metal catalyst, can be used to control the regio- and stereoselectivity of organic reactions. The sections above have discussed some examples of carbonyl coordination, vinyl acetate for instance, and their influence on reactivity and selectivity. Coordination of nitrogen in aminoalkenes to rhodium was demonstrated by isolation of complexes with the double bond

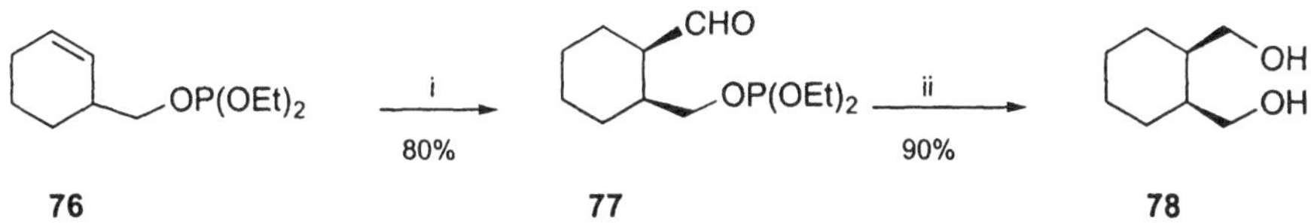

i) $[Rh(OAc)_2]_2$, alkene/Rh= 50:1, PhH, CO/H_2= 1, 37bar, 100°C, benzene, 44h. ii) $LiAlH_4$, Et_2O.

Figure 23. Phosphite directed hydroformylation of homoallylic cyclic alkenes

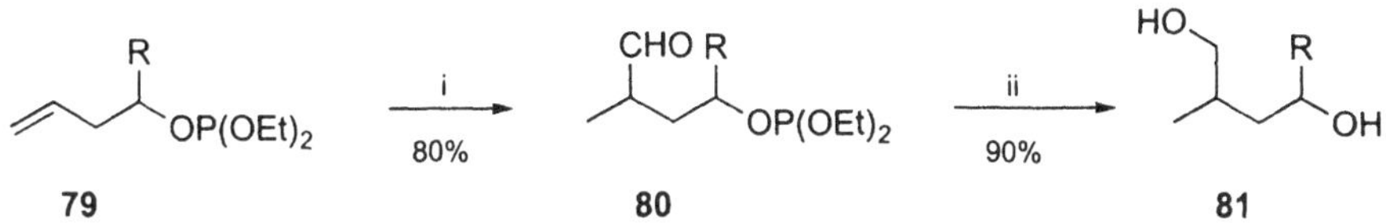

i) $[Rh(OAc)_2]_2$, alkene/Rh= 50:1, PhH, CO/H_2= 1, 37 bar, 50°C, benzene, 5-22h. ii) $LiAlH_4$, Et_2O.

Figure 24. Phosphite directed hydroformylation of homoallylic linear alkenes

Figure 25. Phosphino-alkenes for directed hydroformylation

and the amino group coordinated to rhodium [52]. Other illustrative examples of carbonyl or nitrogen coordination will be discussed below (see for instance Figures 35, 36 and 39)

Introducing phosphorus-containing groups in the substrate has also proved to be a flexible and efficient procedure for controlling the regio- and stereoselectivity of the reaction. Thus, homoallylic alcohols have been transformed into homoallylic phosphites, and the influence of the coordinating heteroatom on the regio- and stereoselectivity of cyclohexenyl derivatives has been studied [53].

By using an unmodified system such as $[Rh(OAc)_2]_2$, the homoallylic phosphite **76** was converted into the formyl derivative **77** with total regio- and stereocontrol (Figure 23). In the open chain homoallylic phosphites **79**, even though the stereoselectivity was low (60:40 for R= Me and 70:30 for R=Ph) regiocontrol was high which led to the formation of the branched aldehydes **80** (Figure 24). In both cases a six-member chelating ring was formed. This ring was shown to exist by treating the phosphate derivative of **79** under similar hydroformylation conditions that led to an equimolar mixture of branched and terminal aldehydes.

Some chelation was also observed to form larger rings. Thus, the hydroformylation of n-pentenylphosphites gave a mixture with a branched-linear ratio of 87:13.

Treatment of formyl-phosphites **77** and **80** with lithium aluminum hydride gave respectively the 1,4-diols **78** and **81** in good yields.

Similarly, the alkene **82** was hydroformylated with complete regio- and stereocontrol by using a Rh/phosphine catalytic system (Figure 25). Under the same conditions **83** was hydroformylated with good regioselectivity but low stereoselectivity (ratio *cis/trans*= 4:1) [54, 55]. Regioselectivity was also complete in the hydroformylation of the open-chain phosphine **84** although in this case the stereoselectivity was negligible.

When the phosphino derivatives **82**, **83** were hydroformylated at higher temperatures (90°C) alcohols were formed as a consequence of consecutive hydroformylation-reduction. This is an interesting result since alcohols are not usually obtained with modified rhodium catalysts.

The regio-and stereoselective introduction of a formyl group into the cyclohexene ring of **85** was a key step in the synthesis of phyllantocin (Figure 26) [56, 57]. The first experiments in the hydroformylation of **85** led to the undesired regioisomer.

An attempt was then made to anchor the phosphorus ligand to the substrate and this required an easily removable group incorporated in the phosphino group. The diphenylphosphinobenzoate group fulfilled these requirements, and the ester **86** was prepared by reacting **85** with *p*-$(Ph_2P)C_6H_4COOH$ using DCC as coupling reagent. Compound **86** gave a very low yield in hydroformylation, but its phosphine oxide derivative gave moderate yields of a mixture of formyl derivatives. It was suggested that this was because the spacer was too long, and pushed the catalysts beyond the olefin. Coupling the *m*-$(Ph_2P)C_6H_4COOH$ to the alcohol **85** gave compound **87** in an 88% yield, which was hydroformylated to give a mixture of aldehydes in a 72% yield, presumably via the intermediate **88**. The selectivity in the major isomer **89** was 77 %. This result is the first and overwhelming example of a long distance regio- and stereocontrolled hydroformylation reaction.

The formyl group in **89** was then epimerized to obtain the compound **90**, in which the formyl group had the same configuration as the natural product. When treated in basic medium the *meta*-(diphenylphosphino)benzoic acid directing group, was recovered.

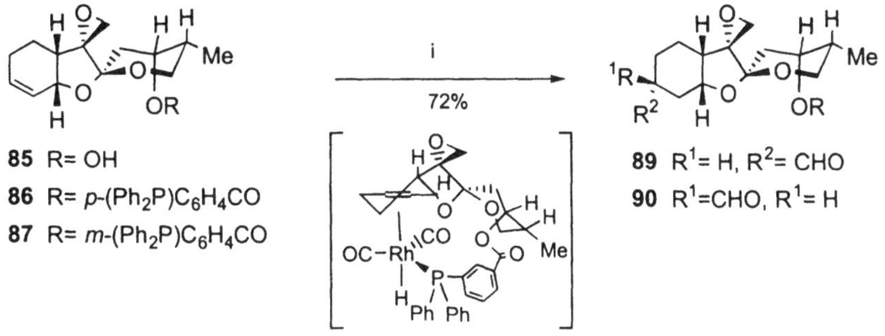

85 R= OH
86 R= *p*-$(Ph_2P)C_6H_4CO$
87 R= *m*-$(Ph_2P)C_6H_4CO$

89 R^1= H, R^2= CHO
90 R^1=CHO, R^1= H

i) 8 mol % $[Rh(OAc)(cod)]_2$, $CO:H_2$= 1, 55 bar, 85°C, benzene, 3h.

Figure 26. Hydroformylation step in the synthesis of phyllantocine

The introduction of the *ortho*-diphenyl-phosphinobenzoate (*o*-DPPB) group led to a considerable improvement in the directed diastereoselective hydroformylation of methallylic [58, 59, 60] and homo-metallylic alcohols [61, 62]. It also supposes the formation of a big chelate ring. In these processes high 1,2- and 1,3-asymmetric induction is obtained.

Protecting groups such as *t*-$BuPh_2Si$ or the free OH do not provide good stereoselectivity. This proves the absence of a directing effect in these cases. The fact that the coordinating properties of the hydroxyl are lower than those of CO may explain this result.

The hydroformylation of **91** (Figure 27) using $[Rh(CO)_2(acac)]$/4 $P(OPh)_3$ as catalyst quantitatively furnished the aldehydes **92, 93** with diastereoselectivities of up to 92:8. An additional phosphorus ligand (monophosphine) was necessary to ensure the presence of two phosphorus

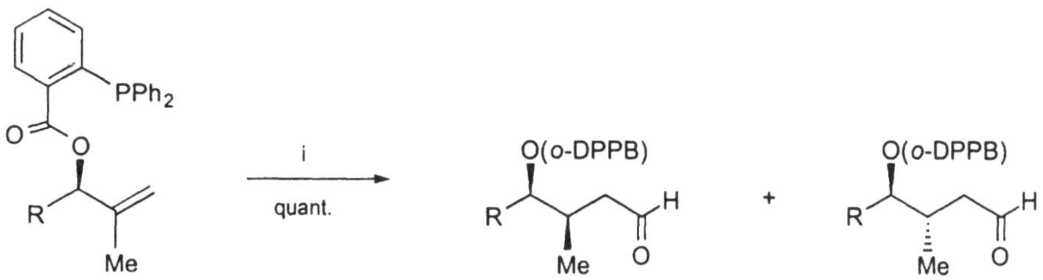

91 **92** >92% **93** <8%

i) 0.7 mol/% $[Rh(CO)_2(acac)]$/4 $P(OPh)_3$, CO/H_2= 1, 20 bar, 90°C, toluene, 24 h.

Figure 27. Stereoselective hydroformylation of metallylic alkenes controlled by the group *o*-DPPB

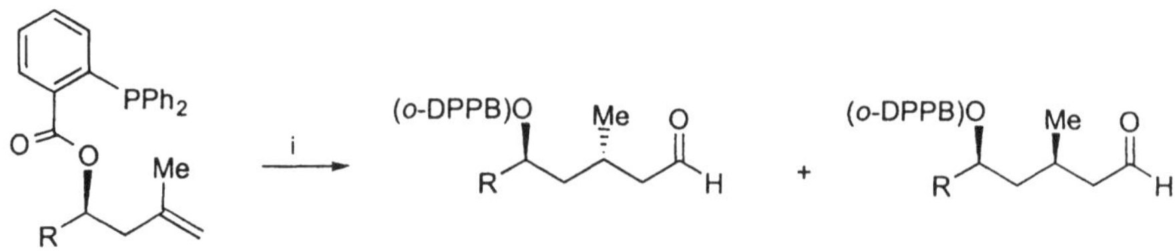

94 **95** >91% **96** <9%

(*o*-DPPB)= *ortho*-diphenylphosphinobenzoate

i) 0.7 mol/% [Rh(CO)$_2$(acac)]/4 P(OPh)$_3$, CO/H$_2$= 1, 20 bar, 50°C, toluene, 72 h.

Figure 28. Stereoselective hydroformylation of homometallylic alkenes controlled by the *o*-DPPB group

ligands coordinated to rhodium. Ligands such as PPh_3, $P(OPh)_3$, P(O-2,6-di-^tBu-$C_6H_3)_3$, $P(OEt)_3$ and P(N-pyrrolyl)$_3$ were tested, and the results were best with $P(OPh)_3$. Temperatures higher or lower than 90 °C decreased the stereoselectivity.

The scope of the procedure was tested in a variety of substrates (different R) and diastereoselectivity was highest in 1,1-disubstituted alkenes. The directing group was compatible with the presence of several groups such as ester, acid, phenyl, etc., and there was no loss in enantioselectivity when the starting compounds were enantiomerically pure.

Compound **94** was also quantitatively converted into aldehydes **95**, **96** with diastereoselectivity up to 91:9 when the same catalytic system was used (Figure 28). In the absence of *o*-DPPB group some diastereoselectivity is also observed, although the disastereomer *cis* is mainly obtained in this case. This result confirms that the *o*-DPPB group acts as directing group by reversible coordination to the catalyst. A significant dependence of the diastereoselectivity on the reaction temperature was observed. The diastereoselectivity was best at 50°C, which mean longer reaction times.

6.6 Consecutive processes under hydroformylation conditions

In the presence of alcohols or amines, the aldehydes generated in the hydroformylaton reaction give acetals or imines. Depending on the hydroformylation catalyst used an additional acid catalyst is required. When the process is intramolecular, (i.e. when alcohols, amines or amides are present in the starting material), it is spontaneous and gives hemiacetals or imines, especially if five or six member rings can be formed [4b]. Moreover the presence of coordinating atoms such as nitrogen leads to a chelate which controls the regioselectivity of the process.

6.6.1 Hydroformylation-acetalization (intramolecular)

The hydroformylation of allylic and homoallylic alcohols has been widely studied in this process because hydroxy-aldehydes can easily lead to hemiacetals with five- or six- member rings. Likewise, the hydroformylation of allylic alcohol to give 4-hydroxybutanol and 1,4-dihydroxybutanol is an important industrial process {63].

The hydroformylation of allylic alcohol **97** usually a mixture of regioisomers **98**, **99**, and propanal (for R=H) which is formed by the isomerization of the double bond and is subsequent tautomerism (Figure 29). The regioselectivity is mainly determined by the ligand. The substitution of the double bond also has a strong influence on the selectivity and reactivity.

The general trend is that reactivity is low when double bond substitution is high, and the formyl group is also introduced into the less substituted carbon in this case.

Aldehyde **99** spontaneously cycles to give the hemiacetal **100**. This leads to the acetal **102** if the reaction takes place in the presence of an alcohol and the catalytic system [Rh(CO)$_2$-zeolite]/PEt$_3$ [64], or in the presence of a carboxylic acid and a rhodium catalyst [65]. Lactone **101** is easily obtained from hemiacetal **100** by oxidation.

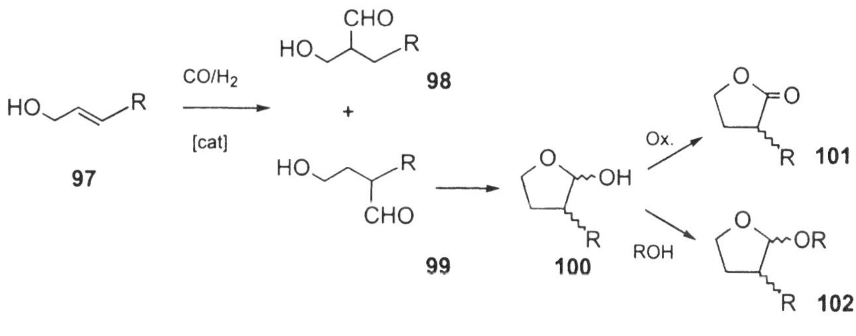

Figure 29. Hydroformylation-intramolecular acetalization of allylic alcohols

An enantioselective version of this reaction has been reported [66]. The hydroformylation of **97** (R=Ph) with the catalytic system [Rh(acac)(CO)$_2$]/ 4 BINAPHOS gives the hemiacetal **100** (R=Ph) with a yield up to 99%, which after oxidation provides the lactone **102** (R=Ph) with 88% ee.

The hydroformylation of the substituted allylic alcohols **103-106** (Figure 30) leads mainly or exclusively to the linear aldehydes which evolve upon cyclization to the respective hemiacetals type **102** [67, 68, 69].

Figure 30. Different allylic alcohols studied in hydroformylation-intramolecular acetalization

A stereoselective version of the hydroformylation of **106** by heteroatom-directing hydroformylation has been discussed above (see Figure 27).

Homoallylic alcohols such as **107** (R=H) give mixtures of regioisomers **108** and **109**. Both can be cycled to provide tetrahydrofuran or tetrahydropyran derivatives (Figure 31). As in the case of allylic alcohols, substitution in the double bond or in the allylic carbon favors the formation of the linear aldehyde. The linear aldehyde **109** provides the hemiacetal **110,** from which acetals [64], lactones [65, 77] and also enolethers such as **111,** can be formed [70, 71, 72, 73].

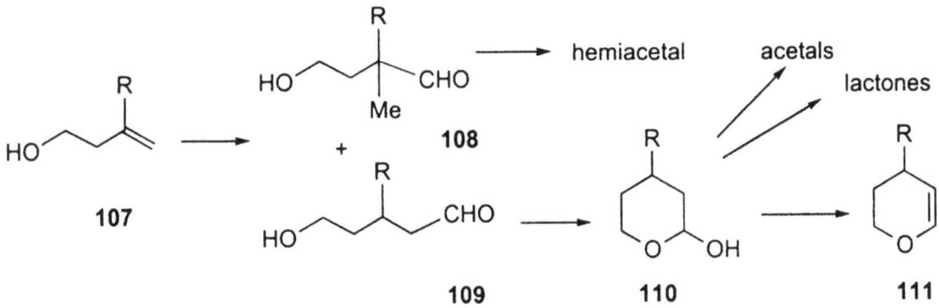

Figure 31. Hydroformylation-acetalization of homoallylic alcohols

6.6.2 Hydroformylation-acetalization (intermolecular)

When alkenes are hydroformylated in the presence of alcohols, dialkoxyacetals or orthoesters, aldehydes are mainly obtained, and there are only small amounts of acetals. Normally, acetals can only be formed in presence of an acid catalyst. Since the catalyst for the hydroformylation process requires the presence of basic phosphine ligands, and the acetalization reaction requires an acid catalyst, the main goal of the process was to find two compatible catalytic systems. The solution was to use weak acids, such as phosphonium or ammoniun salts, or carboxylic acids, in the presence of phosphorus ligands. Strong acids shift favors the equilibrium to the protonated form of the phosphine, thus inhibiting the hydroformylation reaction, and can also react with Rh-H to give dihydrogen and salts (see chapter 9).

If platinum catalysts are used no additives are required, probably because of the Lewis acidity of platinum [74].

Using phosphonium or pyridinium salts, 2,5-dihydrofuran (**113**) is selectively hydroformylated with the [Rh(μ-OMe)(cod)$_2$]/PPh$_3$/PPTS (PPTS =pyridinium *p*-toluensulfonate) catalytic system and 2,2-dimethoxypropane as the solvent, to give the dimethoxyacetal **114** in quantitative yield [75]. Styrene **115** is converted into dimethoxyacetal **116**, with high yield and selectivity. The presence of the weak acid has no influence on the regioselectivity. Interestingly, the aldehyde generated during the hydroformylation of the keto-alkene **117** was selectively acetalized in the presence of a ketone group to give the dimethoxyacetal **118**, (Figure 32).

Catalytic systems such as [Rh(μ-S(CH)$_2$NMe$_2$)(cod)]/PPh$_3$, which are anchored to a sulphonic exchange resin by protonation of the amino group, converts **115** into **116** under hydroformylation conditions and with methanol as the solvent [76]. In a similar reaction the catalytic system [Rh(μ-SCH$_2$NH$_3$)(COOH]$_2$ (OTf)$_2$, in the presence of ethyl orthoformate, converts terpenes into acetals [77].

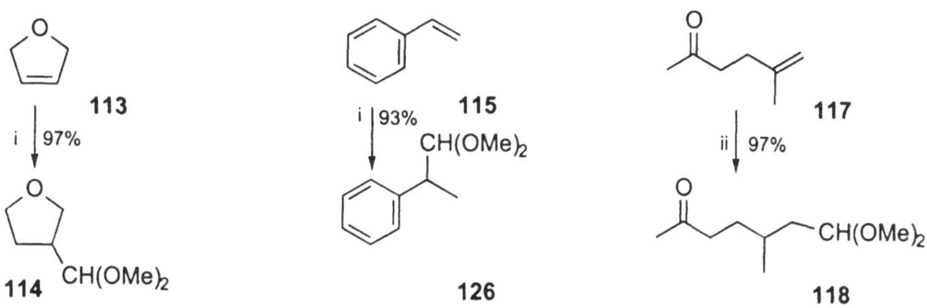

i) 1 mol/% [Rh(μ-OMe)(cod)]$_2$/ 10 PPh$_3$/PPTS, CO/H$_2$= 1, 50 bar, 60°C, 2,2-dimethoxypropane, 24 h.
ii) Identical to i but using P(O-o-^{t}BuC$_6$H$_4$)$_3$.

Figure 32. Consecutive hydroformylation-intermolecular acetalization with the catalytic systems [Rh(μ-OMe)(cod)$_2$]/PPh$_3$/PPTS

As has been shown above (see section 6.4), the hydroformylation of 3,4,6-tri-*O*-acetylglucal gives considerable amounts of the elimination product **121**, which must be obtained by eliminating of acetic acid from **120** (see also Figure 19) [49]. The hydroformylation of allylic esters to give α,β-unsaturated aldehydes by hydroformylation and acid elimination is a well documented process [4e]. *In situ* acetal formation partially avoids this process. Thus, using [Rh(μ-OMe)(cod)]$_2$/P(O-*o*-tBuC$_6$H$_4$)$_3$/PPTS as hydroformylation-acetalyzation catalysts, 3,4,6-tri-*O*-acetyl-D-glucal **119** is converted into the dimethoxy acetal

120 (Figure 34). Only a small amount of the elimination product **121** is formed [78].

Unexpectedly, the 3,4,6-tri-*O*-benzyl-D-glucal gave only the methyl α-glycoside **121** under the hydroformylation-acetalization conditions. In fact, there are two electrophilic reagents in competition for the nucleophilic alkene, the rhodium complex and the proton. When the alkene is deactivated (R=Ac) the coordination of rhodium is preferred and the hydroformylation-acetalization takes place. But when it is not deactivated (R=Bn) the acidic proton reacts faster than rhodium and methanol is added.

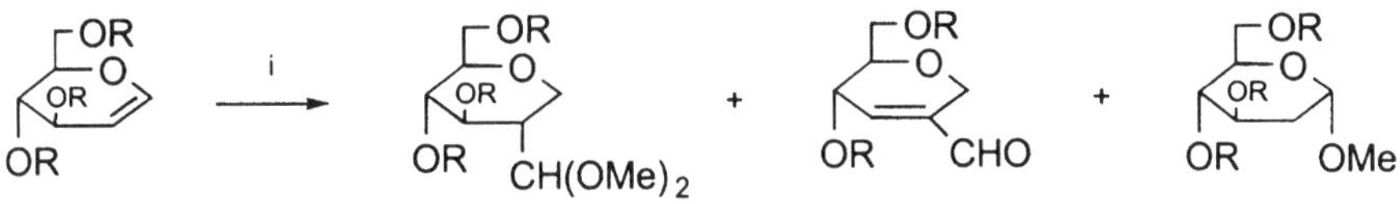

	119	**120**	**121**	**122**
R= Ac		66	9	—
R= OBn		—	—	85%

i) 1 mol/% $[Rh(\mu\text{-}OMe)(cod)]_2$/ $10P(O\text{-}o\text{-}{}^tBuC_6H_4)_3$ / PPTS, CO/H_2=1, 50 bar, 100°C, 2,2-dimethoxypropane, 48h.

Figure 33. Hydroformylation-acetalization of glucal derivatives

6.6.3 Hydroformylation-amination (intramolecular)

When an amino group is present in the substrate (**123**, **125**, **127**), various processes can take place consecutively under hydroformylation conditions to afford, cyclic N,O- (**124**, **126**) [79, 80] or N,N-acetals (**128**) [81] (Figure 34). Imines and enamines can also be formed. The formation of acetals, imines or enamines depends on alcohols being present in the substrate or the solvent, additional amino groups being present in the substrate, the substitution of the amino group, and the reaction conditions. Moreover the presence of coordinative atoms such as nitrogen allows a chelate to be formed which control the regioselectivity of the process.

Because of the chelate control of the process, allylamides such as **129** react under hydroformylation conditions to give mainly the branched aldehyde **131**, together with cyclic derivatives **132** and **133** (Table 5) [82, 83]. Products **132** and **133** are formed from the linear aldehyde through a sequence of reactions involving cyclization to give the enamide **134**, followed by hydrogenation or hydroformylation, respectively [84].

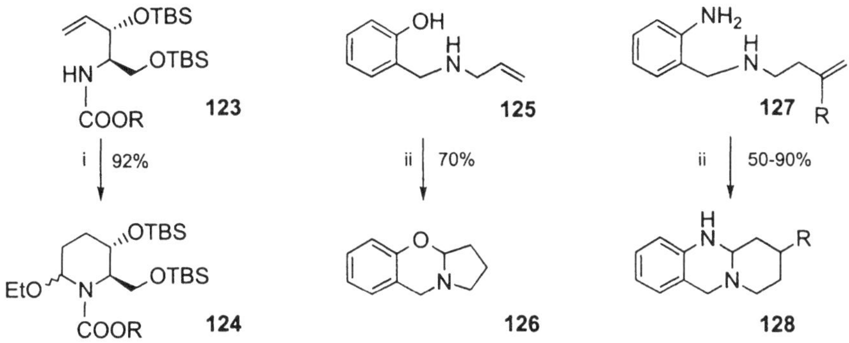

i) 1 mol/% [Rh(acac)(CO)$_2$]/BIPHEPHOS, CO/H$_2$= 1, 4 bar, 65°C, EtOH. ii) [Rh(OAc)$_2$]$_2$/3 PPh$_3$, CO/H$_2$= 1, 28 bar, 80°C.

Figure 34. Consecutive hydroformylation-intramolecular aminoacetalization

Hydroformylation of substrate **130** does not show a similar directing effect and, depending on the catalyst used, gives the terminal aldehyde from which **131** or **132** are formed.

The reaction also works with cyclic amides to afford bicyclic heterocycles [84]. The 3-butenamide **135** (n=1, R=H) also undergoes a consecutive process which affords compounds **136** or the dimer **138** depending on the ligand used (Table 6) [84, 85]. Large excess of PPh_3 affords almost exclusively the linear aldehyde and further to the pyrrolidone **136** by cyclization, water elimination to give **140**, and double bond isomerization.

Cross-coupling reaction between pyrrolidone **137** and the intermediate **140** would give the dimer **138**. In the 4-pentenamide **135** (n=2) the regioselectivity which gives the branched aldehyde is also very high and apparently is controlled by the formation of a chelate. Then the branched aldehyde cyclizes and eliminates water to give the lactam **139**.

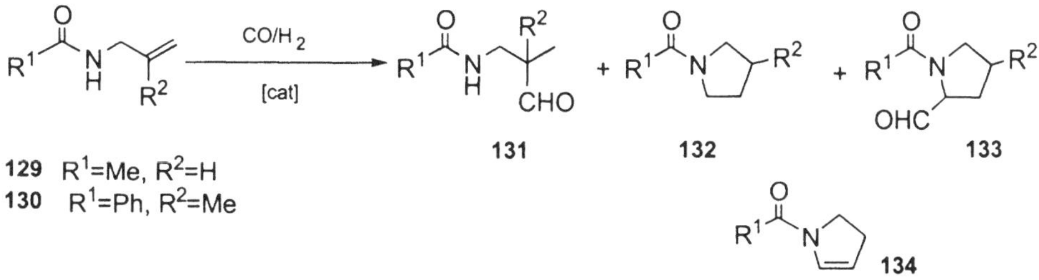

Figure 35. Hydroformylation of N-allylamides

Table 5. Hydroformylation of N-allylamides **129** and **130**.

Substrate	Catalyst	Yield(%)	Ratio **131:132:133**
129	$[Rh(dppb)(NBD)]ClO_4$ [a]	87	71:5:24
129	$Co_2Rh_2(CO)_{12}$ [a]	80	79:21:0
130	$[Rh(dppb)(NBD)]ClO_4$ [b]	87	--:--:>99.5
130	$Co_2Rh_2(CO)_{12}$ [c]	85	--:100:--

[a] CO/H_2=1, 90 bar, 80°C, THF, 18h. [b] CO/H_2=11, 140 bar, 100°C, THF, 18h. [c] CO/H_2=0.3, 92 bar, 100°C, THF, 18h

The bulkiness of the substituent on the amide nitrogen virtually does not have any effect on the regioselectivity, but it exerts a marked effect on the cyclic/acyclic ratio of the products. The effect of the bulky N-substituents is particularly pronounced in the trityl derivative (R=Tr, n=1), since no formation of pyridone **136** was observed. A mixture of open chain linear aldehyde and pyrrolin-2-one **137** was obtained.

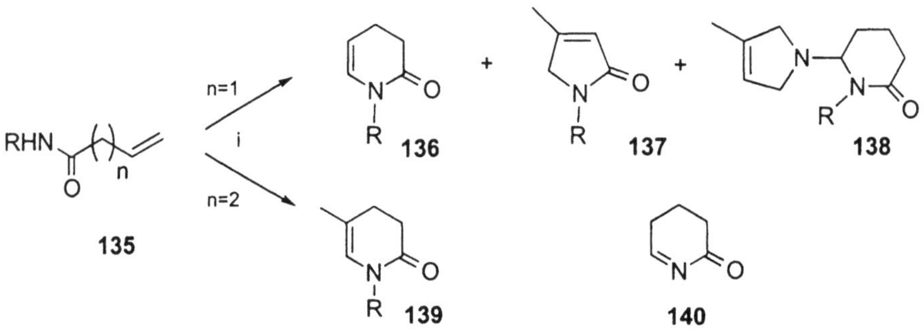

Figure 36. Consecutive processes in the hydroformylation of unsaturated amides.

Table 6. Hydroformylation of unsaturated amide **135**.

n	Catalyst	Yield (%)	Ratio **136:137:138**	**139**
1	$[RhCl(PPh_3)_3$ /20PPh_3 [a]	100	92 : 8 : 0	
1	$[RhCl(PPh_3)_3$ /10$P(OPh)_3$ [a]	90	3 : 3 : 94	
2	$[Rh_4(CO)_{12}]$ [b]	92		100

[a] 1 mol % [Rh], CO/H_2= 3, 90 bar, 100°C, THF, 40h. [b] 1 mol % [Rh], CO/H_2= 1, 90 bar, 100°C, THF, 18h.

Treating the protected amine **141** in the presence of the Rh-BIPHEPHOS catalyst (see Figure 3) under hydroformylation conditions, leads to enamide **142**. The consecutive reactions are hydroformylation to give the linear aldehyde, cyclization to give aminoacetal and the elimination of water. Compound **142** is a key intermediate in the synthesis of prosopinine **143** [86].

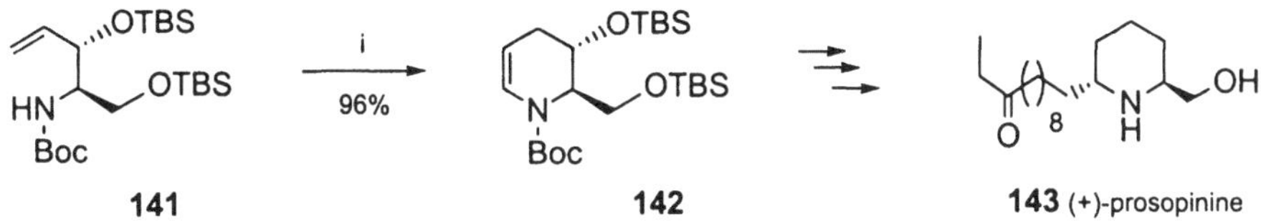

i) 1 mol/% $Rh(acac)(CO)_2$/ 2 BIPHEPHOS, CO/H_2, 4 bar, 65°C, 96%.

Figure 37. Consecutive hydroformylation-acetalyzation-elimination processes in the synthesis of (+)-prosopinine

The hydroformylation of the amidodiene **144** catalyzed by Rh-BIPHEPHOS under standard hydroformylation conditions gave the dehydropiperidine-aldehyde **146** as the sole product. The reaction is extremely chemo- and regioselective. The hydroformylation takes place at the homoallylic olefin moiety exclusively, and yields only the linear aldehyde intermediate **145** [87].

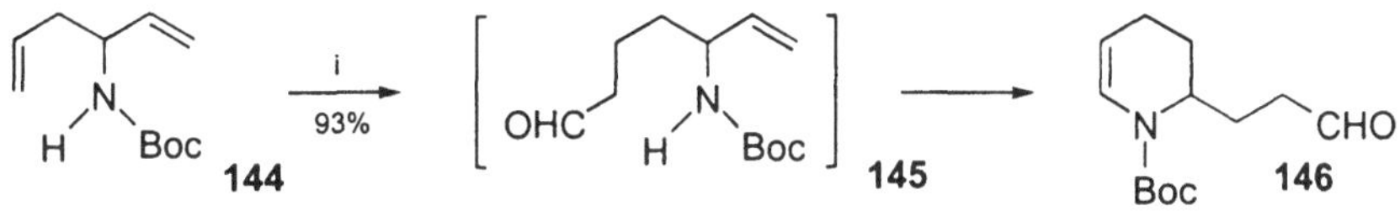

i) 1 mol % $[Rh(acac)(CO)_2]$ / 2BIPHEPHOS, CO/H_2, 4 bar, 65 °C.

Figure 38. Consecutive processes in the hydroformylation of diene **147**

ortho-Propenebenzeneamines **147** are hydroformylated with Rh-PPh_3 catalysts and the resulting product depends on the substitution in the allylic carbon. Thus, when R^1=OH, R^2=H, alkyl, the benzazepine **149** is obtained with regioselectivities up to 91% because of the preferred formation of linear aldehyde and the subsequent formation of the imine. However, when R^1=R=H, only dihydroquinoline **148** is formed, shows that the amino group has a remarkable directing effect since the hydroformylation of 3-phenylpropene gives a ratio b/l 70:30 [88].

N-Alkenyl-1,2-diaminobenzenes are hydroformylated with the Rh-PPh_3

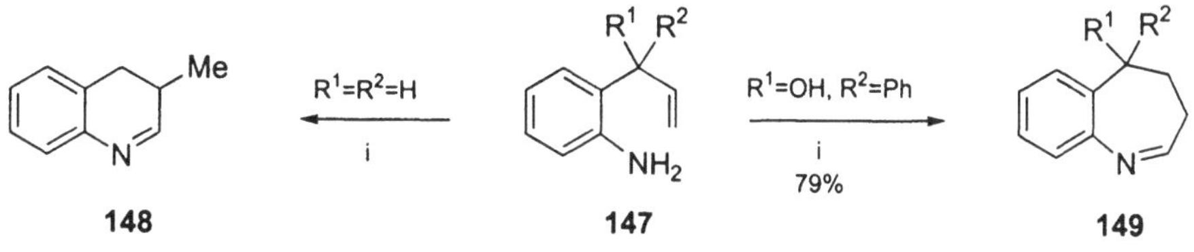

i) 0.5 mol/% $[Rh(OAc)_2]_2$, 4 PPh_3, CO/H_2= 1, 30 bar, 60°C, EtOAc.

Figure 39. Consecutive processes in the hydroformylation of alkene-amine **150**.

catalyst to give fused benzimidazoles [89, 90].

Alkenyl acyclic diamines such as N-allyl-1,3-diaminopropanes (**150**, n=1, R=H) are hydroformylated with Rh-BIPHEPHOS catalysts to give a quantitative yield of product **154**, which is formed from the linear aldehyde **151**, through the intermediate **152**. The homoallylic derivative gives a similar result (**150**, n=2). By changing the auxiliary ligand the reaction can be controlled to give dihydropyrrolidone derivative [91]. When n=4-9, diazabicyclicoalkanes **153** with a macrocycle of 8-13 member rings are formed in yields up to 95% in conditions that do not require high dilution. These aminals can be reduced with DIBAL-H to give macrocycles **154** in good yields [92, 93].

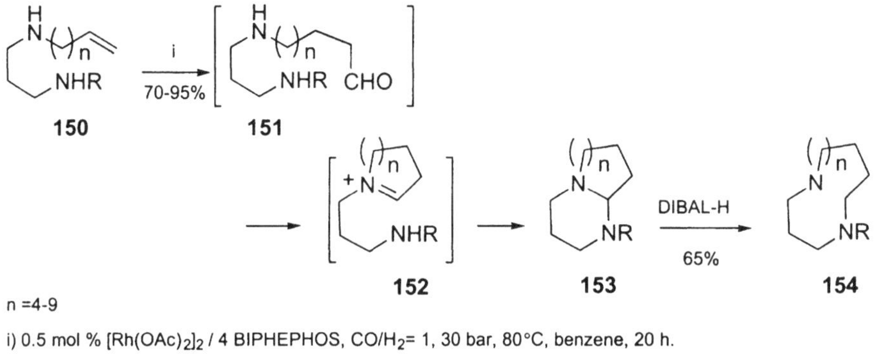

Figure 40. Consecutive processes in the hydroformylation of alkene diamines

6.6.4 Consecutive Hydroformylation-amination-reduction. Hydroaminomethylation

The hydroaminomethylation of alkenes was originally discovered by Reppe [94] and consists of the hydroformylation of an alkene, followed by reaction of

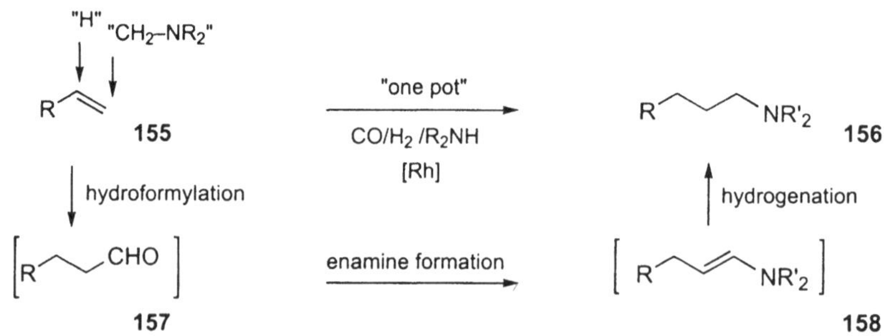

Figure 41. Consecutive processes in the hydroaminomethylation of alkenes.

the intermediate aldehyde with a primary or secondary amine to form an imine or enamine, and a final hydrogenation to give a secondary or tertiary amine (Figure 42) [4b, 95].

The hydroaminomethylation of 1-octene with morpholine in the presence of [Rh(μ-S-^{t}Bu)(CO)$_2$(PPh$_3$)$_2$]/PPh$_3$ gives a mixture of amines **161** and **162**, together with other secondary products (figure 42) [96, 97]. Using an unmodified rhodium catalyst, high pressures and a CO/H$_2$ ratio of 4.5, the same mixture of amines is obtained in quantitative yields that depend on the alkene [96]. No byproducts were detected. The high CO/H$_2$ ratio was chosen to suppress substrate hydrogenation and support catalyst stability.

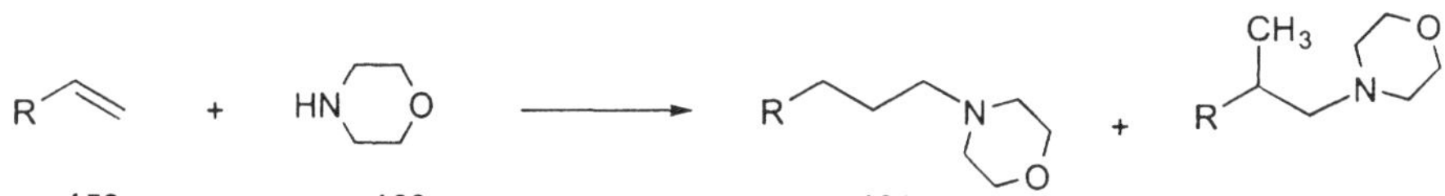

159 **160** **161** **162**

R=Hexyl

	161	162
[Rh(μ-S-Bu^t)$_2$(CO)$_2$(PPh$_3$)$_2$]/PPh$_3$,8 bar, 80°C	42	13%
[RhCl(cod)]$_2$/ P_{CO}= 90 bar, P_{H2}= 20 bar, 80°C,	46	36%

Figure 42. Hydroaminomethylation of 1-octene in the presence of morpholine

This reaction is general and preserves the regioselectivity observed under normal hydroformylation conditions, although variations are observed as a function of the amine. Thus, hydroaminomethylation of styrene leads to the branched derivative **164** with both secondary (R^1= R^2=alkyl) and primary amines (R^1=alkyl, R^2=H), although in this case more drastic conditions are needed. The piperazines give the *iso,iso*-diamine **165** as the major product in 90% yield.

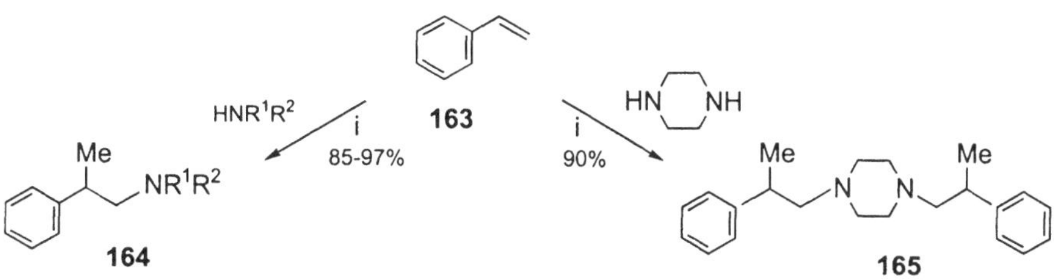

i) 1 mol/% [RhCl(cod)]$_2$, CO/H$_2$= 1, 110 bar, 80°C, dioxane.

Figure 43. Hydroaminomethylation of styrene in the presence of secondary amines and piperazine.

The reaction is compatible with the presence of carbonyl groups in the molecule, which remains unaltered, and only the aldehyde function reacts in the reaction conditions (see also Figure 32).

In the last few years Eilbracht et col. have made a lengthy study of how to apply this methodology to prepare a great number of differently functionalized organic compounds using $[RhCl(cod]_2$ as catalyst [4b]. No phosphorus ligands were used as co-catalysts. Dialkenes **166** are treated under hydroformylation conditions in the presence of secondary amines to give the diamines **167** (Figure 44). Mixtures *n,n, n,iso,* and *iso/iso* are obtained when R^1= R^2= H. Lineal aldehydes are the only products obtained when $R^1$and R^2 are alkyl groups. The use of $[Rh(acac)(CO)_2]$ and BIPHEPHOS exclusively leads to *n,n* products. The group X also has some influence on the regioselectivity but mainly on the reaction that takes place. Thus, when X=O double bond isomerization gives the enol, and when X=NH pyrrolidone is formed. Consequently, in order to obtain triamines the NH function should be protected [98].

Diamines are also obtained from allyl chlorides by initial halogen substitution to give the allylamine and subsequent hydroaminomethylation [99]. With primary amines or diamines this method can also be used in the synthesis of heterocyclic systems.

Different γ- and δ-aminofunctionalized ethers, amines and silanes **169** are obtained from alkenes **168**. Branched derivative is the main product obtained when X=OR, NR. The alkylation of the primary amines leading to secondary or tertiary amines can be controlled by the alkene/amine ratio [100]. Nitro derivatives can be used instead of amines, since they are reduced to amines in the reaction conditions [101]. Selective monoalkylation or dialkylation of nitro compounds is achieved depending on the alkene/nitro compound ratio.

Enamines and enamides such as **170** mainly give the iso aldehyde, which leads to the 1,2-diamines **171** by forming and then reducing intermediate enamines [102].

Hydroformylation of 1,4-dialkenes **172** in the presence of primary amines affords pyrroles **174** or the eight-member heterocycles **173**, depending on the alkene substitution pattern [103]. 1,4-Pentadienes, if substituted only in the 3,3-position (R^1=Me, R=H), undergo intramolecular pyrrole formation to give the bicylic systems **174**. Pyrrole is generated via hydrocarbonylative cyclization leading to the alkyl intermediate **175**. This rhodium intermediate undergoes further insertion of carbon monoxide to form an acyl intermediate, from which pyrrole is obtained by a Pall-Knorr synthesis. In contrast, 1,4-pentadienes with substituents in the double bond (R=alkyl) give an eight-membered heterocycle.

This reaction has been used in the preparation of a series of pharmacologically active amines [104] and diamines [105]. A stereoselective version of this reaction by using the directing group *o*-diphenylphosphinobenzoic acid has been reported [106] (Figure 27).

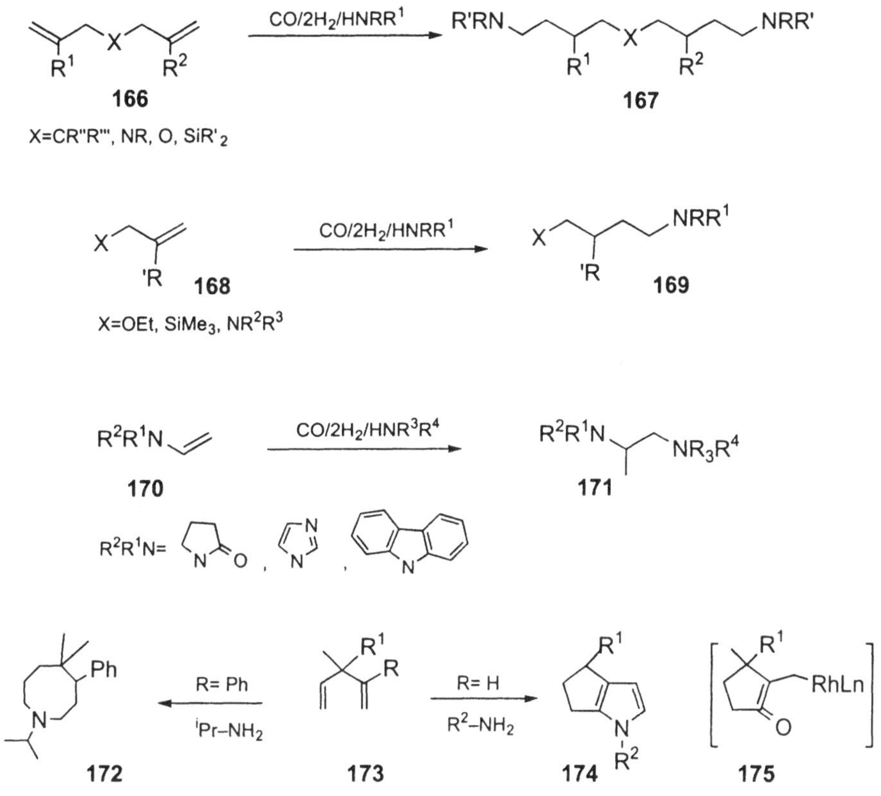

Figure 44. Hydroaminomethylation of differently functionalized alkenes.

6.6.5 Consecutive hydroformylation-aldol reaction

Aldehydes generated in the hydroformylation reaction in the presence of silyl enol ethers, enamines or enolates undergo consecutive aldol reaction. Thus, under usual hydroaminomethylation conditions the siladiene **176** leads to the sylacyclohexane **182**. The process is make up of a sequence of reactions and the proposed mechanism is the following: a) hydroformylation of **176** to give the dialdehyde **177**, b) **177** forms the rhodium enolate **178** (X=O) or the rhodium metalloenamine (X= NR), c) further aldol reaction to give α,β-unsaturated aldehyde **179**, d) double bond reduction to give the aldehyde **180**, from which e) enamine **181** is formed. This enamine is in equilibrium with **181'**. Subsequent hydrogenation takes probably place through conformer **181'** to give **182'** which then evolves to the conformationally more stable **182**. The diastereoselectivity of the process is determined in the last hydrogenation step. The regioselectivity in the 2,5-disubsituted sylacyclohexane is very high in unsymmetrically disubstituted divinylsilanes, and the *trans* isomer is exclusively formed when

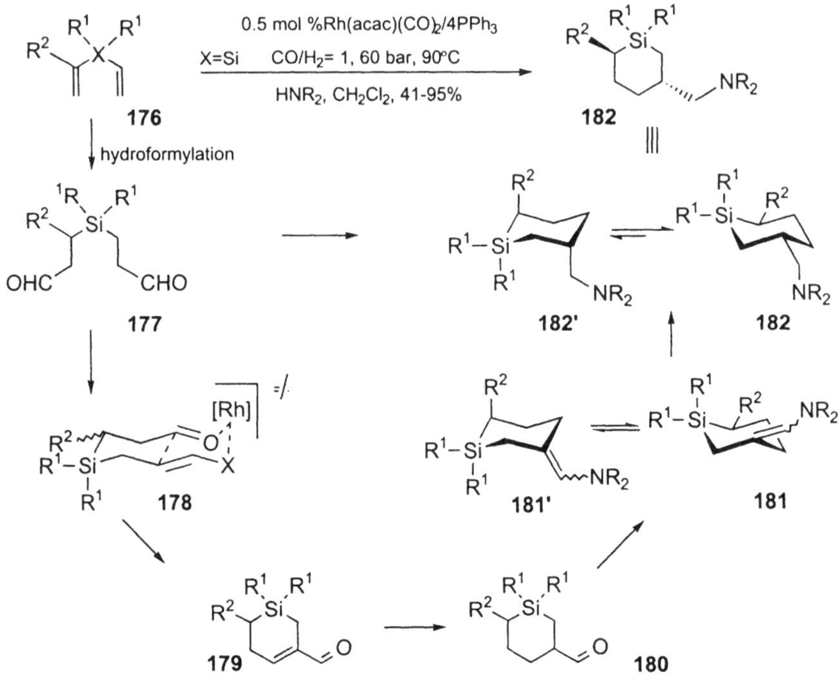

Figure 45. Consecutive hydroformylation-aldol reaction.

R^2=Ph. The reaction provides better yields with cyclic secondary amines than with acyclic ones [107].

Silylenol ethers such as **184** also undergo the hydroformylation-aldol reaction to give the silylated aldol adducts **185** in good yields through a sequence of reactions involving the hydroformylation of the alkene and the intramolecular Mukaiyama type aldol reaction. [108]. Best results were achieved using the trimethylsilyl group.

Different reaction products are obtained if the hydroformylation of the unsturated ketone **183** is carried out in the presence of amines. With secondary amines the hydroaminomethylation of the double bond is observed, leaving the carbonyl group unaffected.

In the presence of benzylamine, ketone **183** is converted into the aminoketone **186** by alkene hydroformylation, imine formation and aldol reaction. Under more drastic hydroformylation conditions the reaction of **183** with benzylamine leads to the amine **187**, which results from a mechanism similar to those above including reductive amination of the ketone moiety.

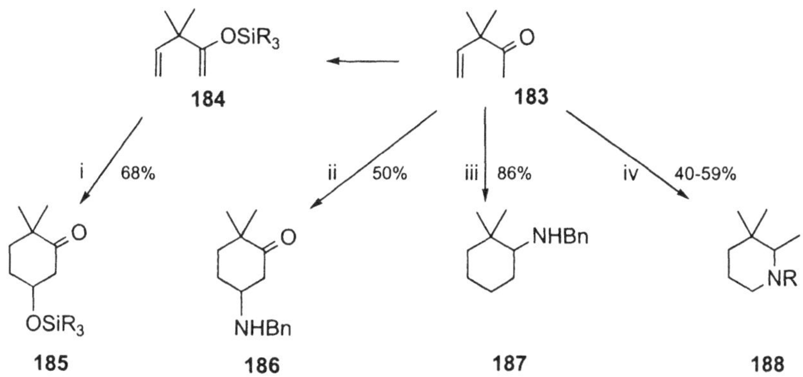

i) $[RhCl(cod)]_2$, CO/H_2=1, 80bar, 90 °C, CH_2Cl_2. ii) $[Rh(acac)(CO_2)]$/BIPHEPHOS, $BnNH_2$, CO/H_2=2, 30 bar, 60 °C, dioxane. iii) $[RhCl(cod)]_2$/ $P(OPh)_3$, $BnNH_2$, CO/H_2=2, 90 bar, 120 °C, dioxane. iv) $[RhCl(cod)]_2$/ $P(OPh)_3$, $^iPrNH_2$, CO/H_2=2, 90 bar, 120 °C, dioxane.

Figure 46. Consecutive hydroformylation-aldol reactions in the presence of amines

Using bulkier primary amines such as isopropyl- or cyclohexylamine, no aldol reaction is observed and, instead, heterocycles of type **188** are generated via alkene hydroformylation, double amine condensation and reductive amination of both carbonyl functions.

6.6.6 Consecutive hydroformylation-Wittig reaction

Consecutive hydroformylation-Wittig reaction is the last reported consecutive reaction involving hydroformylation. As in the case of hydroaminomethylation and hydroformylation-aldol reaction, the last step is also a hydrogenation reaction [109].

The reaction is limited to stabilized ylides, because non stabilizided ylides are too basic and induce rhodium inactivation.

Under hydroformylation conditions and in the presence of $Ph_3P=CHCOR$, compound **189** leads to the oxo derivative **190**. The process involves a sequence of reactions that includes initial hydroformylation to give the aldehyde **191** in a stereoselective way (see also Figure 26), Wittig olefination to give the *trans* conjugated alkene **192** and hydrogenation.

Disubstituted ylides (i.e. $PPh_3=C(Me)COR$) do not undergo hydrogenation and in consequence α,β-unsaturated ketones or esters are obtained. Ylides including the ester function provide low yields. The stereoselectivity of the process is determined by the chelating (directing) group *o*-DPPB, and stereochemistries all-*syn*, *anti-syn*, and all-*anti* can be obtained.

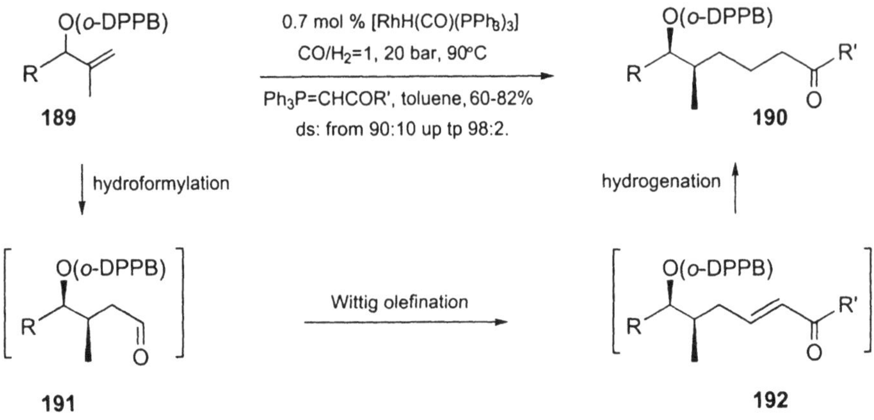

Figure 47. Consecutive hydroformylation-Wittig reaction

6.7 Alkyne hydroformylation

While the hydroformylation of alkenes is an important industrial process which also has enormous potential in organic syntheses, (see previous section in this chapter), the hydroformylation of alkynes has been much less investigated. This is because of the lack of general catalytic systems and the low selectivity of the process, which leads to the formation of undesired products [7, 110]. However, acetylenes are versatile intermediates in organic synthesis, and to improve the efficiency of the hydroformylation should be the main objective of research into this reaction.

The problem of the chemoselectivity is associated with the drastic reactions conditions usually employed in the hydroformylation reaction. Thus, the consecutive hydrogenation of unsaturated aldehyde gives aldehydes or alcohols [111, [112] (Figure 48). No products of double hydroformylation are obtained.

The regioselectivity is low except for the hydroformylation of acetylenic bonds bearing bulky substituents on one side of sp carbons.

For years, the problem of hydroformylating alkynes was circumvented by performing the reaction under water-gas shift conditions or by silylformylation. However, some authors have recently effectively used synthesis gas in alkyne hydroformylation.

– *Using CO/H_2.* Internal alkynes **197** are hydroformylated at room temperature and 1 bar CO/H_2 with the catalytic system [Rh]/BIPHEPHOS to give excellent yields of α,β-unsaturated aldehydes **198** [113].

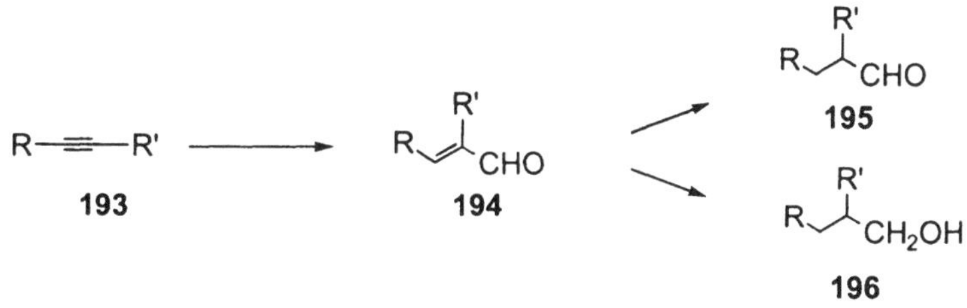

Figure 48. Hydroformylation of acetylenes

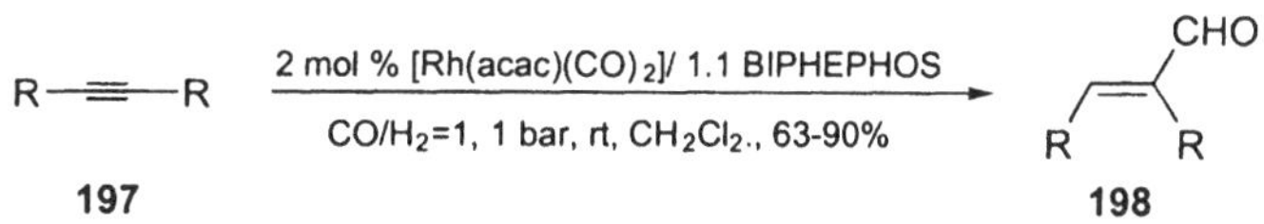

Figure 49. Hydroformylation of acetylenes with CO/H_2.

The hydroformylation of compounds which have a triple bond conjugated with a double bond (**199**) takes place, contrary to what might be expected, in the triple bond to give formyl dienes [114, 115]. The catalytic system $[RhH(PPh_3)_3]$ provides a mixture of diene **200** and cyclopentanone.

However, a zwitterionic rhodium complex $Rh^{zw}/P(OPh)_3$ mainly provides the formyl diene **200**, together with the nonconjugated unsaturated aldehyde **201** as by product. This phosphite provides similar yields to BIPHEPHOS, which, however, gives a more active catalyst.

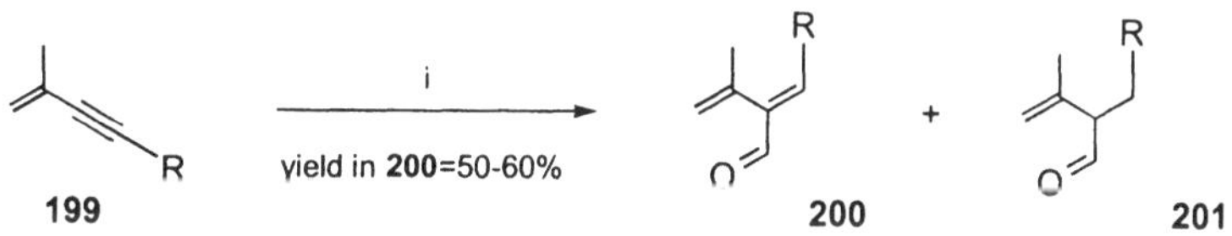

i) 4 mol % $[Rh^+(\eta^6\text{-}C_6H_5\text{-}B^-Ph_3)(cod)]$/ 4 $P(OPh)_3$, CO/H_2=1, 12 bar, 60°C, CH_2Cl_2, 36-48h.

Figure 50. Hydroformylation of alkene-acetylenes.

Terminal acetylenes give complex mixtures. The reaction also takes place in the acetylene that shows that for these substrates the triple bond is more reactive than the double bond.

The hyfroformylation of propargylamines **202** in the presence of the classical catalytic system $[Rh(OAc)_2]_2/PPh_3$ gives 2,4-disubstituted pyrroles **204** in excellent yields, through consecutive hydroformylation, cyclization and double bond isomerization [116, 117].

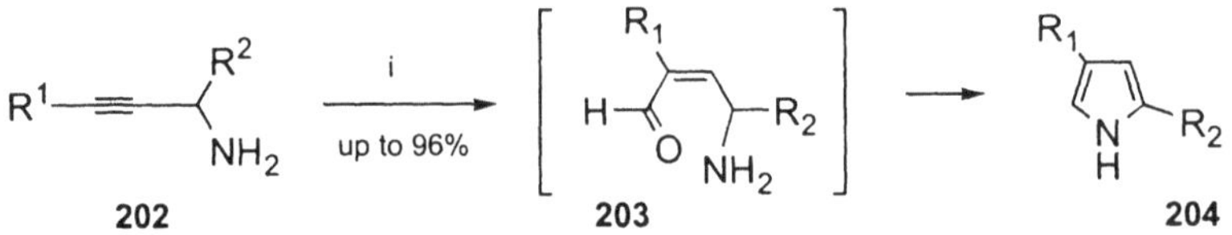

i) $[Rh(OAc)_2]_2/PPh_3$, CO/H_2=1, 30 bar, 70°C, 20h.

Figure 51. Synthesis of pyrrole by hydroformylation of amino-acetylenes.

– *Using CO/H₂O (water-gas shift).* The rhodium-catalyzed carbonylation of alkynes under water-gas shift reaction conditions proceeds through a different mechanism (type of reaction) than that under synthesis gas. Under these conditions alkynes give a regioselective mixture of furanones [118, 119, 120, 121, 122]. Rhodium carbonyl compounds in the presence of triethylamine are used as the catalyst. Triethylamine seems to be necessary for the reaction to initiate because when it was absent the reaction did not occur.

Of special interest is the hydroformylation of functionalized alkynes with amino [123], formyl [124], phenol [125], benzyl alcohol [126] or carboxylate groups [127] which leads to differently fuctionalized heterocycles.

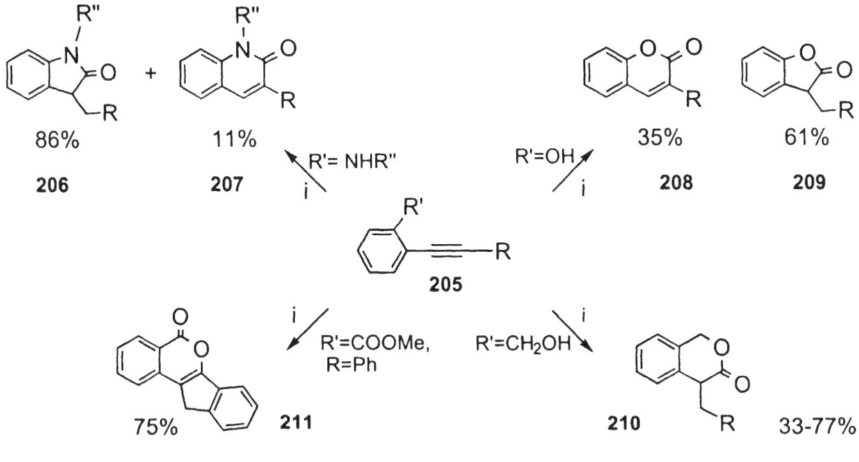

i) $Rh_6(CO)_{16}/NEt_3$, P_{CO}= 100 bar, 180°C, water-dioxane.

Figure 52. Synthesis of heterocycles by hydroformylation of phenylacetylenes.

The reaction gives compounds **206**, **209**, and **210**, and proceeds through the α,β-unsaturated lactones which undergoes an *in situ* hydrogenation to give the saturated lactones.

–Silylformylation. Carbonylation in the presence of hydrosilanes is conceptually comparable to carbonylation under hydrogen pressure [128]. The simultaneous incorporation of a triorganosilyl group and a formyl group into an acetylenic bond via rhodium catalyzed silylformylation [129], is an excellent procedure for synthesizing β-silylenals with high regio- and chemoselectivity [130, 131, 132, 133]. β-Silylenals are precursors of silyl-substituted dienes [134], dienones [135] and α,β-unsaturated ketones [136].

Rhodium carbonyl complexes in the presence of amines are used as catalytic systems, although zwitterionic rhodium complex has also provided excellent yields.

The regioselectivity in the silylformylation of terminal and internal acetylenes seems to be governed in general by the steric bulkiness of the substituents. Thus the bulky silyl group goes to the least hindered carbon, and consequently the formyl group is introduced into the most substituted one [137]. The electronic effect is the major factor that determines the regiochemistry of propiolate derivatives. Hydrosilylation is a competitive reaction.

The substituents of alkynes and hydrosilanes affect the reaction rate. The reactivity of the alkynes decreased in the order phenylacetylene > 1-hexyne > (trimethylsilyl)acetylene >> 3,3-dimethyl-1-butyne, whereas for the hydrosilane was $MePh_2SiH$, Me_2PhSiH >Et_2MeSiH, Et_3SiH > $tBuMe_2SiH$ >> iPr_3SiH. Thus, the rate of silylformylation largely depends on the combination of alkynes and hydrosilanes.

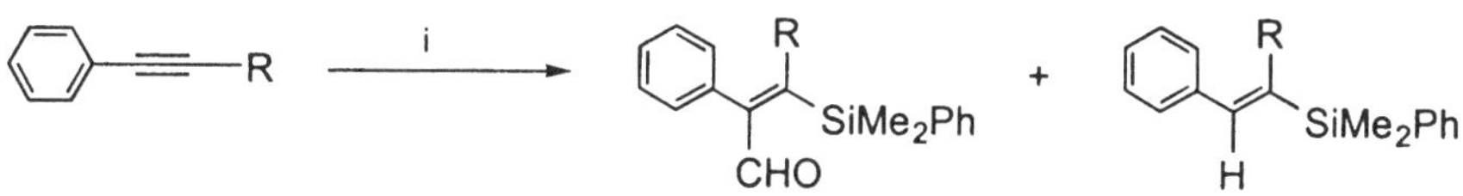

212 213 214

R= H	90% (Z:E= 8:2)	
R= Me	74%	8%

i) $Rh_4(CO)_{12}/Et_3N$, $PhMe_2SiH$, 25°C, benzene, 15h.

Figure 53. Silylformylation of phenylacetylenes.

Regioselectivity in acetylenes with a dimethylsilyloxy moiety is inverted and high, since the transfer of the silylgroup is intramolecular [138]. The reaction works with open chain as well as with cyclic compounds (Figure 54). Thus, differently *ω*-(dimethylsilyloxy)alkynes reacts under hydroformylation conditions to give the corresponding 3-*exo*-(formyl-methylene)oxasilacycloalkanes in excellents yields.

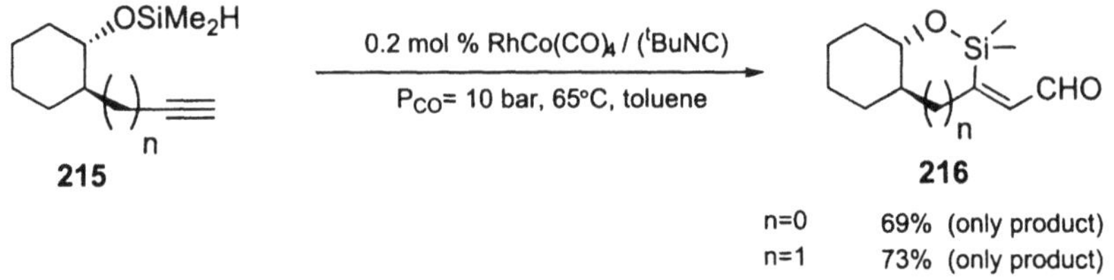

Figure 54. Intramolecular silylformylation.

The reaction of 3-(dimethylsilyloxy)-1-propyne does not proceed at all. But the reaction of of 3-(dimethylsilyloxy)-2-heptyne a 3-exo-(1-formylpentylid-1-ene)1-oxa-2-silacyclobutane. Oxa-silacycles of 4-7-member ring were reported. When starting from cyclic compounds bicyclic compounds are obtained.

Alkynes can also undergoes consecutive process under silylformylation conditions. Thus, rhodium catalyzed silylformylation of alkyne **217** in the presence of primary or secondary amines leads directly to the azadiene **221** by silylformylation and enamine formation [139]. These azadienes undergoes Diels-Alder reaction with dimethyl acetylendicarboxilate to give dihydropyridines **222**.

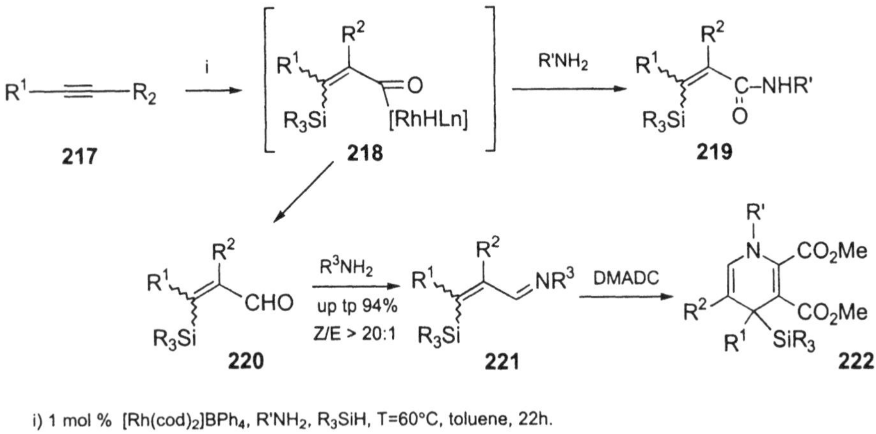

Figure 55. Silylformylation-amination of acetylenes.

6.8 Concluding remarks

The intense investigation of recent years in hydroformylation has allowed the best catalytic precursors to be selected, and new ligands have been found that provides highly active and selective catalysts. The incorporation of a ligand in the substrate now means that there are new ways of controlling the regio- and the stereoselectivity of the reaction. The compatibility of hydroformylation with many organic functionalities and the possibility of performing consecutive reactions with high selectivity also add to the value of the reaction. However, some of the selected examples in this chapter do not use the best catalytic precursors and ligands. Readers should, therefore, take into account all the information that has been accumulated in recent years, since at the present moment appropriate catalyst and reaction conditions can be chosen for hydroformylation of each substrate. We highly recommend that you read the various chapters in this book about the role of ligands and other criteria in selectivity before selecting a catalyst. Although new discoveries are expected in the different aspects of hydroformylation in the coming years, the present situation of knowledge of this reaction allows its use as an efficient synthetic tool in organic synthesis.

References

1 a) Gladiali, S.; Bayón, J. C.; Claver, C. *Tetrahedron: Asymmetry* **1995**, *6*, 1453. b) Agbossou, F.; Carpentier, J. F.; Mortreux, A. *Chem Rev.* **1995**, *95*, 2458. c) Botteghi, C.; Paganelli, S.; Schionato, A.; Marchetti, M. *Chirality* **1991**, *3*, 355. d) Eilbracht, P. in *Methoden der Organischen Chemie Stereoselective Synthesis,* Vol. E21c, sub vol 4, (Eds. Helmchen, G.; Hoffmann, R.W.; Mulzer, I.; Schaumann, E.) Houben-Weyl, Thieme, Sttutgart, **1996**, p. 2488. e) Nozaki, K.; Takaya, H.; Hiyama, T. *Topics in Catalysis* **1997**, *4*, 175. f) Nozaki, K. in *Comprehensive Asymmetric Catalysis*, vol 1, (Eds. Jacobsen, E. N.; Pfaltz, A.; Yamamoto, H.), Springer, Berlin, **1999**, p. 381. g) Consiglio, G. in *Catalytic Asymmetric Synthesis*, (Ed. Ojima, I.), VCH, Winheim, **1993**, p. 273. h) Noyori, R. *Asymmetric Catalysis in Organic Synthesis*, Wiley, New York, **1994**.

2 Eilbracht, P.; Barfacker, L.; Buss, C., Hollmann, C., Kitsos-Rzychon, B. E.; Kranemann, C. L.; Rische, T.; Roggenbuck, R.; Schmidt, A. *Chem. Rev.* **1999**, *99*, 3329.

3 Ojima, Y.; Tzamarioudaki, M.; Li, Z.; Donovan, R. J. *Chem. Rew.* **1996**, *96*, 635.

4 a) Botteghi, C.; Ganzerla, R.; Lenarda, M.; Moretti, G. *J. Mol. Catal.* **1987**, *40*, 129. b) Siegel, H.; Himmele, W. *Angew. Chem. Int. Ed. Engl.* **1980**, *19*, 178. c) Cornils, B.; Herrmann, W. A. *Applied Homogeneous Catalysis with Organometallic Compounds*, VCH, Weinheim, **1996**. d) Beller, M.; Cornils, B.; Frohming, C. D.; Kohlpaintner, C. W. *J. Mol. Catal.* **1995**, *104*, 17. e) Cornils, B.; Herrmann, W. A.; Kohlpaintner, C. W. *Angew. Chem. Int. Ed. Engl.* **1994**, *33*, 2144. f) Botteghi, C.; Marchetti, M.; Paganelli, S. in *Transition Metals for Organic Synthesis*, (Eds. Beller, M.; Bolm, C.), Wiley-VCH, **1998**, p. 25. g) Cornils B. in *New Synthesis with carbon monoxide*, (Ed. Falbe, J.) Springler, Berlin, **1980**, p. 1.

5 Lazzaroni, R.; Settambolo, L. R.; Uccello-Barreta, G. *Organometallics* **1995**, *14*, 4644.

6 Botteghi, C.; Branca, M.; Marchetti, M.; Saba, A. *J. Organomet. Chem.* **1978**, *197*, 161.

7 Brown, C. K.; Wilkinson, G. *J. Chem. Soc., A.* **1970**, 2753.
8 van Leeuwen, P. W. N. M.; Roobeek, C. F. *J. Organomet. Chem.* **1983**, *258*, 343.
9 Billing, E.; Abatjoglou, A. G.; Bryant, D. R. U.S. Patent 4769 498, **1988**.
10 Cuny, C. D.; Buchwald, S. L. *J. Am. Chem. Soc.* **1993**, *115*, 2066.
11 Cuny, C. D.; Buchwald, S. L. *Synlett.* **1995**, 519.
12 Dewar, T. J.; Philips, G.W.; Puckette, T.A.; Stavinoha, J. L.; Vanderbilt, J.J. Eastman Kodak Company, US 4694109, **1987**.
13 Kranenburg, M.; Kamer, P. C. J.; van Leeuwen, P. W. N. M. *Eur. J. Inor. Chem.* **1998,** 155.
14 van der Veen, L. A.; Kamer, P. C. J.; van Leeuwen, P. W. N. M. *Angew. Chem. Int. Ed. Engl.* **1999**, *38*, 336.
15 Beller, M.; Zimmermann, B.; Geissler, H. *Chem. Eur. J.* **1999**, *5*, 1301.
16 Matsui, Y.; Orchin, M. *J. Organomet. Chem.* **1983**, *57*, 246.
17 a) Botteghi, C.; Cazzolato, L.; Marchetti, M.; Paganelli, S. *J. Org. Chem.* **1995**, *60*, 6612. b) Botteghi, C.; Marchetti, M.; Paganelli, S.; Sechi, B. *J. Mol. Catal. A* **1997**, *118*, 173.
18 a) Kleemann, A.; Engel, J. *Pharmazeutische Wirkstoffe*, George Thieme Verlag, Stuttgart, **1987**. b) Elks, J.; Ganellin, G. R.; *Dictionary of Drugs*, Chapman Hall, London **1990**.
19 von Kleemann, A. *Chem. Ztg.*, **1977**, *101*, 389.
20 Takahashi, T.; Machida, K.; Kido, Y.; Nagashima, K.; Ebata, S.; Doi, T. *Chem. Lett.* **1997**, 1291.
21 Botteghi, C.; Soccolini, F. *Synthesis* **1985**, 592.
22 a) Fell, B.; Barl, M. *J. Mol. Catal.* **1977**, *2*, 301. b) Abatjoglou, A. B.; Bryant, D. R.; D'Esposito, L. C. *J. Mol. Catal.* **1983**, *18*, 381.
23 a) Beker, Y.; Eisenstadt, A.; Stille, J. K. *J. Org. Chem.* **1980**, *45*, 2145. b) Delogu, G.; Faedda, G.; Gladiali, S. *J. Organomet. Chem.* **1984**, *268*, 167.
24 a) Ojima, I. *Chem. Rev.* **1988**, *88*, 1011. b) Fuchikami, T.; Ojima, I. *J. Am. Chem. Soc.* **1982**, *104*, 3527.
25 Crameri, Y.; Ochsner, P. A.; Schudel, P. Eur. Pat. Appl. EP52775, **1982**
26 Matsumoto, M.; Tamura, M. *J. Mol. Catal.* **1982**, *16*, 195.
27 Polo, A.; Claver, C.; Castillón, C.; Ruiz, A.; Bayón, J. C.; Real, J.; Mealli, C.; Massi, D. *Organometallics* **1992**, *11*, 3525.
28 Kawabata, Y.; Suzuki, T. M., Ogata, L. *Chem. Lett.* **1978**, 361.
29 Hayashi, T.; Tanaka, M.; Ogata, L. *J. Mol. Catal.* **1981**, *13*, 323.
30 Lai, R.; Ucciani, E. *J. Mol. Catal.* **1978**, *4*, 401.
31 Botteghi, C.; Paganelli, S.; Bigini L.; Marchetti, M. *J. Mol. Catal.* **1994**, *93*, 279.
32 Botteghi, C.; Chelucci, C.; del Ponte, G.; Marchetti, M.; Paganelli, S. *J. Org. Chem.*, **1994**, *59*, 7125.
33 Settambolo, R.; Scamuzzi, S.; Caiazzo, A.; Lazzaroni, R. *Organometallics* **1996**, *17*, 2127.
34 Caiazzo, A.; Settambolo, R.; Pontorno, L.; Lazzaroni, R. In press
35 Caiazzo, A.; Settambolo, R.; Uccello-Barreta, G.; Lazzaroni, R. *J. Organomet. Chem.* **1997**, *548*, 279.
36 Totland, K.; Alper, H. *J. Org. Chem.* **1993**, *58*, 3328.
37 Kurkov, V. P.; U. S. Patent, 4344896 **1982**, *Chem. Abst.* **1982**, *97*, 181731.
38 Yamashita, H.; Roan, B. L.; Sakakura, T.; Tanaka, M. *J. Mol. Catal.* **1993**, *83*, 255.
39 Consiglio, G.; Kollár, L.; Kölliker, R. *J. Organomet. Chem.* **1990**, *396*, 375.

40 Kollár, L.; Consiglio, G.; Pino, P. *J. Organomet. Chem.* **1990**, *386*, 389.
41 Stefani, A.; Consiglio, G.; Botteghi, C.; Pino, P. *J. Am. Chem. Soc.* **1973**, *95*, 6504.
42 Himmele, W.; Siegel, H. *Tetrahedron Lett.* **1976**, 907.
43 Siegel, H.; Himmele, W. *Angew. Chem. Int. Ed. Engl.* **1980**, *19*, 178.
44 Azzaroni, F.; Biscarini, P.; Bordoni, S.; Longoni, G.; Venturini, E. *J. Organomet. Chem.* **1996**, *508*, 5958.
45 Leighton, J. L.; O'Neil, D. N. *J. Am. Chem. Soc.* **1997**, *119*, 11118.
46 Sarraf, S. T.; Leighton, J. L. *Tetrahedron Lett.* **1998**, *39*, 6423.
47 Rosenthal, A. *Adv. Carbohydr. Chem.* **1969**, *24*, 59.
48 Rosenthal, A.; Abson, D.; Field, T. D.; Koch, H. J.; Mitchel, R. E. *Can. J. Chem.* **1967**, *45*, 1525.
49 Fernández, E.; Ruiz, A.; Claver, C.; Castillón, S.; Polo, A.; Piniella, J. F.; Alvarez-Larena, A. *Organometallics* **1998**, *17*, 2857.
50 Takahashi, T.; Ebata, S; Yamada, H. *Synlett.* **1998**, 381.
51 Breit, B.; Zahn, S. K. *Tetrahedron Lett.* **1998**, *39*, 1901.
52 a) Kraft, E. M.; Wilson, L. J. *Organometallics* **1988**, *7*, 2528. b) Kraft, E. M. *Tetrahedron Lett.* ***1989, 30I, 539.***
53 Jackson, W. R.; Perlmutter, P.; Tasdelen, E. E. *J. Chem. Soc., Chem. Commun.* **1990**, 763
54 Jackson, W. R.; Perlmutter, P.; Tasdelen, E. E. *Tetrahedron Lett.* **1990**, *31*, 2461.
55 Jackson, W. R.; Perlmutter, P.; Suh, G. H. *J. Chem. Soc., Chem. Commun.* **1987**, 724.
56 Burke, S. D.; Cobb, J. E. *Tetrahedron Lett.* **1986**, 27, 4237.
57 Burke, S. D.; Cobb, J. E.; Takeuchi, K. *J. Org. Chem.* **1990**, 55, 2138.
58 Breit, B. *Angew. Chem. Int. Ed. Engl.* **1996**, *35*, 2835.
59 Breit, B. *Liebigs Ann. Recueil* **1997**, 1841.
60 Breit, B.; Dauber, M.; Harms, K. *Chem. Eur. J.* **1999**, *5*, 2819.
61 Breit, B. *Chem. Commun.* **1997**,591.
62 Breit, B. *Eur. J. Org. Chem.* **1998**, 1123.
63 Pittman, C. U. Jr.; Honnick, W. D. *J. Org. Chem.* **1980**, *45*, 2132.
64 Wuts, P. G. M.; Obrzut, M. L.; Thompson, P. A. *Tetrahedron Lett.* **1984**, *25*, 4051.
65 Smith, W. E.; Chambers, G. R.; Lindberg, R. G.; Cawse, J. N.; Dennis, A. J.; Harrison, G. E.; Bryant, D. R. In *Catalysis of Organic Reactions*; (Ed. Augustine, R. L.) Dekker, New York, **1985**, p.151.
66 Nozaki, K.; Li. W.; Horiuchi, T.; Takaya, H. *Tetrahedron Lett.* **1997**, *38*, 4611.
67 Trzeciak, A. M.; Wolszcak, E.; Ziolkówsky, J. *New J. Chem.* **1996**, *20*, 365.
68 Anastasiou, D.; Jackson, W. R. *Aust. J. Chem.* **1992**, *45*, 21.
69 Anastasiou, D.; Jackson, W. R.; McCubbin, Q. J.; Trnacek, A. E. *Aust. J. Chem.* **1993**, *46*, 1623.
70 Baker, R.; Herbert, R. H. *Nat. Prod. Rep.* **1984**, *43*, 3309.
71 Jaramillo, C.; Knapp, S. *Synthesis* **1994**, 1.
72 Sirol, S.; Kalck, Ph. *New. J. Chem.* **1997**, *21*, 1129.
73 Kitsos-Rzychon, B.; Eilbracht, P. *Tetrahedron* **1998**, *54*, 10721.
74 Parriello, G.; Stille, J. K. *J. Am. Chem. Soc.* **1987**, *109*, 7122.
75 Fernández, E.; Castillón, S. *Tetrahedron Lett.* **1994**, *35*, 2361.
76 Balué, J.; Bayón, J. C. *J. Mol. Catal.* **1999**, *137*, 193.
77 Soulantica, K.; Sirol, S.; Koinis, S.; Pneumatikakis, G.; Kalck, Ph. *J. Organomet. Chem.* **1995**, *498*, C10.
78 Fernández, E.; Polo, A.; Ruiz, A.; Claver, C.; Castillón, S. *Chem. Commun.* **1998**, 1803.

79 Ojima, I.; Tzamarioudaki, M.; Eguchi, M. *J. Org. Chem.* **1995**, *60*, 7078.
80 Campi, E. N.; Jackson, W. R.; McCubbin, Q. J.; Trnacek, A. E. *Aust. J. Chem.* **1996**, *49*, 219.
81 Campi, E. N.; Habsuda, J.; Jackson, W. R.; Jonasson, C. A. M.; McCubbin, Q. *Aust. J. Chem.* **1995**, *48*, 2023.
82 Ojima, I.; Zhang, Z. *J. Org. Chem.* **1988**, *53*, 4425.
83 Ojima, I.; Zhang, Z. *J. Organomet. Chem.* **1991**, *417*, 253.
84 Ojima, I.; Korda, A. *Tetrahedron Lett.* **1989**, *30*, 6283.
85 Ojima, I.; Korda, A.; Shay, W. R. *J. Org. Chem.* **1991**, *56*, 2024.
86 Ojima, I.; Vidal, E. S. *J. Org. Chem.* **1998**, *63*, 7999.
87 Ojima, I.; Iula, D. M.; Tzamarioudaki, M. *Tetrahedron Lett.* **1998**, *39*, 4599.
88 Anastasiou, D.; Jackson, W. R. *J. Chem. Soc., Chem. Commun.* **1990**, 1205.
89 Anastasiou, D.; Chaouk, H.; Jackson, W. R. *Tetrahedron Lett.* **1991**, *32*, 2499.
90 Anastasiou, D.; Campi, E. M.; Chaouk, H.; Jackson, W. R. *Tetrahedron* **1992**, *48*, 7467.
91 Bergmann, D. J.; Campi, E. M.; Jackson, W. R.; Cubin, Q. J.; Patti, A. F. *Tetrahedron Lett.* **1997**, *38*, 4315.
92 Bergmann, D. J.; Campi, E. M.; Jackson, W. R.; Patti, A. F. Sylik, D. *Tetrahedron Lett.* **1999**, *40*, 5597.
93 Bergmann, D. J.; Campi, E. M.; Roy Jackson, W.; Patti, A. F. *Chem. Commun.* **1999**,1279.
94 a) Reppe, W. *Experientia* **1949**, *5*, 93. b) Reppe, W.; Vetter, H. *Liebigs Ann. Chem.* **1953**, *582*, 133.
95 Rische, T.; Eilbracht, P. *Synthesis* **1997**, 1331.
96 Baig, T.; Kalc, Ph. *J. Chem. Soc., Chem. Commun.* **1992**, 1373.
97 Baig, T.; Molinier, J.; Kalc, Ph. *J. Organomet. Chem.* **1993**, *455*, 219.
98 Eilbracht, P.; Kranemann, C. L.; Bärfacker, L. *Eur. J. Org. Chem.* **1999**, 1907.
99 Rische, T.; Eilbracht, P. *Tetrahedron* **1999**, *55*, 3917.
100 Rische, T.; Barfacker, L.; Eilbracht, P. *Eur. J. Org. Chem.* **1999**, 653.
101 Rische, T.; Eilbracht, P. *Tetrahedron* **1998**, *54*, 8441.
102 Bärfacker, L.; Rische, T.; Eilbracht, P. *Tetrahedron* **1999**, *55*, 7177.
103 Kranemann, C. L.; Kitsos-Rzychon, B.; Eilbracht, P. *Tetrahedron* **1999**, *55*, 4721.
104 Rische, T.; Eilbracht, P. *Tetrahedron* **1999**, *55*, 1915.
105 Rische, T.; Muller, K-S.; Eilbracht, P. *Tetrahedron* **1999**, *55*, 9801.
106 Breit, B. *Tetrahedron Lett.* **1998**, *39*, 5163.
107 Bärfacker, L.; El Tom, D.; Eilbracht, P. *Tetrahedron Lett.* **1999**, *40*, 4031.
108 Hollman, C.; Eilbracht, P. *Tetrahedron Lett.* **1999**, *40*, 4313.
109 Breit, B.; Zahn, S. K. *Angew. Chem. Int. Ed.* **1999**, *38*, 969.
110 Pino, P.; Barca, G. in *Organic Synthesis via Metal Carbonyls*, vol 2, (Eds. Wender, I.; Pino, P.), Wiley-Interscience, New York, **1977**, p. 419.
111 Tkatchenko, I. in *Comprehensive Organometallic Chemistry*, vol. 8, (Eds. Wilkinson, Sir G. Stone, F. G. A., Abel, E.), Pergamon Press, Oxford, **1982**.
112 Botteghi, C.; Salomon, Ch. *Tetrahedron Lett.* **1974**, 4285.
113 Johnson, J. R.; Cuny, G. D.; Buschwald, S. L. *Angew. Chem. Int. Ed. Engl.* **1995**, *34*, 1760.
114 Doyama, K.; Joh, T.; Takahashi, S. *Tetrahedron Lett.* **1986**, *27*, 4497.
115 Doyama, K.; Joh, T.; Shiohara, T.; Takahashi, S. *Bull. Chem. Soc. Jpn.* **1988**, *61*, 4353.
116 Campi, E. M.; Jackson, W. R. *Aust. J. Chem.* **1989**, *42*, 471.
117 Campi, E. M.; Jackson, W. R.; Nilsson, Y. *Tetrahedron Lett.* **1991**, *32*, 1093.

118 Joh, T.; Doyama, K.; Onitsuka, T.; Shiohara, S.; Takahashi, S. *Organometallics* **1991**, *10*, 2493.
119 Laine, R. M.; Crawford, E. J. *J. Mol. Catal.* **1988**, *44*, 357.
120 Cavinato, G.; Toniolo, L. *J. Mol. Catal.* **1996**, *105*, 9.
121 Joh, T.; Fujiwara, K.; Takahashi, S. *Bull. Chem. Soc. Jpn.* **1993**, *66*, 978.
122 Joh, T.; Doyama, K.; Fujiwara, K.; Maeshima, K.; Takahashi, S. *Organometallics* **1991**, *10*, 508.
123 Hirao, K.; Morii, N.; Joh, T.; Takhashi, S. *Tetrahedron Lett.* **1995**, *36*, 6243.
124 Sugioka, T.; Zhang, S.-W.; Mori, N.; Joh, T.; Takahashi, S. *Chem. Lett.* **1996**, 249.
125 Yoneda, E.; Sugioka, T.; Hirao, K.; Zhang, S.-W.; Takahashi, S. *J. Chem. Soc., Perkin Trans. I* **1998**, 477.
126 Yoneda, E.; Kaneko, T.; Zhang, S-W.; Takahashi, S. *Tetrahedron Lett.* **1998**, *39*, 5061.
127 Sugioka, T.; Yoneda, E.; Onitsuka, K.; Zhang, S-W.; Takahashi, S. *Tetrahedron Lett.* **1997**, *38*, 4989.
128 a) Murai, S.; Sonoda, N. *Angew. Chem. Int. Ed. Engl.* **1979**, *18*, 837. b) Chatani, N.; Kajiwata, Y.; Nishimura, H.; Murai, S. *Organometallics* **1991**, *10*, 21. c) Chatani, N.; Murai, S. *Synlett* **1996**, 414.
129 Matsuda, I.; Ogiso, A.; Sato, S.; Isumi, Y. *J. Am. Chem. Soc.* **1989**, *111*, 2332.
130 a) Monteil, F.; Matsuda, I.; Alper, H. *J. Am. Chem. Soc.* **1995**, *117*, 4419. b) Ojima, I., Vidal, E.; Tzamarioudaki, M.; Matsuda, I. *J. Am. Chem. Soc.* **1995**, *117*, 6797.
131 Ojima, I.; Fracchiolla, D. A.; Donovan, R. J.; Banerji, P. *J. Org. Chem.* **1994**, *59*, 7594.
132 a) Doyle, M. P.; Shanklin, M. S. *Organometallics* **1993**, *12*, 11. b) Doyle, M. P.; Shanklin, S. *Organometallics* **1994**, *13*, 1081.
133 Zhou, J.-Q.; Alper, H. *Organometallics* **1994**, *13*, 1586.
134 a) Jung, M. E.; Gaede, B. *Tetrahedron* **1979**, *35*, 621. b) Carter, M. J.; Fleming, I.; Percival, A. *J. Chem. Soc., Perkin Trans. I* **1981**, 2415.
135 Pillot, J.; Dunogues, J.; Calas, R. J. *J. Chem. Res. Synop.* **1977**, 268.
136 Fleming, I.; Perry, D. A. *Tetrahedron* **1981**, *37*, 4027.
137 Matsuda, I.; Fukuta, Y.; Tsuchihashi, T.; Nagashima, H.; Itoh, K. *Organometallics* **1997**, *16*, 4327.
138 Ojima, I.; Vidal, E.; Tzamarioudaki, M.; Matsuda, J. *J. Am. Chem. Soc.* **1995**, *117*, 6797.
139 Bärfacker, L.; Hollmann, Ch.; Eilbracht, P. *Tetrahedron* **1998**, *54*, 4493.

Chapter 7

Aqueous biphasic hydroformylation

Jürgen Herwig and Richard Fischer

Celanese GmbH, Werk Ruhrchemie, Research & Technology, D-46128 Oberhausen, Germany

7.1 Principles of biphasic reactions in water

7.1.1 Why two-phase catalysis? Scope and Limitations

What has to be achieved to combine the major important advantages of homogeneous and heterogeneous catalysis? It seems to be a non-complex problem with a simple solution: find a way of separating the catalyst and the reaction products under mild conditions that is ecologically as well as economically efficient. However, to get there is not that simple. Normally the organometallic catalysts applied in solution are not very stable against high temperatures, water or oxygen. Usually, reaction products, especially bulk chemicals, are separated in two ways. Firstly, by gas phase separation technologies like distillation as applied for e.g. the synthesis of acetic acid by the carbonylation of methanol or the hydroformylation of alkenes to aldehydes. Secondly, by reaction conditions that keep the reaction partners, starting materials, and the products in the gas phase as e.g. for the gas phase process for the production of phthalic anhydride. In the case of the production of terephthalic acid the product is insoluble in the reaction medium acetic acid, in which the Co and Mn based oxidation catalyst are dissolved. This way it is possible to separate the product from the catalyst solution by simple filtration. However, such beneficial circumstances occur rather seldom, as mixtures of liquid reaction media and liquid products are the rule. Thus it is a major step forward if catalytic reactions can be performed in liquid multi-phase systems with clean and fast separation of

P.W.N.M. van Leeuwen and C. Claver (eds.), Rhodium Catalyzed Hydroformylation, 189–202.

reactant and product phase. See Chapter 8 for an overview of the other separation techniques in use for hydroformylation.

In contrast to heterogeneous catalysts, their homogeneous counterparts don't resist high temperatures or harsh work-up conditions mostly due to damage of weak carbon-metal bonds or the loss of ligand stabilization of the central metal atom of the complexes. The avoidance of (traces) of water or air at such conditions is also often a parameter that is critical to the success of the systems. Thus the conditions of the separation of the reaction product from the solution, in which the catalyst has been dissolved, must be as mild as possible. A textbook example of such a modern technology is the application of two- or multi-phase reaction systems. One of the most important developments in 1980-2000 in the area of homogeneous catalysis is the successful development of water-stable as well as highly water-soluble catalyst systems and the consecutive introduction of the aqueous two-phase technology.

What limitations have to be accounted for if multi-phase systems are applied is shown in the following listing of possible disadvantages:

- low reaction rates and the resulting low space-time yields due to insufficient solubility of the reactants or complex interface processes such as mass-transport limitations,
- surfactant effects that might favor the enrichment of the reaction product or side products in the reaction media,
- slow or incomplete phase separation,
- the need of high reaction volumes,
- corrosion effects,
- decrease of the chemoselectivity (e.g. pH-controlled side-reactions, etc.).

7.1.2 Concepts for two-phase hydroformylation

Basis for two-phase catalysis is the presence of two immiscible phases. These phases can consist of:

– water (+ co-solvent)	- organic liquid
– polar organic liquid	- non polar organic liquid
– ionic liquid	- organic liquid
– fluorous organic phase	- organic liquid

The first concept has been realized in the Ruhrchemie-Rhône Poulenc process for propene and butene hydroformylation. This process will be discussed in detail below. The catalyst is retained in the aqueous phase by applying a water-soluble phosphine ligand. The product constitutes the organic solvent.

The second concept is realized in the Shell Higher Olefins Process for the oligomerization of ethylene to linear higher alkenes. Butane-1,4-diol is used as the solvent for the Ni based catalyst. The linear higher alkenes are not soluble in the catalyst phase and can thus easily be separated. This concept is not applied in hydroformylation.

The third concept is used in an IFP process for the dimerisation of alkenes. Hydroformylation has been successfully conducted in ionic liquids in the laboratory [1].

Fluorinated organic liquids, concept four, have not been applied in industry, because of technical and economic disadvantages (expensive ligands and solvents). Hydroformylation has been investigated in fluorinated solvents [2] (see Chapter 10).

7.2 Hydroformylation of propene and butene

7.2.1 Historic overview of two-phase hydroformylation technology

In 1982, Rhône Poulenc and Ruhrchemie AG (now part of Celanese AG) joined to develop a continuous aqueous biphasic hydroformylation process. The work was driven by the fact that Rh was much more efficient in hydroformylation of short chain alkenes than cobalt; Rhodium gives lower raw material consumption, lower cost for catalyst recycling, and a higher l:b ratio. In order to stay competitive Ruhrchemie as a leading oxo producer decided to switch to a Rh based system. Furthermore, a two-phase system could offer several advantages over a homogeneous system e.g. lower catalyst loss, lower heat stress of the catalyst, easy catalyst separation from the products and higher energy efficiency.

Following the laboratory results of several, biphasic catalytic reactions (hydroformylation, hydrocyanation, and diene conversion) performed by E. Kuntz and patented by Rhône Poulenc [3] the process work started. In only two years' time these laboratory results were transferred to a 100,000 t/a plant in Oberhausen by a team managed by B. Cornils. The first aqueous biphasic propene hydroformylation reactor came on stream in 1984.

7.2.2 Ligand developments

The water-soluble ligand used by Kuntz and Cornils is TPPTS [tri(m-sulfonyl)triphenylphosphine trisodium salt], the properties of which are very similar to the parent compound tpp. Sulfonation of triphenylphosphine, neutralization, followed by pH-controlled selective extraction and re-extraction procedures, yields an aqueous solution of the trisodium salt of tris(m-sulfonyl)triphenylphosphine in good yields. The catalyst used is a rhodium complex of TPPTS, which is highly water-soluble as in the order of 1 kg of the ligand "dissolves" in 1 kg of water. The ligand forms complexes with rhodium that are most likely similar to the ordinary triphenylphosphine complexes (i.e. $RhH(CO)(PPh_3)_3$) [7].

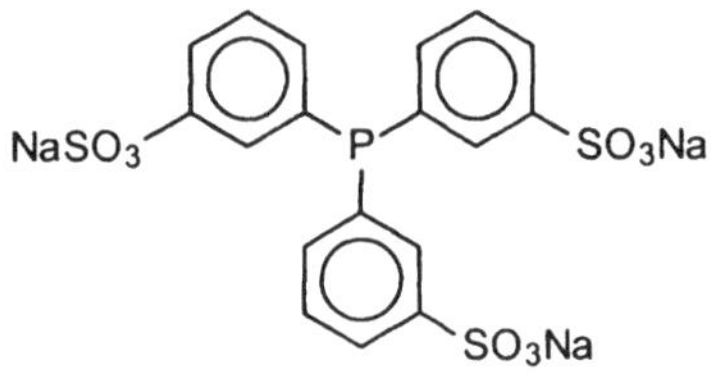

TPPTS

Figure 1. The RCH/RP ligand for two-phase hydroformylation

Until today the ligand synthesis is object of permanent improvement and investigation. Described by a quite simple reaction scheme (see below), the sulfonation technology and the consecutive work up procedure require significant know-how [17].

$$P(C_6H_5)_3 + 3\ SO_3 \quad \rightarrow \quad P(C_6H_4SO_3H)_3$$

Almost two decades of experience with the technically applied system demonstrated that it is of great importance to reduce by-product formation during the ligand synthesis: not only the improved yields but even more so the elimination of specific reaction impurities bear a large potential for further economic savings. Ligand decomposition can lead to the formation of inactive rhodium species. Ligand decomposition can take place via P-C-bond cleavage, followed by formation of propyldi(sulfonylphenyl)-phosphine, which acts as strong electron donor reducing the activivity of the rhodium catalyst. To prevent fast catalyst deactivation [18] it turned out be beneficial to control the P(III)/Rh ratio very carefully and never let it drop below 60/1. In addition it was found that an almost continuously performed addition of fresh ligand solution increases the catalyst life time (catalyst activity and productivity) more than a third compared to the nominal life time [19]. This pseudo-continuous ligand addition mainly compensates the formation of phosphine oxides formed by traces of oxygen brought into the system by the feedstock and the syn gas, which is produced by partial oxidation of heavy oil or natural gas.

Besides the TPPTS-system a number of other sulfonated phosphine ligand systems were investigated and tested on a pilot plant scale. Among them are systems which are derived from biphenyl (BISBIS = sulfonated bis(diphenylphosphinomethyl)biphenyl, varying grades of sulfonation of

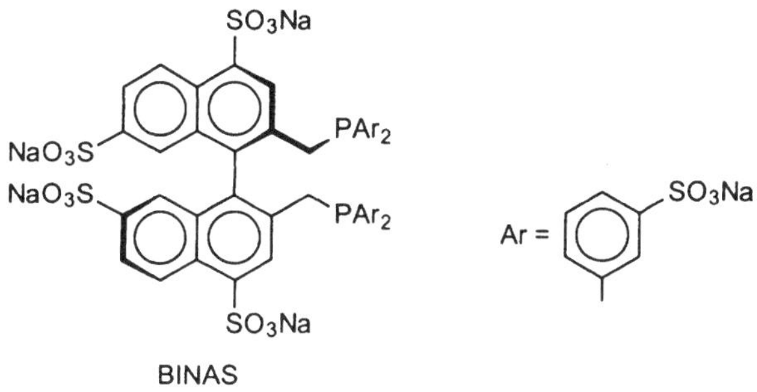

Figure 2. Formula of BINAS

BISBI, see Chapter 4) or binaphthyl structures (BINAS = sulfonated NAPHOS, sulfonation grade between six and eight).

It turned out that rhodium-BINAS is the most active and selective water-soluble hydroformylation catalyst [20]. Its reactivity is up to 10 times higher than TPPTS, even low Rh/P-ratios (P/Rh = 7/1) give l:b ratios of 98/2 (TPPTS: 94/6 with P/Rh = 80/1). Concerning the outstanding performance data, BINAS would be clearly the ligand of choice. Compared to TPPTS, these more complicated systems carry higher manufacturing costs. This disadvantage could be compensated for by the lower amount of ligand needed. More importantly, BINAS shows higher decomposition rates than TPPTS, which was up to now the overriding reason not to change the established system. As a consequence up to now TPPTS is the best choice for technical application in water based two-phase catalyst systems.

7.2.3 Kinetics and catalyst pre-formation

Kinetics.

When two-phase technology is applied, phase transfer phenomena can play a dominant role in determining reaction rates and kinetics. The hydroformylation of alkenes in a two-phase water/organic solvent system is defined by the reaction of two gases, carbon monoxide and hydrogen, with gaseous or liquid alkenes like ethene, propene or higher homologues. Dependent on mass transport and reaction conditions (temperature, partial pressures of the gases, stirring) and reactions partners (type of alkene), the reaction of the alkene with CO and H_2 takes place at the organic-aqueous interface or in the aqueous bulk phase. No published data are available for propene or butene.

One study has been carried out using *octene-1*, which is easy to handle, thus yielding reliable results for the derivation of kinetic data, although solubilities change with the composition of the system. Using the Rh/TPPTS catalyst in an aqueous solution a rate law was determined for the two-phase

To achieve high catalyst lifetimes it is beneficial to keep the reaction temperature below a certain ceiling value like 130 °C to minimize ligand degradation. To stabilize the active catalyst species it turned out to be efficient if ligand is added by a saw-tooth-type dosing during the whole reaction run. This keeps the Rh/P(III) ratio at a level which minimizes the catalyst degradation (see also chapter 9). At the end of a catalyst life - marked by significant and constant drop of productivity - the complete aqueous catalyst solution is taken out of the production system and the reactor is refilled with fresh catalyst solution. The old catalyst will be worked up with full Rh-recovery (>90%) and even recovery of the portion of the not degraded or oxidized TPPTS.

7.2.4 Process description

The Ruhrchemie Rhône-Poulenc-Process applies a catalyst - TPPTS/Rh - which is capable to convert both propene and raffinate-II feedstocks to butyraldehyde and valeraldehyde, respectively. The potential of this process is its flexibility to convert propylene or butenes in the same reactor system. The design of the Ruhrchemie/Rhône-Poulenc (RCH/RP) reaction system and catalyst recovery method is completely different from that of homogeneous hydroformylation processes. In principle, the two-phase RCH/RP process applies a stirred tank reactor, followed by a phase separator and a strip column. The reactor containing the aqueous catalyst solution (Rh/TPPTS) is fed with propylene and syn gas. The oxo crude product is passed on to the phase separator where gaseous components are removed and the liquids separated into two phases, the aqueous catalyst solution and the organic aldehyde phase. The catalyst solution is recycled to the reactor passing a heat exchanger system, where steam is generated. The organic phase containing the crude aldehydes is stripped with syn gas to remove dissolved propylene that is recycled to the reactor. The crude aldehyde mixture is split into iso- and n-aldehyde in an aldehyde distillation unit.

In the past at the Ruhrchemie site, the portion of the not reacted propene, present in the waste gas from the two phase oxo reactors, was converted to butyric aldehyde in a vent reactor, which was based on high pressure cobalt hydroformylation technology. In 1998 the old cobalt technology was eliminated by the implementation of a high-pressure rhodium process giving a more selective process yielding a cleaner reaction product. The rhodium system allows now the possibility to steer the overall l:b ratio within a comfortable range to serve the market needs. The capacity of this high-pressure rhodium-unit is 70,000 tons per year.

The TPPTS ligand easily carries all of the used Rh into the water phase by forming Rh/TPPTS complexes analogous to the complexes known from the simple Rh/TPP system. Due to the sufficient strong coordination of the TPPTS-ligand to the Rh metal-center almost no Rh leaching into the organic phase is observed - a performance prerequisite with respect to the desired economy of the process. Even with the low solubility of propene or butene in water, the high reactivity of the used catalyst system is high enough to provide a very efficient hydroformylation rate. To increase the solubility of propylene in the water phase, plant trials were performed with polyethylene glycol with various chain lengths, acting as an auxiliary-agent for better propylene solubility. Good reaction rates (rate increase of more than 10 %) were achieved, demonstrating a simple tool to increase productivity or to switch to milder conditions. In summary, the major advantages of the TPPTS-system are the simple, automatically performed mild catalyst recycle, the comparable low Rh-inventory (rhodium savings), the high l:b ratio of ~20/1 and the beneficial low energy consumption of the whole process as well as the improved utilization of propylene and synthesis gas (CO/H_2).

The water-based process is normally operated at a *pH-value* between 5 and 6 to provide reaction conditions, which depress unwanted side reactions of the formed aldehydes. It turned out that a careful pH-control is beneficial for various operation parameters of the protic reaction media: catalyst reactivity and selectivity are directly influenced by and thus dependent on the H_3O^+ concentration present in the aqueous phase.

7.2.5 Status of the operated plants

With the RCH/RP process it is possible to hydroformylate alkenes from propene to pentene with satisfying space-time yields. Currently the process is solely used for the conversion of propene and butene-1. In the early 1990s developmental work was started to apply the RCH/RP process also to the conversion of alkenes higher than C3. The overall economics looked attractive and the market was prepared. Consequently, a 12,000 tons unit went on-stream at the Oberhausen site in 1995, to convert raffinate-II (a mixture of butene-1 and butene-2) into C5-aldehydes that are predominantly converted to the corresponding alcohols and acids in further downstream processes at the site.

Compared to propene, butenes show a lower solubility in water. Due to this the observed hydroformylation reaction rate of butene mixtures like raffinate-II (in general 40 % butene-1, 40 % butene-2, 20 % butane isomers) is significantly lower than that found for propene. However, a slight increase in the catalyst concentration is sufficient to achieve satisfying space-time yields for the pro-duction of valeric aldehydes. As a very selective catalyst the Rh/TPPTS-system does not catalyze the hydroformylation of butene-2 providing a simple method for the single conversion of the terminal alkene. The catalyst isomerizes a small fraction of butene-1 to butene-2. The latter is not converted under standard reaction conditions. Thus the butanes and the unconverted butene-2 are separated form the reaction products in a stripping column. Advantageously the RCH/RP technology can be applied to the hydroformylation of C3- as well as C4-alkenes in the same production unit; only minor changes (e.g. reaction temperature, partial pressures) are necessary to switch between the different feedstock streams [21]. This is not the case for alkenes with five or more carbon atoms. Nevertheless the latter feedstocks correspond to approximately 25 % of the worldwide oxo-business.

Today the Ruhrchemie/Rhône-Poulenc Process (RCH/RP) is successfully operated at two locations in the world. In the early 1980s a 100,000 ton scale process was developed and commercialized at the Oberhausen site of Ruhrchemie AG (now Celanese AG). The first unit was started up in 1984. In 1988 and in 1999 two additional lanes went on stream. The actual combined capacity of all three units at Oberhausen (combined with the conversion of not reacted propylene in a high-pressure hydroformylation set up) amounts to more than 500,000 tons of C4-products. The l:b-C4-aldehydes produced are converted further to the corresponding C4-alcohols or after aldol condensation and hydrogenation to 2-ethyl-1-hexanol (2-EH). With an annual volume of 360,000 tons of 2-EH the Oberhausen facility became the world's largest 2-EH operating unit.

In 1996 the RCH/RP-process was licensed to Hanwha Corp. in South Korea. With a total capacity of 120,000 tons the unit started up successfully in 1997. The aldehydes are dedicated mainly for the manufacturing of butanols.

All units are running well without any major problems being observed. The process can be operated for long periods of time at steady and stable conditions. In Table 1 typical process conditions and the composition of the oxo crude product are shown as fifteen years' averages.

7.2.6 Economics

Three major factors attribute to the reduction in the overall manufacturing costs of the RCH/RP two-phase hydroformylation process:

- A highly efficient energy recovery system
- Minimum capital requirements for an RCH/RP oxo unit.
- A high feedstock usage in combination with a classical reactor that converts not reacted off gas.

These are results of the simplified process design, which was only possible by applying a water-soluble catalyst. The catalyst/product separation is achieved by a simple phase separation. Therefore, the exothermal reaction heat of the hydroformylation reaction can easily be recovered by a heat exchanger inside the reactor (high energy efficiency) and, secondly, no further distillation steps and columns are required to separate the catalyst from product (less capital).

Table 1. Reaction conditions and composition of crude product of the RCH/RP process (15 years' average)

Product [Unit]	Typical Value	Variation
n-Butanal[%]	94.5	95 - 91
i-Butanal [%]	4.5	4 - 8
Heavy Ends [%]	0.4	0.2 - 0.8
l:b selectivity	19	13.3 - 32.3
Selectivity towards C4-aldehyde [%]	99	99
Process conditions		
Temperature [°C]	120	110 - 130
Total pressure [bar]	50	40 - 60
Vol.(water)/Vol.(organics)	6	4 - 9
Heat recovery*) [%]	>99	>99
Conversion [%]	95	85 - 99

*) Minus losses by radiation

Both factors together with the reduced fixed costs and the usage of an own technology (no license fees in case of Celanese plants) makes the RCH/RP process ca. 10 % cheaper in manufacturing costs (costs for ligand synthesis already included) compared to classical rhodium process applying a homogeneous phosphine-modified catalyst.

7.3 Reaction of various alkenes

7.3.1 Ethylene to propanal: why not applied?

The major advantage of applying two-phase technology is the easy separation of catalyst phase and product phase. Due to the low solubility of butanal, pentanal, etc. in water and vice versa the separation of the reaction and product layer causes no problem in general, phase separation occurs fast and complete. Propanal, however, derived from the hydroformylation of ethylene, bears a significant miscibility with water. This causes two problems:

a) the formed propanal is "wet", the water in the aldehyde has to be removed be distillation, which is not trivial due to the formation of azeotropic mixtures,

b) together with the water dissolved in the propanal a significant amount of catalyst is transported out of the reactor, a recovery of this portion of the catalyst is not very efficient.

Nevertheless, the hydroformylation of ethylene itself in the water-based Rh/TPPTS system is fast, but the advantage of ethylene, *viz.* its higher solubility in the water phase, is counterbalanced by the unfavorably high solubility of water in propanal. For this reason the hydroformylation of ethylene is performed in homogeneous Rh/TPP systems to avoid the aforementioned problems.

7.3.2 Long-chain alkenes

Concepts for the conversion of long-chain alkenes

Long chain alkenes cannot be hydroformylated economically by the classical Ruhrchemie/ Rhône Poulenc process, because of low solubility of alkenes higher than butene in water [7]. Especially for higher alkenes it would be highly desirable to convert them with a two-phase system because of the higher boiling points of the resulting aldehydes. In a homogeneous system these aldehydes have to be stripped from the catalyst solution at a high temperature which leads to formation of more heavy ends and faster catalyst deactivation. Numerous attempts have been undertaken to hydroformylate higher alkenes with a two-phase system. In principle the difference in polarity between the organic and the catalyst phase has to be decreased compared to the classical RCH/RP system. The following principal approaches have been investigated and will be discussed below:

1. increase the lipophilic character of the catalyst phase,
2. increase the solubility of the catalyst in the product phase,
3. immobilize the catalyst on a support.

Cosolvents of the classical RCH/RP – system

Increasing the lipophilic character will increase the solubility of the alkene in the catalyst phase and thus it will increase the rate of hydroformylation. The limits are the solubility of the cosolvents in the lipophilic phase, and leaching of ligand and rhodium into the organic phase. This means, that choosing the right system for the alkene to be converted is crucial.

To increase the lipophilic character, cosolvents can be added to the classical RCH/RP system. Solvents like methanol, ethanol, polyethylene glycol strongly increase the hydroformylation rate [8]. The potential disadvantages of these cosolvents are leaching into the organic phase and their reactivity towards acetals. Leaching of catalyst and ligand seems to be no problem, provided that certain concentrations of cosolvents are not exceeded. Another disadvantage is a drop in l:b selectivity with increasing amount of cosolvent. This implies that the solvent inters with the catalyst.

Other solvents than water for the two-phase hydroformylation

Because of the increasing lipophilic character of the long chain aldehydes, solvents other than water can be applied in a two-phase system. TPPTS as ligand is almost exclusively soluble in water, but TPPMS (monosulfonated TPP) is soluble in polar organic solvents. Abatjoglou developed a two-phase system with TPPMS as ligand and N-methylpyrrolidone (NMP) as solvent [9]. After the reaction, water is added to obtain phase separation. Prior to recycling of the catalyst phase, the added water is removed by distillation (see Chapter 8).

Ionic liquids have been used for the hydroformylation of higher alkenes. Here the lipophilic character of the catalyst phase can be adjusted by proper choice of the ionic liquid. In the recent past good results in hydroformylation of 1-octene have been reported by P. Wasserscheidt and coworkers. Linear:branched ratios up to 16 with special ionic ligands have been achieved [10]. Bahrmann discovered that certain amines form ionic liquids with TPPTS. In this instance the ligand itself is the ionic liquid [11].

Detergents in the classical RCH/RP system

Anionic detergents like soaps, AOS and alkylsulfonic acids show no increase in hydroformylation rate compared to the classical system. Cationic detergents, on the other hand, can increase the hydroformylation rate up to a factor of 4. This is best explained by the fact that sodium of TPPTS is exchanged by the detergent cation, which increases the solubility of the catalyst in the organic phase. Technical application is hampered by loss of rhodium into the organic phase and foaming [12].

Immobilization of the catalyst

Numerous attempts have been made to immobilize rhodium or its complexes on polymeric or on inorganic supports. Immobilization without ligands leads to systems that mainly hydrogenate the alkene [13]. Recently van Leeuwen and coworkers described the immobilization of Rh - Xantphos complexes in silicagel [14]. The system can be recycled 8 times without significant loss of Rh or activity, but the reaction rate is very low.

Another possibility is the linkage of phosphines to linear polystyrene [15] or other polymers like PEG and polyacrylic acid. These polymers can be separated by membrane filtration or in some cases by thermoregulated phase transfer [16]. In the latter case, the catalyst is soluble at lower temperature and insoluble at higher temperatures. So the catalyst can be separated from the products by phase separation at higher temperatures.

References

1 Chauvin, Y.; Mußmann, L.; Olivier, H. *Angew. Chem.* **1995**, *107*, 2941.
2 Horváth, T.; Rábai, J. *Science*, **1994**, *266*, 72.
3 a) Kuntz, E. D. *CHEMTECH*, **1987**, *17 (9)*, 570. b) Kuntz, E. G., Fr. Pat. 2,314,910, **1977** (to Rhone-Poulenc); *Chem. Abstr.* **1977**, *87*, 101944.
4 Bhanage, B.M.; Studies in hydroformylation of olefins using transition metal cataysts, Ph. D. Thesis, University of Pune, 1995; b) Deshpande, R. M.; Chaudhari, R. V. *Ind. Eng. Chem. Res.* **1988**, *27*, 1996.
5 Frohning, C. D.; Kohlpaintner C. W. in Aqueous-Phase Organometallic Catalysis (Eds.: B. Cornils, W. A. Herrmann), VCH, Weinheim, **1998**, p. 295.
6 Chaudhari, R. V.; Bhanage B. M. in Aqueous-Phase Organometallic Catalysis (Eds.: Cornils, B.; Herrmann, W. A., VCH, Weinheim, **1998**, p. 290; b) (Cornils, B.; Wiebus, E., EP 0,158,246, **1985** (to Ruhrchemie AG).
7 Horváth, I.; Kastrup, R. V.; Oswald, A. A.; Mozeleski, E. J. Catal. Letters,**1989** , 2, 85.
8 Cornils, B.; Herrmann, W. A.; Aqueous Phase Organometallic Catalysis, Wiley-VCH Weinheim, **1998**, p. 306 ff.
9 Abatjoglou, A. G.; Bryant, D. R. US Patent, 4,731,486, **1988** (to Union Carbide Cooperation). *Chem. Abstr.* **1988**, *109,* 57042.
10 Wasserscheid, P.; Salzer, A. XXXIII. Jahrestreffen Deutscher Katalytiker, Tagungsband p. 66ff.
11 Bahrmann, H.; Schulte, M. DE 19756945 (to Celanese). *Chem. Abstr.* **1999**, *131*, 75261.
12 Cornils, B.; Herrmann, W. A. Aqueous Phase Organometallic Catalysis, Wiley-VCH Weinheim, **1998**, p. 306 ff.
13 Greiner, R.; Burger, B. XXXIII. Jahrestreffen Deutscher Katalytiker, Tagungsband p. 119 ff.
14 Sandee, A. J.; van der Veen, L. A.; Reek J. N. H.; Kamer, P. C. J.; Lutz, M.; Spek, A. L.; van Leeuwen, P. W. N. M. *Angew. Chem. Int. Ed. Engl.* **1999**, *38*, 3231.
15 Bayer, E.; Schurig, V. *Chemtech* **1976**, 212.
16 Jin, Z.; Zheng, X.; Fell, B. *J. Mol. Catal. A.: Chem.* **1997**, *116*, 55.
17 Gärtner, R.; Cornils, B.; Springer, H.; Lappe, P., DE 3.235.030, to Ruhrchemie AG (1982).
18 Kulpe, J. A. Dissertation, **1989**, Technische Universität München; b) Kohlpaintner, C. W. Dissertation, **1992**, Techniche Universität München.
19 Frohning, C. D.; Kohlpaintner, C. W. in Aqueous-Phase Organometallic Catalysis (Eds.: Cornils, B.; Herrmann, W. A. VCH, Weinheim, **1998**, p. 304.
20 Herrmann, W. A.; Kohlpaintner, C. W.; Bahrmann, H.; Konkol, W. *J. Mol. Catal.* **1992**, *73*, 191; b) Herrmann, W. A.; Kohlpaintner, C. W.; Manetsberger, R. B.; Bahrmann, H. ; Kottmann, H. *J. Mol. Catal.* **1995**, *97*, 65; c) Bahrmann, H.; Bach, H.; Frohning, C. D.; Kleiner, H. J.; Lappe, P.; Peters, D.; Regnat, D.; Herrmann, W. A. *J. Mol. Catal.* **1997**, *116*, 49; d) Herrmann, W. A.; Kohlpaintner, C. W. *Angew. Chem. Int. Ed. Engl.* **1993**, *32*, 1524.
21 a) Frohning, C. D.; Kohlpaintner, C. W. in Aqueous-Phase Organometallic Catalysis (Eds.: Cornils, B.; Herrmann, W. A., VCH, Weinheim, **1998**, p. 302; b) Bahrmann, H.; Frohning, C. D.; Heymanns, P.; Kalbfell, H.; Lappe, P.; Peters, D.; Wiebus, E. *J. Mol. Catal.* **1997**, *116*, 35.

Chapter 8

Process aspects of rhodium-catalyzed hydroformylation

Peter Arnoldy

Shell International Chemicals, Shell Research and Technology Centre Amsterdam, Badhuisweg 3, 1031 CM Amsterdam, The Netherlands.

8.1 Introduction

This chapter treats industrial applications of rhodium-catalyzed hydroformylation technology. The field of hydroformylation, based on homogeneous catalysts in all instances, is now dominated by technologies based on rhodium and cobalt. The development of processes with modified rhodium catalysts was initially driven by the desire to replace of some of the older Co-based technologies by simpler processes with higher selectivities. Later, rhodium processes have also been developed for new applications in which Co-based technologies play no role. This chapter is centered around the major problem of Rh-catalyzed hydroformylation, which is slowing down the speed of commercial implementation, i.e., the high catalyst cost related largely to the price of rhodium. Process development is directed to full Rh containment, thus avoiding Rh loss.

Some economic aspects, including rhodium catalyst cost, are treated in section 8.2. Catalyst performance aspects are treated in sections 8.3 (activity, selectivity) and 8.4 (stability, loss routes for Rh and ligand). In 8.5 and 8.6, several commercial processes are described. Four generic, industrially used process types are described in 8.5, *viz.* processes using a stripping reactor, a liquid recycle, a two-phase reaction, and an extraction after a one-phase reaction. In 8.6, interesting, current developments in a few petrochemical product areas are shortly discussed.

P.W.N.M. van Leeuwen and C. Claver (eds.), Rhodium Catalyzed Hydroformylation, 203–231.

8.2 Economics

Apprehension of processes and their commercial viability starts with a good understanding of their economics. An important concept is the manufacturing costs sheet including all cost elements, resulting in the overall manufacturing costs of a chemical product. For a typical (hydroformylation) process on petrochemical scale (10-500 kt/a), all cost elements are relevant: variable costs (alkene, syn gas, utilities, catalyst), depreciation of capital investment, and capital-related fixed costs (labor, maintenance). Catalyst performance has a critical impact on the overall manufacturing costs, influencing many of the cost elements via its *selectivity* and *activity*, and determining catalyst cost directly via its *stability* and the extent of Rh and ligand losses.

Direct catalyst cost seems to be only a small cost element. Catalyst costs of about 1-3% of total costs are typical, especially in areas with mature products and optimized processes, with some technology competition present. Note however, that firstly other variable cost elements (alkene and syn gas feed) can constitute higher overall costs, but they can be influenced to a lesser extent because they are at least consumed stoichiometrically. Secondly, an increase of catalyst consumption to about 10% of total costs by lack of control easily results in a totally unacceptable cost structure. For homogeneous processes, catalyst consumption/cost is generally a major issue, because of both limited stability/lifetime and the occurrence of physical losses, related to imperfect separation of catalyst from products (see 8.4). Catalyst cost determines most of the development effort in the area of Rh-catalyzed hydroformylation. Figure 1 gives the development of the Rh price since 1972 [1,2]. In 1999, circa 16,100 kg Rh has been produced globally, of which 94% of Rh has been consumed for automobile exhaust catalysts [3]. Rh prices sank in the nineties because of recession and reduced Rh demand for exhaust catalysts (due to the introduction of Pd-only catalysts in the US). Recently, Rh prices have started to rise again, *e.g.* because of the return to Rh-containing exhaust catalysts which give a better NO_x reduction at high speed.

In the overall catalyst cost, also ligand consumption can play a role, especially if (i) ligand structure/synthesis becomes more complicated, (ii) ligands are not chemically stable, and (iii) high ligand/Rh molar ratios have to be applied. A typical petrochemical ligand cost is ca. $10-100/kg, but more complicated syntheses can lead to prices around $1000/kg. Ligand costs rapidly escalate using expensive precursors or multi-step synthesis routes. Also the small manufacturing scale and production of large waste volumes pushes up the costs.

Going from large-scale, petrochemical processes to smaller-scale fine-chemical processes, the structure of the manufacturing cost sheet will change. Feedstocks will be more expensive and minimization of capex (capital expenditure), labor, and catalyst cost is somewhat less important. The latter, in combination with the generally smaller total financial turnover per product and the often shorter product lifecycle, will lead to less process development. Processes can be carried out in batch rather than continuously, generally using generic types of separation of catalyst and product (like distillation), in spite of some incurring Rh losses. Even a "Rh throw away" process (once-through in Rh) has been proposed for small-scale production of expensive chemicals [4]!

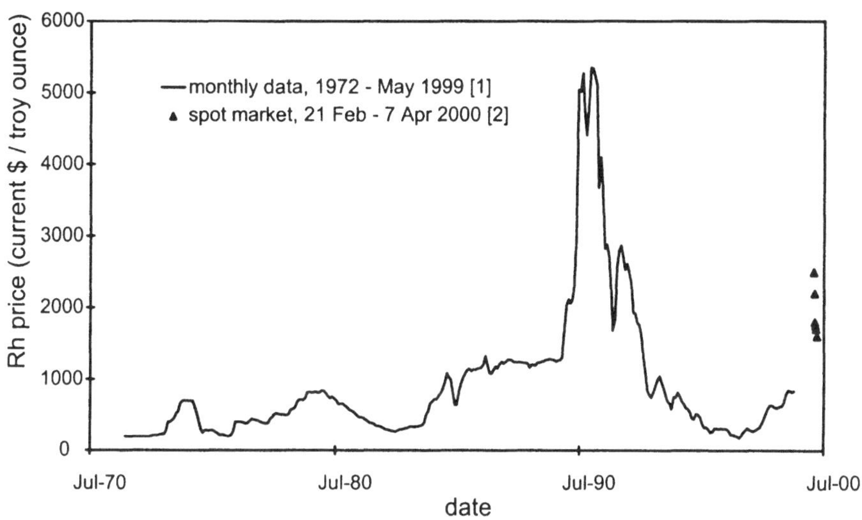

Figure 1. Development of the rhodium price over the years

The following calculation shows how product value determines the need for effective Rh recycle:

Using a Rh price of $30,000/kg and a Rh concentration in the reactors of 300 ppmw, the Rh value in the reactors is 9 $/kg reactor content. Assuming 50%w product in the reactors, a typical petrochemical product value of $1/kg and the need to limit Rh cost to 1% of manufacturing costs, leads to the need to recycle Rh not less than 1800 times before it can be wasted. Using the same assumptions but taking a fine chemical product value of $100/kg results in need for only 18 recycles. Going back to the petrochemical example, 1% Rh cost is equivalent to a physical Rh loss with crude product of only 0.33 ppmw, assuming that there are no other Rh loss pathways! Or in chemical terms, at a product mol weight of 72 (butanal), a turnover is required of 4.3 million mol product/mol Rh!

8.3 Catalyst selectivity and activity

8.3.1 Catalyst selectivity

While catalyst stability affects directly the catalyst cost, other catalyst performance elements (activity and selectivity) also have their impact. Catalyst selectivity (alkene loss to aldehyde byproduct, paraffin, alcohol, heavy ends) determines alkene variable costs, but also capital costs, via process simplifications and increased production of desired product in the same facilities. Chemoselectivity to aldehydes is high for all Rh catalysts. By-products can include aldehyde isomers, low-reactive alkene isomers, alcohols, alkanes, and heavy ends. Some aldehyde isomers (generally branched) have a significant value (e.g., isobutanal, some branched detergent alcohol constituents).

Of all by-products, the formation of *heavy ends* constitutes the biggest problem, generally not so much from a *product* point of view, but rather from a *process* point of view. Heavy-ends accumulation can pose serious process problems, since it can lead to a forced Rh-containing bleed. Figure 2 gives a survey of heavy ends formation reactions, as they can take place in reactors and other hot places like distillation bottoms.

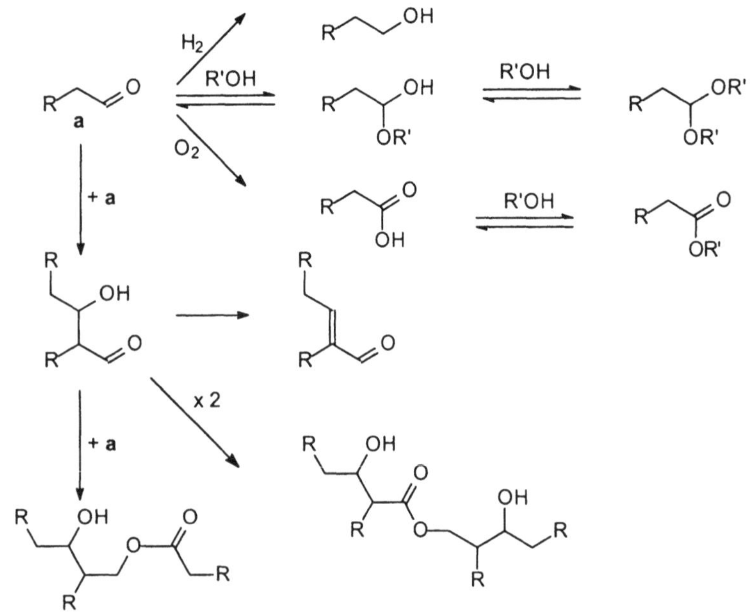

Figure 2. Heavy ends formation

Central in heavy-ends formation is the high reactivity of aldehyde products, in aldol condensation, Tischenko reaction [5], acetalization and oxidation reactions. Aldol condensation is base-catalyzed, acetal formation is acid-catalyzed, so it is important to keep the process medium more or less neutral. Acetals formation can only take place if alcohol by-product is produced (or if a hydroxy group is part of the feedstock/product).

Activity for *double-bond isomerization* is a mixed blessing. For most Rh catalysts, isomerization is slow with respect to hydroformylation, resulting in some loss of reactive 1-alkene to inert 2-alkene. For some Rh catalysts (few diphosphine ones and especially diphosphite-based ones), however, the rate of isomerization is enhanced to such an extent that internal alkenes can be converted to terminal aldehydes (see Chapters 3 and 4).

8.3.2. Catalyst activity

Catalyst activity improvements can be used only up to a certain extent for reduction of capex for high-pressure reactors, but are generally translated into reduction of the Rh concentration (limiting catalyst cost) or temperature (improving selectivity). Extreme reduction of reactor size does not always pay out. The highly exothermic hydroformylation reaction (28-35 kcal/mol [6]) requires sufficient cooling area. While normally sparged/bubble columns and CSTR's are used with enough volume available for internal cooling, a small internal volume would necessitate external cooling, which would add new capex elements. Also the risk of a thermal runaway and the need to avoid mass transfer limitations (especially for CO) favors moderate reaction rates.

Typically, reaction rates of around 0.5-2 $mol.l^{-1}.h^{-1}$ are applied commercially, and translate at 300 ppmw Rh and a density of 800 kg/m^3 into a turnover frequency (TOF) of circa 215-860 $mol.mol^{-1}.h^{-1}$. The following may serve as an example: assuming 300 ppmw Rh in the reactors, a productivity of 2 $mol.l^{-1}.h^{-1}$, a product molecular weight of 72, 50%w product in the reactors, a plant capacity of 100 kt/a, a stream factor of 8000 h/a, and a gas hold-up of 20%, an impressive reactor volume of 208 m^3 is required. The latter is probably split over more reactors, in parallel, or - preferentially - in series in order to benefit from reduced back-mixing. With such volumes, also the Rh inventory cost is quite high: at 30,000 $/kg Rh, it is $1.25 million, which is a significant working capital and risk element.

Several other chapters indicate how activity can be tuned *via* variation of type and concentration of ligand, as well as by other process parameters. *Kinetic equations* can be specific for each catalyst system, since they are dependent on the rate-determining step (see, e.g., Chapter 4).

8.4 Catalyst stability; degradation routes, losses and recovery

As has become clear from the above, catalyst cost (Rh and ligand) should be controlled very carefully. Especially *Rh containment* is the dominant theme in development of Rh-based processes. Besides direct catalyst losses, also catalyst deactivation can take place, which leads to a forced bleed of deactivated catalyst. Rh can be recovered from such bleeds as well as from crude product. Below, a survey is given of potential degradation/ deactivation, loss and recovery routes [7]. The chemistry of the degradation reactions of the ligands will be discussed in Chapter 9, while in this chapter we will concentrate on the process aspects.

8.4.1 Rhodium loss routes

Three different loss routes can be distinguished:

- *Chemical Rh loss via Rh plating*. Rh, as one of the noble metals, has a strong tendency to plate out as zero-valent metal, which can be present as aggregates of colloid particles or as a film on wall surfaces. Mechanism of such "plating" most likely goes via gradual growth of Rh-metal clusters, till sufficient size is reached for adhesion to other particles or the wall. Rh-Rh interactions are strong, as can be seen from the presence of Rh dimers or $Rh_6(CO)_{16}$ under normal hydroformylation process conditions [8, 9]. The unmodified Rh carbonyl catalyst is stabilized against plating by CO only as the ligand, leading to a need for high CO pressures (total pressures of ca. 200 bar). In the presence of phosphorus ligands, these take over the stabilization of Rh, and pressures can be reduced to 1-50 bar. Rh plating can be suppressed by low Rh concentration, high ligand/Rh ratios, use of chelating ligands (diphosphines, diphosphites), and low temperature (high activity). Plating can take place inside, but also outside the reactors, especially under harsh distillation conditions (significant temperature, no CO pressure at all, air ingress if vacuum distillation would be applied oxidizing protecting P-ligands).
- *Physical Rh loss, with crude product ("Rh leaching")*. Depending on the process, this can be due to entrainment of catalyst in a gaseous or liquid product phase, or some solubility in the liquid product phase In view of the small Rh concentrations, this Rh will be more difficult to recover and only at additional cost, e.g., using efficient adsorption beds for Rh concentration.
- *Physical Rh loss, via a bleed* (from reactor/catalyst recycle loop). A bleed stream is generally required because of (i) accumulation of heavy ends in

the reactor system (when sufficient separation of catalyst and heavy ends is not possible) or (ii) because of catalyst deactivation by interaction of Rh with poisons, like ligand degradation products (see 8.4.2) or external poisons such as sulfur species and dienes.

8.4.2 Ligand loss routes

Obviously, ligand will be lost if coordinated to Rh, via the physical Rh loss routes mentioned above. Also free ligand can be lost, e.g., if the volatility is similar to that of heavy ends (trimers/tetramers), in the event they are removed by evaporation. But *ligand degradation* is often the root cause for the occurrence of catalyst/ligand losses. Below a survey is given of degradation pathways of the industrially used phosphorus ligands. Since phosphines and phosphites show very different degradation pathways, they are treated separately in the following.

8.4.2.1 Phosphine degradation

- Oxidation can occur with molecular oxygen and hydroperoxides (thermally or Rh-catalyzed), or with oxygenates like water, carbon dioxide and allyl alcohol (always Rh-catalyzed). Oxygen can be present in syn gas feed, but will mainly enter the process via air ingress during vacuum distillation. Peroxides can easily form by contact of alkenes with air.
- Sulfidation can occur with simple sulfur components as can be present in syngas or propene (H_2S, COS), but also as a result of breakdown and reduction of the sulfonate groups of sulfonated aryl-ligands).
- P-C bond cleavage [7, 10-16] results in formation of Rh-coordinated aryl- and diarylphosphide fragments. P-C splitting has several consequences, a.o. (i) formation of inactive rhodium compounds, (ii) formation of side products derived from the aryl group stemming from the ligand, (iii) formation of diarylalkylphosphines causing a lower activity, and (iv) scrambling of aryl groups at phosphorus if different types of aryl groups are present, resulting in formation of different phosphines with potentially undesirable properties.
- Reaction of nucleophilic phosphines with electrophiles forming quaternary phosphonium salts, either thermally (with α,β-unsaturated aldehydes present in heavy ends) or Rh-catalyzed (alkenes).

8.4.2.2 Phosphite degradation

Phosphites probably undergo oxidation and sulfidation reactions like phosphines, albeit at much lower rate due to their lower nucleophilicity. Decomposition of phosphites can be much faster than that of phosphines due to their hydrolytic properties. Below some specific phosphite degradation chemistry is summarized.

- P-O bond splitting via hydrolysis with water (at least available in trace amounts due to heavy ends formation) can be very fast, non-catalyzed or acid-catalyzed, depending on pH [17]. Hydrolysis produces hydroxyphosphites; further hydrolysis, with/without additional oxidation reactions, can result in formation of phosphorus acids, like H_3PO_3, H_3PO_4 and α-hydroxyalkyl phosphonic acids. Since acids catalyze hydrolysis and can be formed as a product of hydrolysis, hydrolysis kinetics show an autocatalytic behavior, making efficient acid removal a process requirement. The development of cyclic mono-phosphites [18, 19] and especially rigid diphosphites [20] has led to a strong suppression of hydrolysis activity.
- Reaction with aldehyde to phosphonium salt adducts containing 2 or 3 mol aldehydes/mol phosphite. Reaction with water gives α-hydroxyalkyl-phosphinate esters, acetal phosphonate, and eventually phosphinic acids [19, 21] (see Chapter 9).
- P-O-bond splitting via alcoholysis. Alcoholysis can occur via Rh-coordinated alkoxy species (ex aldehyde/Rh-hydride, or alcohol). Alcoholysis of diphosphites leads to a mono-phosphite, which poisons catalyst activity [17, 22].
- Hydrogenolysis of P-O bonds [19], probably Rh-catalyzed.
- H-C bond splitting can take place via oxidative addition to Rh. This so-called orthometallation has been found for aryl phosphites [23]. In the case of diphosphites this reaction is catalyzed by Rh clusters and results in catalyst deactivation [9].
- Alkyl phosphites cannot be used as ligands, since they rearrange producing an alkylphosphonate esters via the Arbusov reaction [7].

8.4.3 Catalyst recovery processes

In spite of all efforts to contain Rh catalyst in the reactor and a primary catalyst recycle loop and to maintain its catalytic performance, it is generally required to take a bleed from the system. Such a bleed can be taken continuously (in combination with continuous make-up of fresh catalyst). More preferably (making maximum use of active Rh in reactors), bleeding is delayed as long as possible by taking measures to compensate for performance loss. These involve a gradual change to more severe process conditions (temperature, Rh concentration), or the addition of fresh ligand [24] or scavengers of ligand degradation products (such as maleic anhydride [25]). At one point in time, however, a catalyst change is required, thus marking the end of a catalyst life cycle. Catalyst (Rh and ligand) should be recovered as completely as possible, from a point of view of both economy and sustainability. There seems to be a preference for on-site catalyst recovery, using methods that can be process-specific. In such cases, significant additional capex is involved. Alternatively, use can be made of the services of a noble metal reclaimer. Standard treatments of Rh-containing bleed streams are:

- distillation, in order to concentrate Rh by removing heavy ends and poisons (preference for short-contact time distillation);
- oxidation of ligand and ligand degradation products, enabling bare Rh extraction into an acidic aqueous layer, from which active Rh catalyst can be re-formed via addition of fresh ligand and reduction; and
- combustion of Rh-containing waste.

8.5 Process concepts

The selection of the optimum process for commercialization is very feedstock- and product-specific. The volatility (molecular weight) and polarity of alkene feedstocks and products play an important role. Fortunately, Rh-catalyzed hydroformylation can be carried using a wide variety of ligands, allowing for extensive ligand variation and optimization. Such *ligand design* should be done with several objectives in mind; besides the normal wish list of activity, selectivity, and stability (of Rh catalyst and free ligand), it is important that the ligand fits to the process concept, which sets requirements on the ligand volatility and polarity. Also ligand cost and availability play an important role. Attention should be paid to efficient ligand manufacture as well as to measurement of physical and toxicological properties (for notification, needed in the case of introduction of new commercial chemicals).

The current industrial processes for rhodium hydroformylation can be divided in four categories/types. Each type will be described below, using industrial applications as example, in the chronological order of commercial implementation. The general theme in all processes is good Rh containment. The major principles for separation of catalyst from products are evaporation (process types I and II) and extraction (process types III and IVA/B). The schemes that we present are not complete; facilities like alkene and syn gas feed purification with adsorption beds, catalyst feed and recovery systems, and upgrading of crude product (generally via distillation and further reactions like hydrogenation and aldol condensation) are neglected. More separation principles than Types I-IV are possible (using, e.g., solid supports, membranes, crystallization), but will not described in here (see Chapter 10).

8.5.1 Type I: Stripping reactor process/Rh containment in reactor

The development of processes with modified Rh catalysts started mainly in order to achieve advantages with respect to existing Co-based technology (simpler process; lower selectivities to paraffin/alcohol/heavy ends; less diethylketone in ethene hydroformylation; higher butanal linearity in propene hydroformylation) [26, 27]. In view of the considered risk of Rh loss [28], a logical start was to start with the conversion of the lightest 1-alkenes available (ethene and propene) thus enabling a process in which the Rh catalyst was contained as much as possible, i.e., by keeping Rh within the reactor. For these light feeds, the volatile aldehyde products can be stripped with excess gas from the liquid phase in a so-called stripping or sparged reactor [29]. Such a process was independently developed by several companies (Celanese (commercialized in 1974), Union Carbide/Davy Powergas/Johnson Matthey (commercialized in 1975 for ethene, in 1976 for propene) and BASF, and was called "the low-pressure oxo process" (LPO) [6, 27, 30-34]. The catalyst used is always Rh/triphenylphosphine (TPP) [26], solvents are the heavy ends formed (trimers and tetramers, which have been claimed to stabilize the Rh catalyst [5]) as well as the ligand excess. A limitation of the process is the coupling of reaction and separation conditions; stripping requires a combination of high temperature (lower selectivity: lower linearity, more paraffin), low pressure (lower activity) and high gas rate (capex, utility costs). From this perspective, the concept seems to be most suited for ethene, just acceptable for propene, and unacceptable for butene and higher alkenes. The concept is essentially limited by the volatility of especially heavy ends (trimers, tetramers). Their removal rate via the gas phase should at least equal their rate of formation, and this becomes rapidly more demanding with increasing carbon number. Insufficient heavy ends volatility leads to either (i) heavy ends accumulation and need for a bleed of heavy ends containing active catalyst (see 8.3.1), or (ii) application of lower pressures or higher temperatures, with unacceptable consequences for catalyst performance.

Figure 3 shows the stripping reactor scheme for propene hydroformylation in Union Carbide's "gas recycle process" [27, 30-32, 34, 35]. Reaction conditions are about 100 °C, 15-20 bar, 300 ppmw Rh, a molar TPP/Rh ratio of 100-200. Mixing in the reactor is ensured by gas sparging, but can be enhanced by mechanical stirring. A large syn gas compressor is required to generate the gas flows required for stripping, via syn gas recycling (syn gas recycle/feed ratios of ca. 10). Propene conversion per pass is low (ca. 30%); unconverted propene is recycled with the syn gas and overall propene conversions are in the range of 85-90%, probably dependent on propene quality.

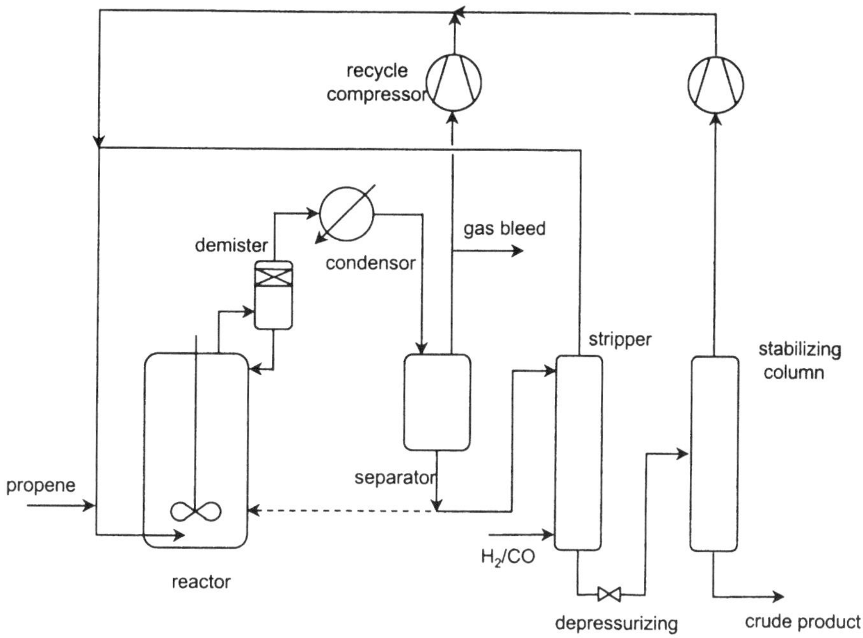

Figure 3. Process type I: stripping reactor (gas recycle process Union Carbide)

There is no need for staging via use of more reactors in series, because of the presence of propene recycle. Butanal linearity is ca. 92%, circa 2% paraffin is produced, but no butanols. A syn gas bleed is required to get rid of inert gases (methane, nitrogen, propane), and it is inevitable that some propene is lost via this bleed. Due to syn gas recycling, H_2/CO ratios above 10 are achieved. A large reactor freeboard and a de-mister pad are required to ensure that no liquid (containing Rh!) is entrained with the gas. Crude product is condensed at high pressure and subsequently degassed and stabilized by removal of residual propene. Level control is maintained via gas flow and crude product recycling. BASF has developed a similar process, using a slightly higher temperature (110 °C) and lower TPP concentration, resulting in a smaller gas recycle, but a somewhat lower linearity (86%) [32].

8.5.2 Type II: Liquid recycle process/use of distillative separation

This process has been developed for propene hydroformylation by Mitsubishi Chemical and Union Carbide/Davy Process Technology [32, 33, 36], as improvement of the stripping reactor concept. Product is now removed from the reactors via the liquid phase and carefully evaporated from the catalyst. The uncoupling of reactor and product/catalyst separation

Figure 5 gives a description of this process. The reactor is mechanically stirred, to minimize mass transfer limitations for both gas/liquid- and liquid/liquid phase transfer. In spite of this, the reaction still seems to be mass-transfer limited and restricted to the liquid/liquid interface [43], which may be caused by high catalyst activity and temperature combined with the low solubility of propene. The reactor cooling is integrated with the butanals distillation reboiler. The temperature is ca. 120 °C, the pressure is ca. 50 bar (H_2/CO ratio of 1.0). The water phase contains ca. 300 ppmw Rh at a TPPTS/Rh molar ratio of 50-100. The water/organic phase ratio in the reactors is high (ca. 6) making water the continuous phase and giving an overall high Rh concentration in the reactors.

The two-phase liquid is taken from the reactor (without cooling) to a decanter vessel, in which excess syn gas is separated and the two liquids separate by simple settling. The catalyst-containing aqueous phase is recycled back to the reactor and is cooled against condensate to absorb some of the reaction heat and produce low-pressure steam. The product phase (containing all heavy ends) is stripped at high pressure with syn gas feed to recover unconverted propene. The relatively high temperature of 120 °C seems to be required because of the limited solubility of propene in water, the desire for energy integration, and the need for high propene conversion per pass (ca. 95% in one reactor). This higher temperature might have an impact on catalyst stability, although a high ligand/Rh ratio is applied; no data are available on Rh loss via plating or catalyst deactivation. High CO pressure seems required, maybe for catalyst stability reasons, maybe because of low CO solubility in water [44]. Heavy ends are minimized by pH control (pH 5.5-6.2); the presence of some CO_2 (1-3%) in the syn gas seems to be beneficial to minimize heavy ends [45].

The major advantages of this process are: (i) the high level of heat integration (making this process a steam exporter rather than a steam importer), (ii) simple separation of catalyst and product/heavy ends (without "thermal stress" of catalyst via distillation), (iii) better selectivities (99% aldehydes, 95% linearity; no paraffin) and (iv) lower sensitivity to some poisons (the preference of some poisons for the organic product layer). The higher product linearity seems to be hardly an advantage in view of the value and market size for isobutanal derivatives. The higher reactor pressure and the heat integration might increase the capex somewhat.

8.5.4 Type IV: Extraction *after* one-phase reaction

In this concept, catalysis is carried out in a homogeneous, one-phase reaction system, while a two-phase catalyst extraction is done afterwards. A one-phase reaction seems to be preferred, when the type III two-phase reaction leads to (i) too low reaction rates because of low substrate solubility or mass transfer limitations, or (ii) inefficient phase separation, because of emulsification. The two separate phases form after reaction just by cooling or, more typically, by addition of a solvent. In order to have efficient extraction after the reaction, the distribution coefficients over two liquid phases of catalyst and products should be sufficiently different. Generally one phase is aqueous, and the other one is apolar. Given the product (a)polarity, ligand design will result in the opposing ligand (a)polarity. Two type IV concepts will be described: use of apolar catalyst for polar product (type IVA), and use of polar catalyst for apolar product (type IVB).

8.5.4.1 Type IVA: apolar Rh catalyst, polar product

The Type IV concept has been commercialized for the first time in the Rh-catalyzed hydroformylation of polar feed to polar product: allyl alcohol (AA) conversion to 4-hydroxybutanal (for production of 1,4-butanediol (BDO)) [46]. This process was developed by Kuraray and Daicel Chemical Industries [47, 48] and commercialized in 1990 by Arco (Lyondell since 1998) [49], and is described in Figure 6.

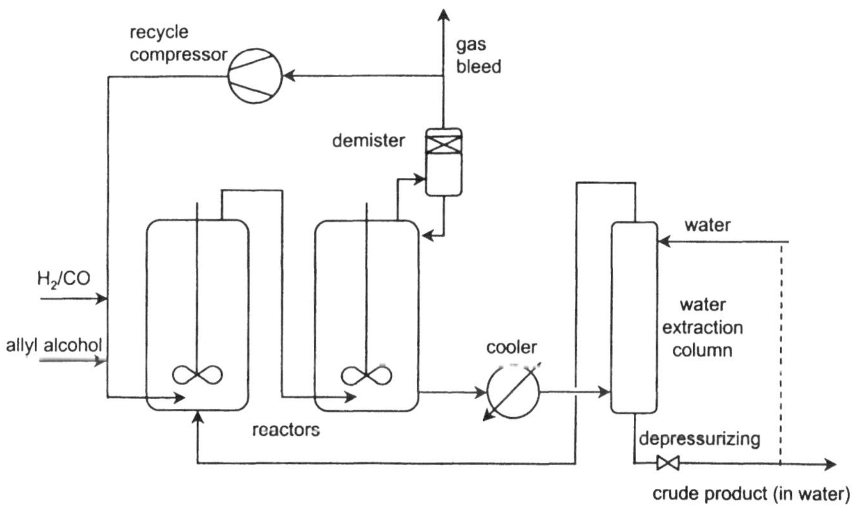

Figure 6. Process type IV: one-phase reaction, extractive separation; type IVA: apolar catalyst, polar product (Kuraray's 1,4-butanediol process)

The catalyst used is Rh/TPP (apolar), to which 1,4-bis(diphenylphosphino)butane (dppb) is added in small amounts (molar ratio Rh/TPP/DPPB ca. 1:150:0.2). The presence of dppb elegantly avoids rapid deactivation of the Rh catalyst by (i) poisoning by acyl intermediates or methacrolein (formed by dehydration of branched aldehyde by-product), and

(ii) ligand oxidation [50-53]. Very mild process conditions (ca. 60-65 °C, 2-2.5 bar) can be used, because of the high reactivity of AA compared with unfunctionalized alkenes. These mild conditions will help to suppress heavy ends (acetals) formation. Reaction takes place in one phase using an apolar catalyst in an aromatic solvent like toluene. Multi-stage extraction with water (ca. 30 °C, water/organic phase volume ratio of ca. 1:1) leads to recovery of the hydrophilic product in the aqueous phase, while essentially all catalyst is left in the organic phase and can be recycled. Rh losses *via* the aqueous phase are negligible (10 ppbw). Note that heavy ends as well as TPPO (formed by oxidative addition of AA's C-O bond to Rh [52]) are less polar and will accumulate in the apolar catalyst recycle, leading to need for some bleed from the catalyst recycle. The reaction is very sensitive to CO pressure. A too low a CO pressure results in low selectivities (production of more propanal and propanol via AA isomerization and hydrogenation, respectively) and catalyst deactivation. A too high a CO pressure leads to lower linearity (< 70% rather than ca. 87%) as well as Rh catalyst loss with the aqueous product phase. In order to achieve the optimum CO pressure, good control of reaction rate and avoidance of mass transfer limitation (good mixing) are important. The H_2/CO ratio of syngas feed is ca. 4; CO starvation is probably prevented by high syngas recycling rates, but addition of CO in a second stage is an alternative option. AA concentrations should be kept low: high AA conversions are preferred and back-mixing in the reactors will help. Typical AA conversions are ca. 98%, propanal and propanal selectivities are ca. 7 and 3 mol%, respectively. Recently Arco has been able to achieve a linearity improvement (up to ca. 95%) by use of a group 8 metal (Fe or Ru) as an additive [54]. There still seems to be scope in Rh-catalyzed hydroformylation for BDO production: Lyondell has announced to build 126 kt/a new capacity in the Netherlands in 2001, in addition to their current 55 kt/a capacity in the US, which would be the largest single stream BDO unit in the world [55].

8.5.4.2 Type IVB: polar Rh catalyst, apolar product

The Type IV concept can also be used in the field of higher alkenes, but in this instance the product phase is the apolar one and a polar catalyst has to be applied. In this field, technology developed by Union Carbide seems most advanced (see below). Recently, Sasol announced the commercialization in 2001 of a Rh-catalyzed hydroformylation ("low-pressure oxo") technology, licensed from Kvaerner (previously Davy McKee), for conversion of Sasol's Fischer-Tropsch 1-alkenes to detergent alcohols, on 120 kt/a scale [56]. On the basis of the advanced status of Union Carbide's higher alkene technology, it seems probable that Sasol will apply the latter in South Africa. Sasol's Fischer-Tropsch streams consist of alkenes diluted in a soup of paraffins, aromatics and oxygenates (e.g., alcohols) [57], and hydroformylation of the full stream is feasible since product alcohols have a higher boiling point and can be readily separated from the other components by distillation. The alkenes are > 90% 1-alkenes, so are well suited for hydroformylation by Rh/phosphine catalysts; a significant part of the 1-alkenes is mono-methyl-branched, but these branches are well distributed over the alkene chain [57], so probably, they hardly affect the hydroformylation rate.

For use in their higher alkene process, Union Carbide use mono-sulfonated ligands, most probably ω-diphenylalkane-α-sulfonates (alkane = butane (DPBS) or propane (DPPS; DPBS seems preferred) [58]. These ligands cannot give undesired aryl exchange reactions, and are probably not expensive, since they can be conveniently produced from diphenylphosphide and 1,4-butane- and 1,3-propane-sultone (carcinogenic), respectively [59]. Several production routes for the sultone precursors have been described [60]. The presence of an alkyl group on the phosphorus leads to lower activity than found for the Rh/TPP system, resulting in need for higher temperature (ca. 110 °C) and Rh concentration (ca. 300-400 ppmw), but a lower ligand/Rh molar ratio (ca. 15). Pressure is low (ca. 7 bar) and H_2/CO ratios are high (ca. 4). Figure 7 gives a process scheme [33, 58, 61]. A series of well-mixed reactors is used to obtain high alkene conversion (ca. 95%). Selectivity to aldehydes is ca. 95%, with 2-alkenes being the major by-product.

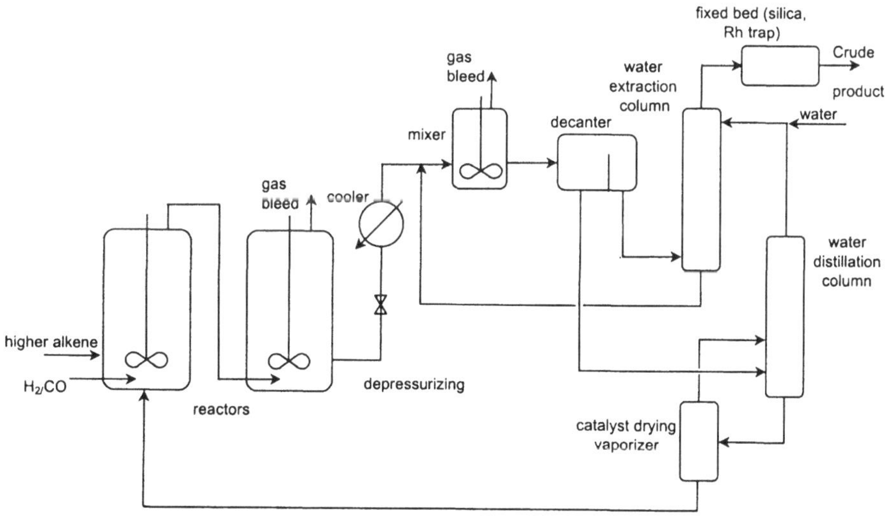

Figure 7. Process type IV: one-phase reaction, extractive separation; type IVB: polar catalyst, apolar product (Union Carbide's higher alkene hydroformylation process)

Aldehyde linearity is high (ca. 90%). Sufficient N-methyl-pyrrolidone (NMP; ca. 40%w) and some water (1-2%w) are applied to achieve a one-phase system in the reactors. After reaction, water is added in a mixer (phase ratio 1:1 v/v), followed by efficient phase separation in a settler, with virtually all catalyst in the NMP/water layer. The crude product layer is subjected to a multi-stage water extraction to remove residual NMP and catalyst, and a final treatment over a silica-bed to reduce Rh leach levels from 0.2 ppmw to 0.02 ppmw. The recycle catalyst layer (in NMP/water) is dried in two steps, to evaporate water and achieve the low water concentrations required for one-phase reaction, and then recycled to the reactors. Water is recycled, from evaporators, via water extraction, to the mixer. The flexibility of this process with respect to alkene carbon number seems excellent: good performance has been found for C_6-C_{14} alkenes [61].

8.6 Survey of commercialized processes and new developments

Table 1 gives a survey of industrially applied Rh-catalyzed hydroformylation processes, with indications of type of catalyst and process. The list is probably not complete; especially smaller-scale applications might have been missed, because of lack of open literature data. Below, some interesting, new developments in a few product areas are shortly discussed, when they have special process/catalyst features and/or involve new petrochemical applications.

8.6.1 Hydroformylation of butenes

Conversion of linear butenes is mainly of interest for production of the plasticizer alcohol 2-propyl-1-heptanol *via* pentanal. Linear butenes are generally available as a Raffinate-1 stream (after butadiene extraction of cracker C_4's) or a Raffinate-2 stream (after additional isobutene removal, generally *via* MTBE production). The latter contains 50-65% 1-butene, the remainder being 2-butenes plus some butanes [24]. The Ruhrchemie-Rhone Poulenc process has been developed for butene conversion by Hoechst (now Celanese) and is very similar to the process for propene (see Chapter 7 and 8.5.3) [24]. The same catalyst, Rh/TPPTS, is used. To compensate for lower alkene solubility, 5-10 °C higher temperature and slightly higher Rh concentrations are applied. Occurrence of double-bond isomerization to 2-butene results in a lower selectivity to aldehydes. The fact that only 1-butene is hydroformylated, leaving all 2-butene, makes this process less elegant than the propene process.

Table 1. Survey of commercial applications of Rh-catalyzed hydroformylation processes

alkene	products	D[#]	year	process	ligand	kt/a	Reference
C_6-C_{14} 1-alkenes	higher alcohols	M	1970	II	none	23[a]	Ch. 8.6.3
ethene	propanol[g]	C,U	1974	I	TPP	400	Ch. 8.5.1
propene	butanol, isobutanol, 2-ethyl-1-hexanol, neopentyl glycol	B,C,U	1974	I	TPP	4000 [b]	Ch. 8.5.1
do	do	M,U	1978	II	TPP	[b]	Ch. 8.5.2
do	do	R	1984	III	TPPTS	600	Ch. 7, 8.5.3
1,2-diacetoxy-3-butene	vit-A[c]	B	70's	II	none	3[d]	[32,34]
1,4-diacetoxy-2-butene	do	H	70's	II	TPP	[d]	[32,34]
1-hexene, 1-octene	carboxylic acids	C	1980	II	TPP	18	Ch. 8.6.3
branched internal octenes	isononanol	M	1987	II	TPPO	30	Ch. 8.6.2
3-methyl-3-butene-1-ol	3-methyl-1,5-pentanediol[e]	K	1988	II	L*[f]	3	[32,49,94,95]
Allyl alcohol	1,4-butanediol	K	1990	IVA	TPP + dppb	180	Ch. 8.5.4.1
7-octenal	1,9-nonanediol	K	1993	IVB	TPPMS	2-3	[32,49,96,97]
do	do	K	> 1993	II	L*[f]	do	[32,98,99]
1-butene	2-propyl-1-heptanol[h]	Ho	1995	III	TPPTS	40	Ch. 7, 8.6.1
1-butene/2-butenes	2-propyl-1-heptanol	U	1996	II	diphosphite	80	Ch. 8.6.1, [100]
higher 1-alkenes	detergent alcohols	Kv (U?)	2001	IVB	DPBS	120	Ch. 8.5.4.2

[#] Developed by: M=Mitsubishi, C=Celanese, U=Union Carbide, B=BASF, R=Ruhrchemie-RhonePoulenc, H=Hoffman-LaRoche, K=Kuraray, Kv=Kvaerner, Ho=Hoechst. [a] Probably obsolete. [b] Sum for Type I and Type II. [c] 4-acetoxy-2-methyl-2-butenal; vitamin-A precursor. [d] 3 kt/a sum for B and H. [e] Together with β-methyl-δ-valerolactone. [f] L*=bulky monophosphite. [g] Plus propanoic acid, methyl methacrylate. [h] Plus pentanoate esters.

Union Carbide has focused on the development of diphosphite ligands (see also Chapter 3), for conversion of butenes including 2-butene to linear pentanal [20, 62, 63]. Compared to cyclic mono-phosphites [18, 19], they have the advantages of substantial double-bond isomerization activity, higher ligand stability and higher regioselectivity to linear aldehydes, albeit at somewhat slower (but still significant) hydroformylation rate [20,64]. The ligand applied commercially could well be 6,6'-[3,3'-bis(tert.butyl)-5,5'-dimethoxy-1,1'-biphenyl-2,2'-diyl]-bis(oxy)bis-dibenzo[d,f][1,3,2]dioxaphosphephin, or a related structure with additional tert.butyl groups instead of the methoxy groups [20, 34, 64, 65]. The activity for 2-butene is lower than for 1-butene, resulting (at ca. 250 ppmw Rh) in temperatures of 85-90 °C rather than the 70-75 °C required for pure 1-butene. Pentanal linearity is so high (ca. 97%) that no aldehyde purification is required before aldol condensation. Related to the mild conditions, heavy ends make is likely to be low and therefore their accumulation can probably be avoided in this type II liquid recycle process. Catalyst/(by)product separation occurs via atmospheric distillation (ca. 110 °C). Reactor pressures are low (circa 10 bar). The H_2/CO ratio in the feed is probably slightly below 1, in order to have a high CO pressure enabling lower paraffin make and suppression of inhibition by poisoning mono-phosphite. This high CO pressure, however, leads to CO inhibition and a negative order in CO pressure, resulting in potentially unstable reactor operation (cycling, runaway). This problem is tackled by (i) reactor cooling with high cooling flow and small temperature differential, and (ii) running at high conversion, especially of H_2 [66, 67]. Ligand/Rh molar ratios are 1.1 to 4, some ligand excess being required for protection against Rh plating, the more so because some ligand degradation takes place (formation of poisoning mono-phosphites and strong phosphorus acids which can catalyze hydrolytic ligand degradation). Some water is present for destruction of the poisoning mono-phosphite, by hydrolysis to phosphorus acids (the latter hydrolysis being much faster than that of the diphosphites themselves) [22]. The phosphorus acids (like α-hydroxypentylphosphonic acid) have to be efficiently removed, by addition of epoxides [17, 68], but preferentially by water extraction [69, 70]. A phosphate buffer can be applied to increase the acid uptake capacity of the water phase and stabilize the pH at 4.5-7.5 [71]. Heterocyclic nitrogen components (like benzoimidazole or benzotriazole) stabilize Rh clusters formed during distillation [72], but also assist in trapping acids [66,69,70]. In water extraction, the heterocyclic nitrogen components stay in the organic phase. A group VIII metal (Ru preferred, 1-5 mol/mol Rh) can be added to suppress Rh cluster formation and its catalysis of orthometallation [9].

8.6.2 Branched higher alkenes to mainly plasticizer alcohols

Plasticizers (mainly for PVC) are dialkylphtalates formed by reaction of phtalic anhydride with plasticizer alcohols (C_7-C_{11}). Market size (1995) for plasticizer alcohols is ca. 3900 kt/a including 2400 kt/a 2-ethyl-1-hexanol production ex n-butanal [73]. In the plasticizer alcohols market, branched alcohols dominate. Most of the plasticizer alcohols are produced from relatively cheap propene/butenes feedstocks. Besides the direct hydroformylation of propene and butenes (giving 2-ethyl-1-hexanol and 2-propyl-1-heptanol), propene and butenes are also oligomerized to C_9 and C_8 branched alkenes, respectively, which are then converted producing C_9-C_{10} alcohols (isononanol, isodecanol), generally using Co-based hydroformylation technology (with/without phosphine modification). Rh-based technology has difficulty to enter this market, because (i) most Rh catalysts are not suited for conversion of branched alkenes, (ii) the products are cheap, (iii) no high selectivity to linear alcohols is required, and (iv) heavy ends are more difficult to remove.

Mitsubishi Chemical has commercialized a Rh-based process on 30 kt/a scale in 1987, for conversion of branched octenes into isononanol (INA) using distillative separation of catalyst (type II) [36, 74, 75]. The catalyst is Rh/triphenylphosphine-oxide (TPPO; Rh/TPPO molar ratio ca. 20). TPPO is claimed to be not just inert, but to be an activity promoter. Hydroformylation of branched octenes takes place at ca. 130 °C and 200 bar, with low Rh concentrations (10-100 ppmw). Aldehyde yields are high (95-98%; the rest being mainly useful alcohol). TPP (1-10 mol/mol Rh) is added after reaction and before distillation, in order to prevent Rh plating during vacuum distillation (at ca. 130 °C bottoms temperature). Part of the TPP is oxidized during distillation via air ingress, part is oxidized on-purpose after distillation with air or hydroperoxide (formed by air oxidation of the alkene feed). Heavy ends will be formed in significant amounts in this process due to high temperatures in reaction and distillation; their accumulation results in a forced catalyst-containing bleed, from which Rh has to be recovered in separate facilities. Although the INA yield is high, this process seems to be not very attractive from a Rh containment point of view (catalyst cost, capital to recover catalyst).

8.6.3 Linear higher alkenes to mainly detergent alcohols

In the area of hydroformylation of linear higher alkenes, applications as detergent alcohol dominate (mainly C_{12}-C_{16}) [76]. The market size for detergent alcohols was ca. 1200 kt/a in 1995 [73]. Linearity (generally in the range 40-100%) relates to the need for detergency (micelles) and biodegradability. Figure 8 gives a survey of routes to detergent alcohols and detergents [76, 77]. Most detergent alcohols are produced from ethene and higher alkenes, but part is produced from natural oils and fats, by transesterification with methanol followed by ester hydrogenation. Higher alkenes can be produced in various ways. 1-Alkenes are produced via ethene oligomerization, or from syn gas via the Fischer-Tropsch process, but presently are produced in insufficient amounts to satisfy the need for surfactant alcohols. Internal alkenes can be produced by an isomerization-metathesis sequence (Shell process) or by dehydrogenation of alkanes (e.g., UOP's Pacol-Olex process). Commercially virtually all hydroformylation of higher alkenes takes place using Co catalysts (Co-oxo and Shell process using phosphine-modified Co catalyst), which convert both terminal and internal alkenes: they are able to rapidly isomerize the double bond, leading to formation of aldehydes/alcohols, with linearities up to 80%.

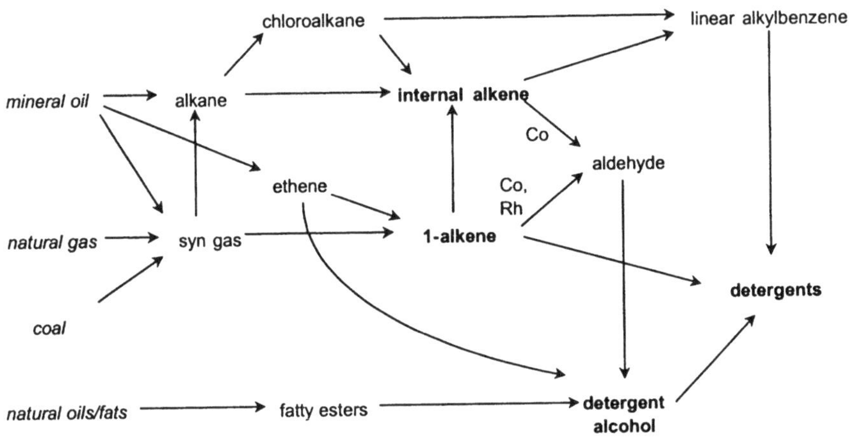

Figure 8. Routes to linear higher alkenes, detergent alcohols and detergents

For Rh-catalyzed hydroformylation of higher alkenes, preferably phosphine-based catalysts are used for the conversion of 1-alkenes, enabling linearities of 80-90%. The low-volatile nature of the higher aldehydes and their heavy ends clearly disadvantage any process with distillative catalyst/product separation (type II). Mitsubishi Chemical has reported the start-up of such a unit in 1970 (23 kt/a), using the unmodified Rh catalyst and C_6-C_{14} 1-alkenes [36], but that unit is probably obsolete. A better type II process is run by Celanese since 1980; they convert C_6-C_8 1-alkenes to linear aldehydes on 18 kt/a scale (for production of carboxylic acids), using Rh/TPP and distillative separation under vacuum [78, 79]. Probably, C_8 is the maximum alkene carbon number to be used in a type II process.

The use of a process of the extractive type (type III or IVB) seems most logical for these higher alkenes, since this enables efficient separation under mild conditions of aldehydes and heavy ends from polar catalyst. For *type III* processes, the major problem is the low solubility of higher alkenes in water (the presence of salts in the aqueous layer making it even worse) [73,80], thus limiting the reaction essentially to the interface. Combinations of the following basic approaches can be used: (i) increase the alkene solubility in the polar phase, (ii) increase the interfacial area, (iii) use a phase-transfer catalyst, and (iv) enhance the reaction rate at the interface. To achieve these objectives, either additives are used (generally selecting the Rh/TPPTS catalyst) or the phosphine structure is modified. Most approaches, however, seem to fail in the two-phase system. Once the problems around catalyst activity are solved, significant problems are still present around loss of catalyst and additives. Type IVB processes seem to be most convenient. In 8.5.4.2, the process developed by Union Carbide has been described. Many groups are exploring similar approaches [73,81].

8.6.4 1,4-Butanediol

1,4-Butanediol (BDO) is a key intermediate in the petrochemical industry (ca. 600 kt/a) [54, 82]. It is used mainly as monomer for the production of the polyester polybutyleneterephtalate (PBT). But significant amounts are also converted into tetrahydrofuran (THF) via dehydration (ca. 250 kt/a) which finds applications as monomer for polytetramethyleneglycol (PTMEG) and as solvent. BDO can also be dehydrogenated to γ-butyrolactone (GBL) (ca. 100 kt/a), the main outlet of which is the production of N-methyl-2-pyrrolidone (NMP) and N-vinylpyrrolidone (monomer for PVP). THF and GBL can also be produced independently from BDO, by hydrogenation of maleic anhydride (MALA). Figure 9 gives a survey of routes to BDO and THF. The major commercial route to BDO is the Reppe process, converting acetylene with formaldehyde to 2-butyne-1,4-diol, followed by hydrogenation. Smaller amounts are produced via acetoxylation of butadiene to 1,4-diacetoxy-2-butene and conversion of 1,4-dichloro-2-butene. A last commercial route is Rh-catalyzed hydroformylation of allyl alcohol (see 8.5.4.1), obtained via isomerization of propene oxide. The old acetylene route is still going strong, due to the fact that a lot of acetylene extraction capacity is just available. But there is also a lot of promise in routes using cheaper feedstocks like propene or butane. Butane can be converted, via selective oxidation to MALA as intermediate, into BDO, THF and GBL, or combinations thereof. New propene-based Rh-hydroformylation routes to BDO are being explored, using allyl acetate or acrolein. In view of the high cost of propene oxide, an alternative for allyl

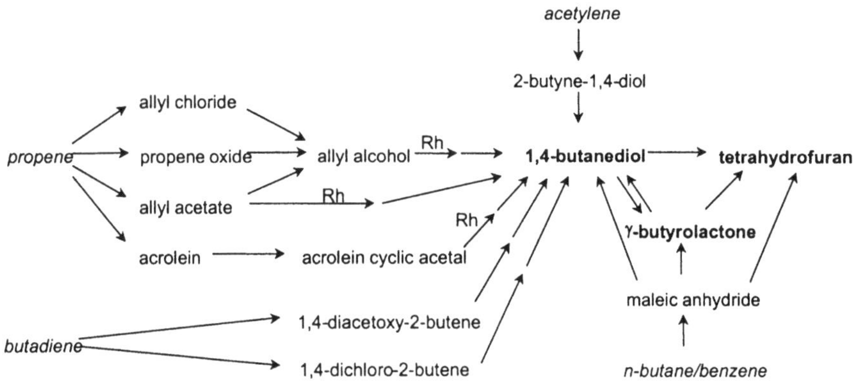

Figure 9. Routes to 1,4-butanediol, tetrahydrofuran and γ-butyrolactone

alcohol production could be the hydrolysis of *allyl acetate*. Direct hydroformylation of allyl acetate via Rh-catalyzed hydroformylation suffers from poisoning by acetic acid. It might therefore be preferable to first hydrolyze allyl acetate to crude allyl alcohol, and hydroformylate that crude mixture. This still requires control of acetic acid traces, via extraction with slightly alkaline water [83]. Alternatively, *acrolein* can be used. This cannot be hydroformylated directly, but is converted to a cyclic acetal (using well-available 1,3-diols like 2-methyl-propane-1,3-diol (by-product) or 2,2,4-trimethyl-pentane-2,4-diol). This acetal now contains an isolated terminal double bond and can be hydroformylated. Subsequently, BDO is formed by combined hydrolysis/hydrogenation. The best results can be obtained with diphosphines with C_4 bridge: at only 8 ppm Rh, full conversion can be obtained with selectivity to linear aldehyde of > 99% [84].

8.6.5 Nylon monomers

Significant work is being carried out to develop new routes to nylon starting from butadiene, including (generally Rh-catalyzed) hydroformylation steps. Nylon-6 and nylon-6,6 are formed from the monomers ε-caprolactam and adipic acid/hexamethyleen-1,6-diamine, respectively [85]. Figure 10 gives a survey of routes to these monomers. ε-Caprolactam is predominantly produced from cyclohexanone oxim, via multi-step routes starting with benzene or toluene. Adipic acid is mainly produced by cyclohexanol/cyclohexanone oxidation. Hexamethylene-1,6-diamine is produced mainly from adiponitrile. Adiponitrile, on its turn, is produced from butadiene mainly via hydrocyanide addition or by electrochemical coupling of acrylonitrile. There is potential for new routes to especially ε-caprolactam and adipic acid, using cheap butadiene.

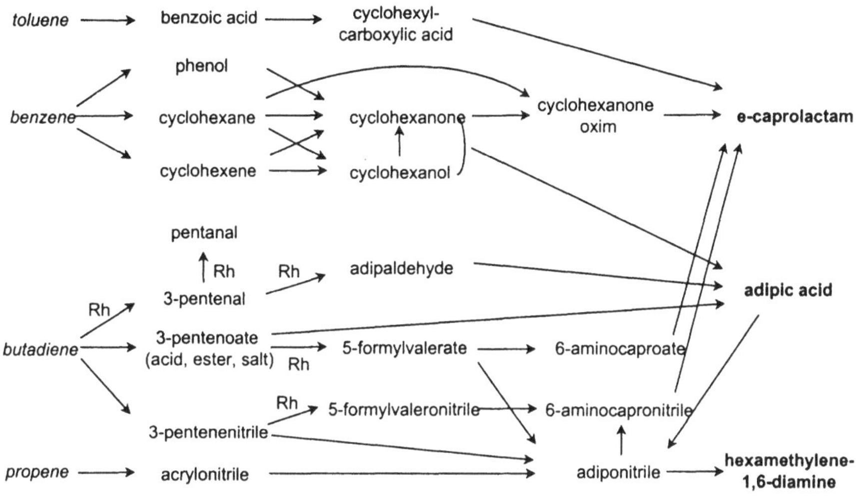

Figure 10. Routes to nylon monomers

One route uses partial hydrogenation of adiponitrile to 6-aminocapronitrile, followed by hydrolysis and ring-closure to ε-caprolactam (BASF). Another uses butadiene double carbonylation to adipic acid (Rhodia). And a third category consists of exploratory routes in which hydroformylation (generally Rh-catalyzed) plays a role: converting (i) butadiene directly to adipaldehyde, (ii) 3-pentenenitrile to 5-formylvaleronitrile, or (iii) 3-pentenoate esters to 5-formylvalerates. The direct hydroformylation of butadiene is probably the most difficult case, with, a.o., low reactivity and selectivity [86, 87]. Rh-catalyzed hydroformylation of 3-pentenenitrile and 3-pentenoate esters is explored using generally diphosphite ligands since these enable the double-bond isomerization to the terminal position. Hydroformylation of methyl 3-pentenoates so far seems most promising. Typical numbers are an aldehyde selectivity of ca. 90% (losses to paraffin and 2-alkene isomer) and an aldehyde linearity of ca. 90%, at conversions of ca. 90%, but reactions are generally very slow (10-20 h) [88-90].

Two process variants have been described. A type II process [91] resembles the Union Carbide process for butenes hydroformylation (see 8.6.1). Rapid oxidation of the diphosphite ligands by air ingress in vacuum distillation columns is mitigated by addition of an excess of sacrificial ligand (tri-orthotolylphosphine). In a type IVA process, very much like the Kuraray process (see 8.6.4.1), apolar and polar solvents are applied to effect the desired combination of one-phase reaction followed by phase separation, with most Rh/diphosphite catalyst in the apolar layer [92,93].

References

1 London Fix, historical rhodium prices, http://www.kitco.com/cgi-bin/yearly_graphs.cgi.
2 New York spot prices, http://www.specialtymetals.com/prices.html.
3 Hillard, H. E., http://minerals.usgs.gov/minerals/pubs/commodity/platinum.
4 Solodar, A. J.; Sall, E. D.; Stout, L. E., Jr.,. in "*Catalysis of Organic Reactions*" (Malz, R. E., Ed.), Chem. Ind. *68*, Dekker, New York, **1996**, *68*, 119.
5 Pruett, R. L.; Smith, J. A. *U.S. Pat.* 4.148,830, 1979 (to Union Carbide); *Chem. Abstr.* **1971**, *75*, 109844.
6 Smith, F. J. *Plat. Met. Rev.* **1975**, *19*, 93.
7 van Leeuwen, P. W. N. M., Appl. Catal. **2000**,
8 Moser, W. R.; Papile, C. J.; Brannon, D. J.; Duwell, R. J.; Weininger, S. J., *J. Mol. Catal.* **1987**, *41*, 271.
9 Abatjoglou, A. G.; Bryant, D. R.; Maher, J. M., U.S. Pat. 5,756,855, **1998**; U.S. Pat. 5,929,289, **1999** (to Union Carbide); Eur. Pat. Appl. 697391, **1996** (to Union Carbide); *Chem. Abstr.* **1996**, *124*, 316213.
10 Billig, E.; Jamerson, J. D.; Pruett, R. L., *J. Organomet. Chem.* **1980**, *192*, C49.
11 Jamerson, J. D.; Pruett, R. L.; Billig, E.; Fiato, F. A., *J. Organomet. Chem.* **1980**, *193*, C43.
12 Abatjoglou, A. G.; Billig, E.; Bryant, D. R., *Organometallics* **1984**, *3*, 923.
13 Abatjoglou, A. G.; Bryant, D. R., *Organometallics* **1984**, *3*, 932.
14 Moser, W. R.; Papile, C. J.; Weininger, S. J., *J. Mol. Catal.* **1987**, *41*, 293.
15 Abatjoglou, A. G.; Billig, E.; Bryant, D. R.; Nelson, J. R., ACS Symp. Series **1992**, *486*, "Phosphorous Chemistry", pp. 229-239.
16 Herrmann, W. A.; Kohlpainter, C. W., *Angew. Chem.* **1993**, *105*, 1588.
17 Babin, J. E.; Maher, J. M.; Billig, E., U.S. Pat. 5,364,950, **1994** (to Union Carbide); *Chem. Abstr.* **1994**, *121*, 208026.
18 Dennis, A.; Harrison, G. E.; Wyber, J.P., U.S. Pat. 4,482,749, **1984** (to Davy McKee); *Chem. Abstr.* **1984**, *100*, 174267; U.S. Pat. 4,496,768, **1985** (to Davy McKee); *Chem. Abstr.* **1984**, *100*, 191375.
19 Billig, E.; Abatjoglou, A. G.; Bryant, D. R.; Murray, R. E.; Maher, J.M., U.S. Pat. 4,599,206, **1986**; U.S. Pat. 4,717,775, **1988** (to Union Carbide); *Chem. Abstr.* **1988**, *109*, 233177.
20 Billig, E.; Abatjoglou, A. G.; Bryant, D. R., U.S. Pat. 4,668,651 and 4,769,498, **1988** (to Union Carbide); *Chem. Abstr.* **1989**, *111*, 117287.
21 Ramirez, F.; Patwardhan, A. V.; Heller, S. R., *J. Am. Chem. Soc.* **1964**, *86*, 514.
22 Maher, J. M.; Babin, J. E.; Billig, E.; Bryant, D. R.; Leung, T. W, U.S. Pat. 5,288,918, **1994** (to Union Carbide); *Chem. Abstr.* **1994**, *121*, 56996.
23 Trzeciak, A. M.; Ziolkowski, J. J., *Z. anorg. allg. Chem.* **1989**, *577*, 255.
24 Frohning, C. D.; Kohlpainter, C. W., in "*Aqueous-Phase Organometallic Catalysis*" (Cornils, B; Herrmann, W. A., Eds.), Wiley-VCH, Weinheim **1998**, pp. 294-306.
25 Konkol, W.; Bahrmann, H.; Herrmann, W. A.; Kohlpainter, C. W., Ger. Pat. 4135050, **1993** (to Hoechst); *Chem. Abstr.* **1994**, *120*, 79891.
26 Pruett, R. L.; Smith, J. A., U.S. Pat. 3,527,809, **1970** (to Union Carbide); *Chem. Abstr.* **1976**, *84*, 30433.
27 Brewester, E. A. V.; Pruett, R. L., U.S. Pat. 4,247,486 **1981** (to Union Carbide); *Chem. Abstr.* **1981**, *95*, 24271.
28 Cornils, B.; Payer, R.; Traenckner, *Hydrocarbon Proc.* **1975**, *54*, June 1975, 83.
29 Hershman, A.; Robinson, K. K.; Craddock, J. H.; Roth, J. F., *Ind. Eng. Chem., Prod. Res. Dev.* **1969**, *8*, 372.

30 Brewester, E. A. V., *Chem. Eng.* **1976**, *83 (24)*, 90.
31 Fowler, R.; Connor, H.; Baehl, R. A., *Hydrocarbon Proc.* **1976**, *55*, Sept. 1976, 247; *Chemtech* **1976**, 772.
32 Frohning, C. D.; Kohlpainter, C. W., in "*Applied Homogeneous Catalysis with Organometallic Compounds*" (Cornils, B; Herrmann, W. A., Eds.), VCH Weinheim **1996**, pp. 29-104.
33 Billig, E.; Bryant, D. R., in "*Kirk-Othmer Encyclopedia of Chemical Technology*", 4th ed., Vol.17, **1996**, pp. 902-919.
34 Kieboom, A. P.; Moulijn, J. A.; Sheldon, R. A.; van Leeuwen, P. W. N. M., in "*Catalysis: an Integrated Approach*" (*Stud. Surf. Sci. Catal. 123*; van Santen, R.A.; van Leeuwen, P. W. N. M.; Moulijn, J. A.; Averill, B. A., Eds.), 2nd ed. **1999**, pp. 29-80.
35 Bryant, D. R.; Billig, E., U.S. Pat. 4,277,627, **1981** (to Union Carbide); *Chem. Abstr.* **1978**, *89*, 163066.
36 Onoda, T., *Chemtech* **1993**, 34.
37 Kuntz, E. G., Fr. Pat. 2314910, **1977** (to Rhone-Poulenc); *Chem. Abstr.* **1977**, *87*, 101944.
38 Cornils, B.; Hibbel, J.; Konkol, W.; Lieder, B.; Much, J.; Schmidt, V.; Wiebus, E., Ger. Pat. 3234701, **1984** (to Ruhrchemie); *Chem. Abstr.* **1984**, *100*, 194022.
39 Kuntz, E. G., *Chemtech* **1987**, 570.
40 Gaertner, R.; Cornils, B.; Springer, H.; Lappe, P., Ger. Pat. 3235030, **1984** (to Ruhrchemie); *Chem. Abstr.* **1984**, *101*, 55331.
41 Bexten, L.; Cornils, B.; Kupies, D., Ger. Pat. 3431643, **1986** (to Ruhrchemie); *Chem. Abstr.* **1986**, *105*, 117009.
42 Albanese, G.; Manetsberger, R.; Herrmann, W. A.; Schwer, C., Eur. Pat. Appl. 704451, **1996** (to Hoechst); *Chem. Abstr.* **1996**, *125*, 11135.
43 Wachsen, O.; Himmler, K.; Cornils, B., *Catal. Today* **1998**, *42*, 373.
44 Chaudhari, R. V.; Bhanage, B. B., in "*Aqueous-Phase Organometallic Catalysis*" (Cornils, B; Herrmann, W. A., Eds.), Wiley-VCH, Weinheim **1998**, pp. 283-294.
45 Cornils, B.; Konkol, W.; Bach, H.; Daembkes, G.; Gick, W.; Wiebus, E.; Bahrmann, H., Ger. Pat. 3415968, **1985** (to Ruhrchemie); *Chem. Abstr.* **1986**, *104*, 209147.
46 Pittman, Jr., C. U.; Honnick, W. D., *J. Org. Chem.* **1980**, *45*, 2132.
47 Harano, Y., U.S. Pat. 4,465,873, **1984** (to Kuraray and Daicel Chemical Industries); *Chem. Abstr.* **1984**, *101*, 9088.
48 Matsumoto, M.; Miura, S.; Kikuchi, K.; Tamura, M.; Kojima, H.; Koga, K.; Yamashita, S., U.S. Pat. 4,567,305, **1986**; Eur. Pat. Appl. 129802, **1987** (to Kuraray and Daicel Chemical Industries); *Chem. Abstr.* **1985**, *102*, 184723.
49 Nojori, N.; Misono, M., *Appl. Catal. A General* **1993**, *93*, 103.
50 Matsumoto, M.; Tamura, M., U.S. Pat. 4,215,077, **1980** (to Kuraray); *Chem. Abstr.* **1979**, *91*, 174826.
51 Matsumoto, M.; Tamura, M., *J. Mol. Catal.* **1982**, *16*, 195.
52 Abatjoglou, A. G.; Bryant, D. R., *Arab. J. Sci. Eng.* **1985**, *10*, 429.
53 Yoshimura, N.; Tamura, M., in "*Successful Design of Catalyst*" (Inui, T., Ed.) **1988**, Elsevier, pp. 307-314.
54 Pitchai, R.; Gaffney, A. M.; Nandi, M. K.; Han, Y.-Z., U.S. Pat. 5,276,210, **1994** (to Arco Chemical); *Chem. Abstr.* **1993**, *119*, 225573.
55 *Eur. Chem News*, 22-28 Nov. 1999, p. 20; 17-23 Jan. 2000, p. 33.
56 *Chem. Eng. News*, **1999**, 18 Oct., p. 20.
57 Betts, M. J.; Dry, M. E.; Geertsema, A.; Rall, G. J. H., Int. PCT Pat. Appl. WO 97/01521 **1997** (to Sastech, Sasol Chemicals Europe); *Chem. Abstr.* **1997**, *126*, 173362.

58 Abatjoglou, A. G.; Bryant, D. R.; Peterson, R. R., Eur. Pat. Appl. 350922, **1989;** U.S. Pat. 5,180,854, **1993** (to Union Carbide); *Chem. Abstr.* **1990**, *113*, 80944.
59 Paetzold, E.; Kinting, A.; Oehme, G., *J. Prakt. Chem.* **1987**, *329*, 725.
60 Breslow, D. S.; Skolnik, H., "*Multi-Sulfur and Sulfur and Oxygen Five- and Six-Membered Heterocycles*", Interscience, New York, **1966**; Part 1, pp. 76-95; Part 2, pp. 775-789.
61 Abatjoglou, A. G.; Peterson, R. R.; Bryant, D. R., in "*Catalysis of Organic Reactions*" (Malz, R. E., Ed.), Chem. Ind. *68*, Dekker, New York, **1996**, pp 133-139.
62 Billig, E.; Abatjoglou, A. G.; Bryant, D. R., U.S. Pat. 4,748,261, **1988**; U.S. Pat. 4,885,401, **1989** (to Union Carbide); *Chem. Abstr.* **1987**, *107*, 7392.
63 Bryant, D. R.; Nicholson, J. C.; Briggs, J. R.; Packett, D. L.; Maher, J. M., U.S. Pat. 5,886,235, **1999** (to Union Carbide); *Chem. Abstr.* **1999**, *130*, 254055.
64 van Rooy, A.; Kamer, P. C. J.; van Leeuwen, P. W. N. M.; Goubitz, K.; Fraanje, J.; Veldman, N.; Spek, A. L., *Organometallics* **1996**, *15*, 835.
65 Cuny, G. D.; Buchwald, S. L., *J. Am. Chem. Soc.* **1993**, *115*, 2066.
66 Nicholson, J. C.; Bryant, D. R.; Nelson, J. R, U.S. Pat. 5,763,679, **1998** (to Union Carbide); *Chem. Abstr.* **1998**, *129*, 55721.
67 Nicholson, J. C.; Bryant, D. R.; Nelson, J. R.; Briggs, J. R.; Packett, D. L.; Maher, J. M., U.S. Pat. 5,874,639, **1999** (to Union Carbide); *Chem. Abstr.* **1999**, *130*, 184071.
68 Billig, E.; Bryant, D. R., U.S. Pat. 5,763,670, **1998** (to Union Carbide); *Chem. Abstr.* **1998**, *129*, 55719; U.S. Pat. 5,767,321, **1998** (to Union Carbide); *Chem. Abstr.* **1998**, *129*, 82975.
69 Bryant, D. R.; Nicholson, J.C., U.S. Pat. 5,763,671, **1998** (to Union Carbide); *Chem. Abstr.* **1998**, *129*, 69096; U.S. Pat. 5,789,625, **1998** (to Union Carbide); *Chem. Abstr.* **1998**, *129*, 166637.
70 Bryant, D. R.; Nicholson, J. C.; Briggs, J. R.; Packett, D. L.; Maher, J. M., U.S. Pat. 5,917,095, **1999** (to Union Carbide); *Chem. Abstr.* **1999**, *131*, 60316.
71 Bryant, D. R.; Leung, T. W.; Billig, E.; Eisenschmid, T. C.; Nicholson, J. C.; Briggs, J. R.; Packett, D. L.; Maher, J. M., U.S. Pat. 5,874,640, **1999** (to Union Carbide); *Chem. Abstr.* **1999**, *130*, 184072.
72 Leung, T. W.; Bryant, D. R.; Shaw, B. L., U.S. Pat. 5,731,472, **1998** (to Union Carbide); *Chem. Abstr.* **1998**, *128*, 245454.
73 Bahrmann, H., in "*Aqueous-Phase Organometallic Catalysis*" (Cornils, B; Herrmann, W. A., Eds.), Wiley-VCH, Weinheim **1998**, pp. 306-321.
74 Tano, K.; Sato, K.; Okoshi, T, Ger. Pat. 3338340, **1984** (to Mitsubishi Chemical); *Chem. Abstr.* **1984**, *101*, 170705.
75 Miyazawa, C.; Hiroshi, M.; Tsuboi, A.; Hamano, K., Eur. Pat. Appl. 272608, **1988** (to Mitsubishi Chemical); *Chem. Abstr.* **1988**, *109*, 151931.
76 Wilne, C., in "*Alpha Olefins Applications Handbook*" (Lappin, G. R.; Sauer, J.D., Eds.), Marcel Dekker, New York, **1989**, pp. 139-199.
77 Wagner, J. D.; Lappin, G. R.; Zietz, J. R., in "*Kirk-Othmer Encyclopedia of Chemical Technology*", 4th ed., Vol. 1, **1991**, pp. 893-913.
78 Celanese, *Chem. Eng.* **1981**, *88* (22), 68.
79 Unruh, J. D.; Strong, J. R.; Koski, C. L., in "*Alpha Olefins Applications Handbook*" (Lappin, G. R.; Sauer, J. D., Eds.), Marcel Dekker, New York, **1989**, pp. 311-327.
80 Hanson, B. E., in "*Aqueous-Phase Organometallic Catalysis*" (Cornils, B; Herrmann, W. A., Eds.), Wiley-VCH, Weinheim **1998**, pp. 181-188.
81 Fell, B., *Tenside Surf. Det.* **1998**, *35*, 326.
82 Hort, E. V.; Taylor, P., in "*Kirk-Othmer Encyclopedia of Chemical Technology*", 4th ed., Vol.1, **1991**, pp. 209-216.

83 Chen, S. C.; Chu, C. C., Lin, F. S.; Chou, J. Y., U.S. Pat. 5,426,250, **1995** (to Darien Chemical).

84 Ichikawa, S.; Ohgomori, Y.; Sumitani, N.; Hayashi, H.; Imanari, M., *Ind. Eng. Chem. Res.* **1995**, *34*, 971.

85 Weissermel, K., Arpe, H.-J., "*Industrial Organic Chemistry*", 2nd ed., VCH, Weinheim, **1993**, pp. 235-262.

86 Ohgomori, Y.; Suzuki, N.; Sumitani, N., *J. Mol. Catal.* **1998**, *133*, 289.

87 Fell, B.; Hermanns, P.; Bahrmann, H., *J. prakt. Chem.* **1998**, *340*, 459.

88 Hansen, C.B.; Teunissen, A.J.J.M., Eur. Pat. Appl. 712828 **1995** (to DSM, du Pont); *Chem. Abstr.* **1996**, *125*, 114182.

89 Wissing, E.; Teunissen, A. J.; Hansen, C. B.; van Leeuwen, P. W. N. M.; van Rooy, A.; Burgers, D., Int.`PCT Pat. Appl. WO 96/16923 **1996** (to DSM, Du Pont); *Chem. Abstr.* **1996**, *125*, 114185.

90 Burke, P. M.; Garner, J. M.; Tam, W.; Kreutzer, K. A., Teunissen, A. J. J. M.; Snijder, C. S.; Hansen, C.B., Int. PCT Pat. Appl. WO 97/33854 **1997** (to DSM, Du Pont); *Chem. Abstr.* **1997**, *127*, 294939; U.S. Pat. 5,874,641 **1999** (to DSM, Du Pont); *Chem. Abstr.* **1999**, *130*, 198123.

91 Borman, P. C.; Gelling, O. J., Eur. Pat. Appl. 839787 **1998** (to DSM); *Chem. Abstr.* **1998**, *128*, 323141.

92 Bunel, E. E., Int. PCT Pat. Appl. WO 99/06345 **1999** (to du Pont, DSM); *Chem. Abstr.* **1999**, *130*, 167872.

93 Argyropoulos, J. N.; Bryant, D. R.,; Morrison, D. L.; Stockman, K.E., Int. PCT Pat. Appl. WO 99/38832, **1999** (to Union Carbide); *Chem. Abstr.* **1999**, *131*, 131509.

94 Tokitoh, Y.; Yoshimura, N., U.S. Pat. 4,663,468, **1987** (to Kuraray); *Chem. Abstr.* **1986**, *104*, 148739.

95 Tokitoh, Y.; Yoshimura, N., U.S. Pat. 4,808,737, **1989;** Eur. Pat. Appl. 223103, **1989** (to Kuraray); *Chem. Abstr.* **1987**, *107*, 154896.

96 Matsumoto, M.; Yoshimura, N.; Tamura, M., U.S. Pat. 4,510,332, **1985** (to Kuraray); *Chem. Abstr.* **1984**, *100*, 22329.

97 Tokitoh, Y.; Yoshimura, N., U.S. Pat. 4,808,756, **1989** (to Kuraray); *Chem. Abstr.* **1987**, *107*, 96332.

98 Omatsu, T.; Tokitoh, Y.; Yoshimura, N., Eur. Pat. Appl. 303060, **1989** (to Kuraray); *Chem. Abstr.* **1989**, *111*, 38870.

99 Tokuda, Y.; Ruda, K.; Suzuki, S., Jap. Pat. 11071318 **1999** (to Kuraray); *Chem. Abstr.* **1999**, *130*, 222989.

100 *Chem. Week* **1994**, Jul. 27, p. 26; *Eur. Chem. News* **1994**, Oct. 3, p. 28.

Chapter 9

Catalyst preparation and decomposition

Piet W. N. M. van Leeuwen

Institute of Molecular Chemistry, University of Amsterdam, Nieuwe Achtergracht 166, 1018 WV, Amsterdam, the Netherlands

9.1 Introduction

Catalyst preparation or setting up the catalytic system is closely related to the decomposition of the catalysts. In this chapter we will deal with both aspects. Rhodium hydroformylation catalysts can decompose in a variety of ways: metal deposition, ligand decomposition, reactions with impurities, dimer formation, and reaction of the metal center with the ligand. Sometimes these side-reactions lead to temporary deactivation only and the catalyst activity can be restored. Without attempting to be complete, we have collected several examples of catalyst decomposition. In the area of homogeneous catalysis there has always been less attention, certainly in chemistry journals, for the stability, decomposition, and regeneration of the catalyst as there is in the area of heterogeneous catalysis. This does not mean that stability of the catalyst is not an issue, on the contrary. For the many industrial applications that have come on stream in the last three decades, catalyst stability has been a key issue, next to selectivity and activity. In industry a considerable effort has been devoted to the study of catalyst stability even though few accounts were published in patents or journals. Many examples that we will mention stem from the work of the Bryant group at Union Carbide Corporation.

9.2 Catalyst preparation

The "catalyst" for hydroformylation is a rhodium(I) hydride species that is clearly distinct from the species that are active for hydrogenation. The

P.W.N.M. van Leeuwen and C. Claver (eds.), Rhodium Catalyzed Hydroformylation, 233–251.

hydrogenation catalysts are cationic $Rh(I)^+$ or neutral Rh(I)Cl species. Carbonylation of alcohols also requires an anionic Rh(I) species, e.g. $[Rh(CO)_2I_2]^-$. In modern catalysis we want to know the exact nature of our catalysts and we would like to control its formation and stability.

Often rhodium(I) salts are used as the precursor for hydroformylation catalysts. Under the reaction conditions (H_2, CO, ligands, temperature > 25 °C) these salts are converted to a rhodium hydride complex, although there are several papers that seem to invoke cationic rhodium species as the catalysts. A catalyst that does contain another rhodium species is the one reported by Stanley, containing bimetallic rhodium species [Chapter 10]. Anions that have been used include halides, conjugate bases of weak acids, thiolates, alkoxides, etc. Chlorides have a particular deleterious effect on the activity (i.e. they are not converted into hydrides under mild conditions). Chloride may be present in phosphites or phosphines as a result of the synthesis; especially when the ligands are used in a large excess this may be the cause for low activities. It has been reported that addition of bases such as amines has a strong "promoting" effect on such systems:

$$L_3RhCl + H_2 + CO + R_3N \leftrightharpoons L_3Rh(CO)H + R_3NH^+Cl^-$$

Rhodium salts of weaker acids are smoothly converted into rhodium hydride without the addition of base:

$$L_3RhA + H_2 + CO \leftrightharpoons L_3Rh(CO)H + HA$$

A = acetate, 2,4-pentanedionate

When a large excess of carboxylic acid is present, rhodium carboxylate is not converted into hydride [1]. A preferred dionate is 2,2,6,6-tetramethyl-3,5-heptanedionate (dipivaloylmethane) especially because its solutions are stable upon storage [2]. The use of stock solutions is recommended because very small amounts of rhodium need to be applied when active catalysts are being studied. In batch reactions the formation of rhodium hydride species may be slow compared to catalysis. If so, one should pre-heat the catalyst system to allow the formation of the catalyst before adding the alkene.

As will be discussed in section 9.3 purification of the reagents is important as impurities may lead to decomposition of the catalytic precursor.

9.3 Catalyst decomposition

9.3.1 Metal plating or cluster formation

Formation of a metallic precipitate is the most common mechanism for decomposition of a homogeneous catalyst. This is not surprising, since often reducing agents such as dihydrogen and carbon monoxide are the reagents used. Certainly this is true for hydroformylation. Precipitation of the metal or metal cluster may be preceded by decomposition of the ligand, leaving carbon monoxide as the only stabilizing ligand.

A typical example is the loss of carbon monoxide and dihydrogen from a rhodium hydride tetracarbonyl, the hydroformylation catalyst discussed in Chapter 2 (Figure 1).

$$4\ HRh(CO)_4 \longrightarrow 2\ Rh_2(CO)_8 + H_2 \longrightarrow Rh_4(CO)_{12} + 4\ CO$$

Figure 1. Decomposition of rhodium hydride carbonyl

High temperatures and low pressures accelerate the reaction. The catalyst $HRh(CO)_4$ must lose one molecule of CO before coordination of the alkene substrate can take place (this is neither the catalyst's resting state nor the rate-determining step; for the sake of simplicity it is represented this way). Thus, loss of CO is an intricate part of the catalytic cycle, which includes the danger of a complete loss of ligands giving precipitation of the metal or one of the carbonyl clusters. Precipitation of cobalt metal is quite common in cobalt catalyzed hydroformylation, but because of the high cost of rhodium (a thousand times more expensive than cobalt) precipitation of rhodium has to be avoided under all circumstances. Addition of a phosphine or phosphite ligand stabilizes the rhodium carbonyl species forming $HRh(CO)_{4-n}(PR_3)_n$ complexes as discussed in Chapters 3-8. In most instances when ligands are present, other complexes are formed upon decomposition rather than rhodium carbonyl clusters or metal (*vide infra*).

Phosphorus free catalysts are used successfully both in industry and the laboratory [3-5] and care should be taken to maintain conditions at which no rhodium cluster or metal is formed. The precipitate observed first is that of $Rh_4(CO)_{12}$.

9.3.2 Oxidation of phosphorus ligands

Phosphorus ligands are very prone to oxidation and thermodynamics show that even water and carbon dioxide may oxidize alkyl phosphines to the corresponding oxides. These reactions may be catalyzed by the transition metal in the system. Therefore, oxygen and hydroperoxides have to be thoroughly removed from the reagents and solvents before starting our catalysis. In spite of this common knowledge oxidation of phosphine ligands has frequently obscured the catalytic results. Purification of the alkene feed is often neglected. Since the alkene may be present in a thousand-fold excess of the ligand, careful removal of hydroperoxides in the alkene is an absolute must. Hydroperoxides are the ideal reagents for oxidizing phosphines. Percolation over neutral alumina is usually sufficient for a hydroformylation reaction. Treatment over sodium or sodium-potassium on a support will also remove alkynes and dienes that may influence the catalyst performance. Distillation from sodium may give isomerization of alkenes as an undesirable side-reaction.

Practical recommendations and safety aspects. During the set-up of the experiments air has to be excluded, although the reaction of oxygen with electron-poor phosphorus compounds is much slower than that with components of Ziegler like catalysts for instance. A simple calculation may be useful. Suppose one uses an autoclave having a volume of 100 mL containing 10 mL of solvent and a catalyst concentration of 1 mM, and a tenfold excess of phosphine ligand (10 mM). The autoclave contains 0.1 mmole of phosphine. Thus, assuming a stoichiometry of one dioxygen for two phosphine molecules we need as little as 6 mL of air to oxidize the phosphine inventory! Flushing the tubing, pressurization and depressurization of the autoclave with syn-gas, or evacuation, and cannula and syringe techniques suffice to exclude oxygen in the system.

One can circumvent the problems caused by ligand oxidation by adding an excess of the ligand, unless one is interested in the kinetics in relation to the type of complexes present in the system. This and the addition of triethylamine lead to a fast and practical way of doing hydroformylation reactions. With these precautions the reaction is a highly accessible one for people less experienced in catalysis. High-pressure equipment remains a prerequisite (10 bar); we recommend the use of autoclaves, regularly tested by the workshops, even though glassware for carrying out reactions at 5 bar might be available.

The safety issues for working with high pressures, phosphorus compounds, and syn-gas are well recognized. For high pressures and syn-gas there will usually be government regulations concerning the construction, maintenance, testing, and use of the equipment. For syn-gas one finds a detection system in many laboratories and a regular rehearsal on how to act when the alarm rings. Depressurization of the autoclaves ($<$ 250 mL, $<$ 20 bar) is done in the fume cupboards using a tube ending in the exhaust channel. The rate of depressurization should be such that in the exhaust gas the MAC value of CO is not exceeded.

The most commonly used phosphines such as tpp, tppms, and tppts have been well tested for their toxicity owing to their use on industrial scale. No particular dangers have been reported. The same holds for many phosphites, which are also marketed as anti-oxidants. Apart from these compounds the phosphorus compounds and their intermediates should be handled with the same care as the toxic analogs that are known. Toxicity levels may be comparable to common agrochemicals. Below we show one example of such a group of phosphite ligands that turned out to be toxic (Table 1, Figure 2). Most of the ligands we are using in the laboratory have not been tested.

Table 1. Toxicity data for a few phosphites [6]

Structure	LD_{50} (mg/kg, for mice, Intraperitoneal)
R= CH_2OH (Figure 2)	> 500
R= Et (Figure 2)	1.1
R= Pr (Figure 2)	0.39
R= i-Pr (Figure 2)	0.22
Parathion (as P=S)	5.9
DFP (as P=O)	6.0

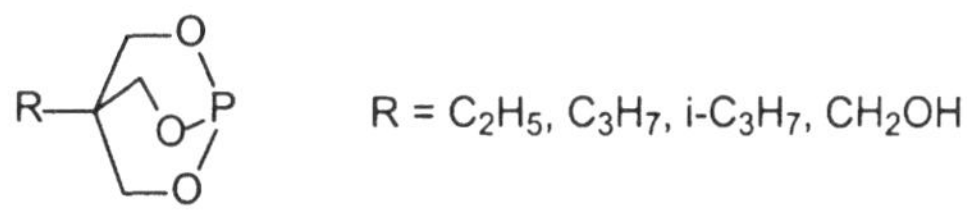

Figure 2. Structure of some toxic phosphites

9.3.3 Phosphorus-carbon bond breaking in phosphines

Oxidation of free phosphines was mentioned above as a reaction leading to phosphine loss. Here we will discuss three further ways of phosphine decomposition: oxidative addition of phosphines to low-valent metal complexes, nucleophilic attack on coordinated phosphines, and aryl exchange via phosphonium species. Interestingly in all cases the metal serves as the catalyst for the decomposition reaction!

In his review [7] and feature article [8] (entitled "I wonder where the ligand went"!) Garrou emphasizes the first mechanism, oxidative addition of the phosphorus-carbon bond to low-valent metal complexes. More recently experimental support for the other two mechanisms has been reported. In Figure 3 the three mechanisms are briefly outlined.

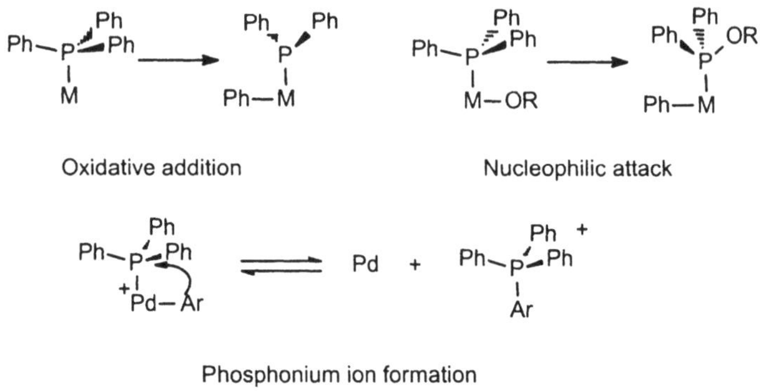

Figure 3. Phosphorus-carbon bond breaking

The latter mechanism has only been observed for palladium compounds as yet, although it would also be feasible for rhodium(I) compounds giving an anionic rhodium species and a phosphonium fragment. The actual phosphorus carbon bond breaking occurs upon the reverse reaction, when the phosphonium ion adds oxidatively to a low-valent metal.

Oxidative addition of P-C bond to a low-valent metal. Oxidative addition of C-Br or C-Cl bonds is an important reaction in cross-coupling type catalysis, and while the reaction of a P-C bond is very similar, the breaking of carbon-to-phosphorus bonds is not a useful reaction in homogeneous catalysis. The reverse reaction, making of a carbon-to-phosphorus bond via palladium or nickel catalysis, is becoming increasingly more important for the synthesis of new phosphines [9]. P-C bond breaking is an undesirable side-reaction that occurs in systems containing transition metals and phosphine ligands and it leads to deactivation of the catalysts. The oxidative addition of a phosphine to a low-valent transition-metal can be most easily understood by comparing the Ph_2P- fragment with a Cl- or Br- substituent at the phenyl ring; electronically they are very akin, c.f. Hammett parameters and the like. The phosphido anion formed during this reaction will usually lead to bridged structures, which are extremely stable. Decomposition of ligands during hydroformylation has been reported both for rhodium and cobalt catalysts [10-12].

Thermal decomposition of $RhH(CO)(PPh_3)_3$, in the absence of H_2 and CO, leads to a stable cluster shown in Figure 4 containing μ_2-PPh_2 fragments [13] as was studied by Bryant's group at Union Carbide. Presumably clusters of this type also form in hydroformylation plants on the long run. Recovery of rhodium from these inert clusters is a tedious operation. Reaction of the cluster mixture with reactive organic halides such as allyl chloride has been described to give allyldiphenylphosphine and rhodium chloride, which can be easily extracted into a water layer. [14].

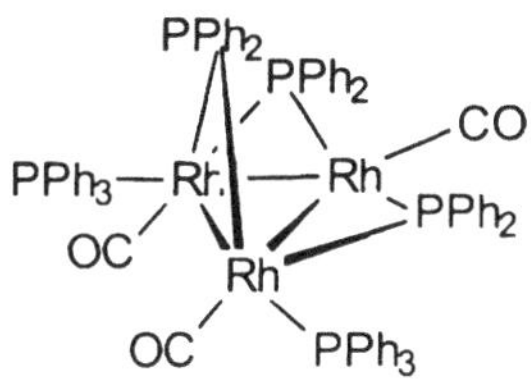

Figure 4. Rhodium cluster formed from decomposition of $RhH(PPh_3)_3CO$

Under hydroformylation conditions also other products are found such as benzaldehyde, benzene, and diphenylpropylphosphine. The mechanism for their formation is outlined in Figure 5.

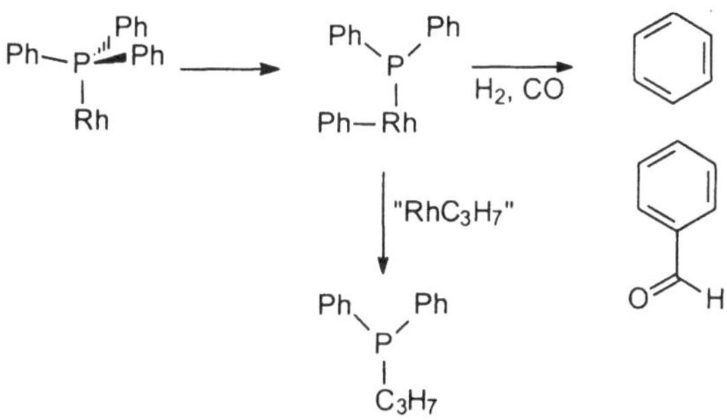

Figure 5. Formation of side-products during ligand decomposition

The phenyl group formed ends up as its hydrogenation or carbonylation product. Instead of cluster formation the phosphido group may give a reductive elimination starting from a rhodium propyl species, which gives diphenylpropylphosphine as the product. The disadvantage of this phosphine is that it is a stronger electron-donor than tpp and it leads to a less active rhodium catalyst. Thus at some stage it has to be removed, which can be done by utilizing its higher basicity by selective phosphonium salt formation.

Benzene and benzaldehyde byproducts were also observed in the Ruhrchemie-Rhône Poulenc process using the trisulfonated analog of triphenylphosphine [15], but the decomposition was reported to be much slower for tppts as compared to tpp. This is an accidental favorable aspect of the RC-RP process; it is unexpected because water might oxidize the phosphine, or it might carry out a nucleophilic attack, thus initiating the second mode of decomposition (vide infra).

Cluster or bimetallic reactions have also been proposed in addition to monometallic oxidative addition reactions. The reactions do not basically change. Several authors have proposed a mechanism involving orthometallation as a first step in the degradation of phosphine ligands, especially in the older literature. orthometallation does take place as can be inferred from deuteration studies, but it remains uncertain whether this is a reaction necessarily preceding the oxidative addition (Figure 6):

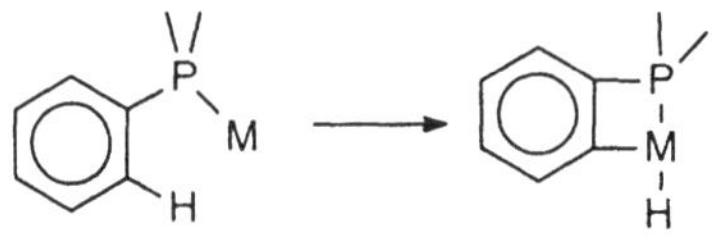

Figure 6. Orthometallation

Subsequently the phosphorus-to-carbon bond is broken and the benzyne intermediate inserts into the metal hydride bond. Although this mechanism has been popular with many chemists there are many experiments that contradict this mechanism. A simple para-substitution of the phenyl group would answer the question whether orthometallation was involved as is shown in Figure 7:

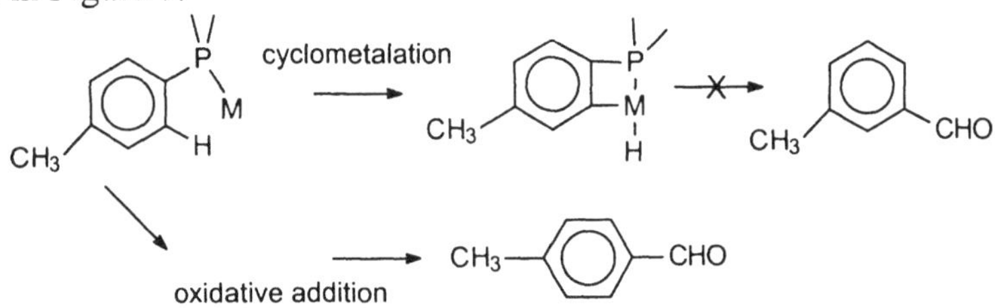

Figure 7. Disproof of orthometallation as the mechanism for P-C cleavage

Decomposition products of tolylphosphines should give 3-methyl substituents if the orthometallation mechanism is operative. For palladium catalyzed decomposition of triarylphosphines this is not the case [16]. Likewise, Rh, Co, and Ru hydroformylation catalysts give aryl derivatives not involving C-H activation [17, 18]. Several rhodium complexes catalyze the exchange of aryl substituents at triarylphosphines [18]:

$$R'_3P + R_3P \text{ — (Rh) —> } R'R_2P + R'_2RP$$

These authors propose as the mechanism for this reaction a reversible oxidative addition of the aryl-phosphido fragments to a low valent rhodium species. A facile aryl exchange has been described for complexes $Pd(PPh_3)_2(C_6H_4CH_3)I$. Again the authors [19] suggest a pathway involving oxidative additions and reductive eliminations. The mechanism outlined below, however, can also explain the results of these two studies.

Phosphido formation has been observed for many transition metal phosphine complexes [7, 8]. Upon prolonged heating and under an atmosphere of CO and/or H_2 palladium and platinum also tend to give stable phosphido bridged dimers or clusters [20, 21].

Nucleophilic attack. Current literature underestimates the importance of nucleophilic attack as a mechanism for the catalytic decomposition of phosphines, especially with nucleophiles such as acetate, methoxy, hydroxy and hydride. For examples of nucleophilic attack at coordinated phosphorus see references [20-25]. A very facile decomposition of alkylphosphines and triphenylphosphine (using palladium acetate, one bar of hydrogen and room temperature) has been reported [20] using acetate or hydride as the nucleophile.

A detailed reaction proving the nucleophilic attack was shown for platinum complexes [25]. The alkoxide coordinated to platinum attacks phosphorus while the carbon atom coordinated to platinum migrates to phosphorus. Thermodynamically the result seems more favorable, but mechanistically this "shuffle" remains mysterious (see Figure 8). Coordination to platinum makes the phosphorus atom more susceptible to nucleophilic attack, and the harder atoms (P and O) and softer ones (C and Pt) recombine as one might expect. The same mechanism was proposed by Matsuda [22] for the decomposition of triphenylphoshine by palladium(II) acetate. In this study the aryl phosphines are used as a source for aryl groups that are converted into stilbenes via a Heck reaction. Even alkyl phosphines underwent P-C bond cleavage via palladium acetate.

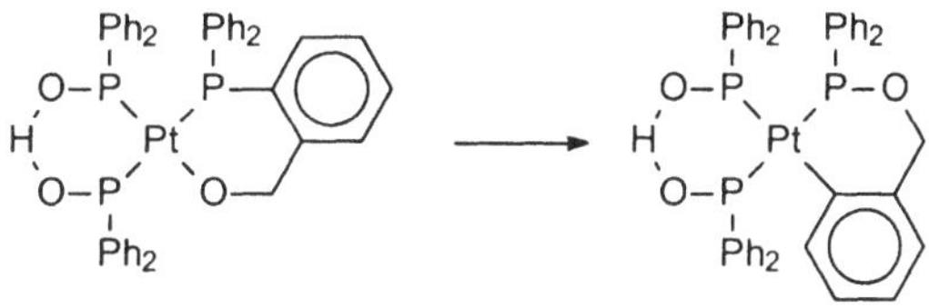

Figure 8. Intramolecular nucleophilic attack

A catalytic decomposition of triphenylphosphine has been reported [26] in a reaction involving rhodium carbonyls, formaldehyde, water, and carbon monoxide. Several hundreds of moles of phosphine can be decomposed this way per mole of rhodium per hour! The reactions that may be involved are shown in Figure 9.

Related to this chemistry is the hydroformylation of formaldehyde to give glycolaldehyde, which would be an attractive route from syn-gas to ethylene glycol. The reaction can indeed be accomplished and is catalyzed by rhodium arylphosphine complexes [27], but clearly phosphine decomposition is one of the major problems to be solved before formaldehyde hydroformylation can be applied commercially.

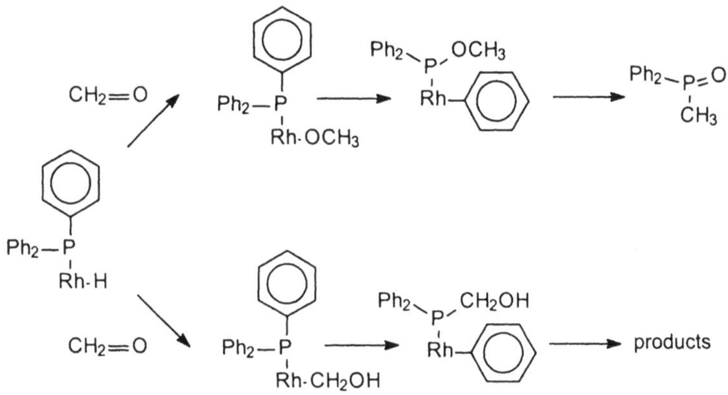

Figure 9. Catalytic decomposition of tpp by formaldehyde and hydrogen

The interest in cross-coupling reactions in the last decade has lead to a large number of reports dealing with the involvement of the ligands in these reactions [e.g. 28-30]. The mechanism has not been elucidated for all cases. Norton [28] proposes an intramolecular nucleophilic attack of a palladium bonded methyl group to a coordinated aryl ligand and explicitly excludes the intermediacy of phosphonium species, as deuterated phosphonium salts present in solution and having the same composition did not participate in the reaction. The mechanism involving phosphonium species will be discussed in the next section.

Aryl exchange via phosphonium intermediates. More recently a variation of the above mechanisms was reported by Novak [30]. Formally the mechanism also involves nucleophilic attack at coordinated phosphines, but after the nucleophilic attack the phosphorus moiety reductively eliminates from the metal as a phosphonium salt. To obtain a catalytic cycle the phosphonium salt re-adds to the zerovalent palladium complex (Figure 10).

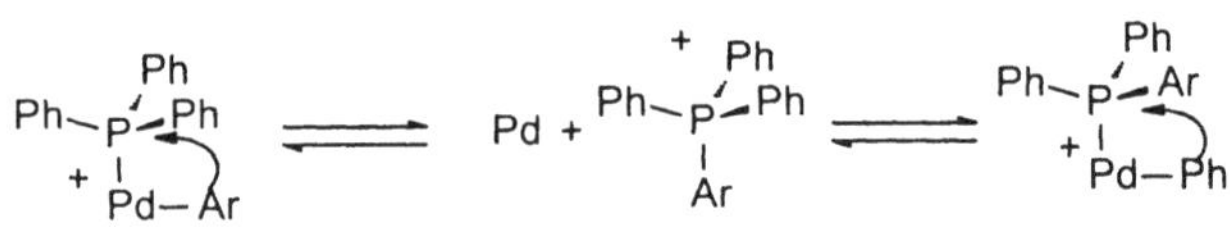

Figure 10. Aryl exchange via phosphonium formation

Scavenging of free phosphines by electrophiles such as protons, other metals, conjugated enones, etc. presents a potential route to phosphine loss in catalytic systems. As yet, the participation of phosphonium intermediates has not been reported for rhodium hydroformylation catalysts, but they could be easily conceived, especially when dienes or enones are also present.

9.3.4 Decomposition of phosphites

Phosphites are easier to synthesize and less prone to oxidation than phosphines. They are much cheaper than most phosphines and a wide variety can be obtained commercially as they are used as anti-oxidants. Disadvantages of the use of phosphites as ligands include several side reactions: hydrolysis, alcoholysis, trans-esterification, Arbusov rearrangement, O-C bond cleavage, P-O bond cleavage. Figure 11 gives an overview of these reactions. In hydroformylation systems at least two more reactions may occur, namely nucleophilic attack to aldehydes, and oxidative cyclizations with aldehydes.

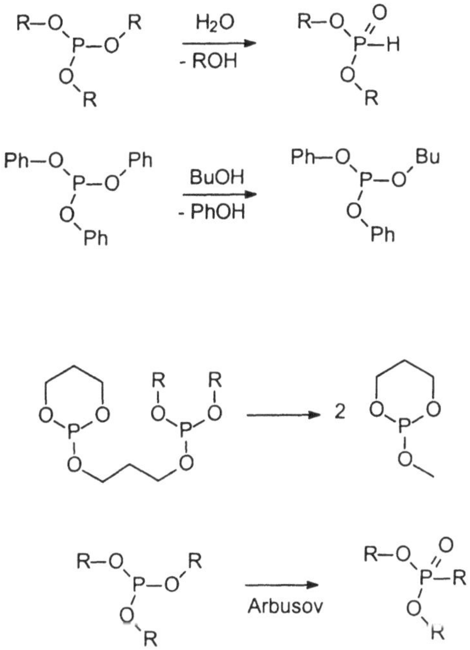

Figure 11. Decomposition pathways of phosphites

Phosphites have been extensively studied for their use as ligands in rhodium-catalyzed hydroformylation (see Chapter 3). The first publication on the use of phosphites is from Pruett and Smith, from Union Carbide [31]. The first exploitation of bulky monophosphites was reported by van Leeuwen and Roobeek [32]. They found that very high rates can be obtained for internal and terminal alkenes, but selectivities were low for linear alkenes. The bulky phosphites not only gave higher rates than less bulky phosphites, but they are also more resistant to hydrolysis. Bryant and coworkers [33] introduced even more stable, bulky phosphites by the

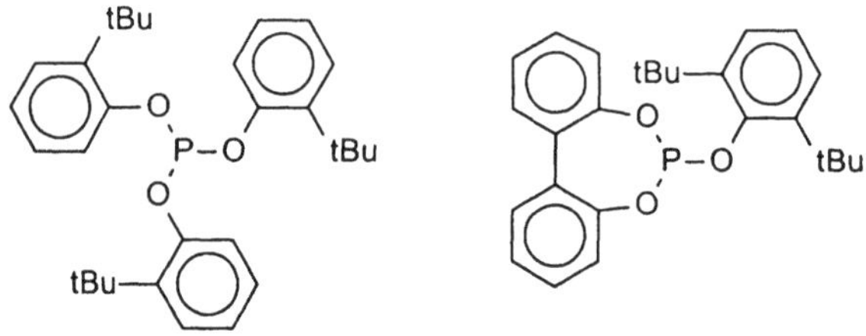

Figure 12. Bulky phosphite ligands

introduction of bisphenols instead monophenols for the phosphite synthesis (see Figure 12). High selectivities are only obtained when diphosphites are used.

Diphosphites came into focus after the discovery of Bryant and coworkers at Union Carbide Corporation that certain bulky diphosphites lead to high selectivities in the rhodium catalyzed hydroformylation of terminal and internal alkenes [34] (see Figure 13). A plethora of diphosphites has been tested and recorded in many patents authored by coworkers of several companies [35-37], indicating their importance for the near future in this field. The patents included in the references of this chapter serve only as examples.

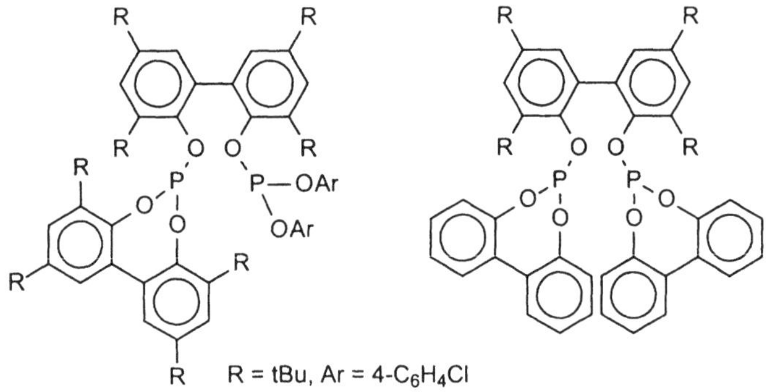

Figure 13. Examples of Union Carbide's diphosphites

The advantages of the diphosphite system for propene hydroformylation as compared to the commercial triphenylphosphine system are that less ligand and less rhodium are required and that higher rates can be obtained. Also high selectivities for conversion of internal alkenes to linear products have been reported.

It should be noted that all phosphites reported are aryl phosphites (sometimes the backbones may be aliphatic) and that the favored ones often contain bulky substituents. One of the reasons that aliphatic phosphites are used only sparingly is that they are susceptible to the Arbusov rearrangement while the aryl phosphites are not. Acids, carbenium ions, and metals catalyze the Arbusov rearrangement. Many examples of metal catalyzed decomposition reactions have been reported [38, 39] (see Figure 14).

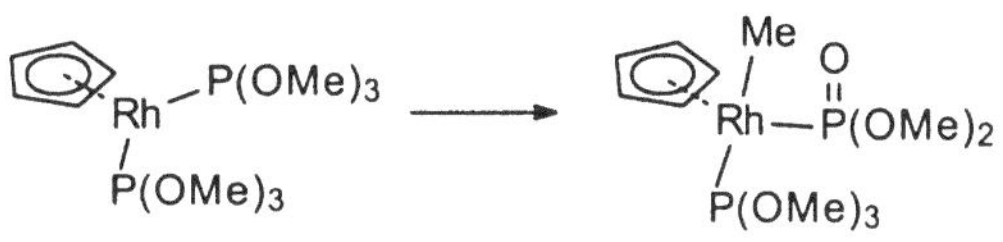

Figure 14. Metal catalyzed Arbuzov reaction

Thorough exclusion of moisture can easily prevent hydrolysis of phosphites in the laboratory reactor. In a continuous operation under severe conditions traces of water may form via aldol condensation of the aldehyde product. Weak and strong acids and strong bases catalyze the reaction. The reactivity for individual phosphites spans many orders of magnitude. When purifying phosphites over silica columns in the laboratory one usually adds some triethylamine to avoid hydrolysis on the column.

Bryant and coworkers have extensively studied decomposition of phosphites [40]. Stability involves thermal stability, hydrolysis, alcoholysis, and stability toward aldehydes. The precise structure has an enormous influence on the stability. Surprisingly it is the reactivity toward aldehydes that received most attention. Older literature [41] mentions several reactions between phosphites and aldehydes of which we show only two in Figure 15.

The addition of a phosphite to an aldehyde giving a phosphonate is the most important reaction [40]. The reaction is catalyzed by acid and since the product is acidic, the reaction is autocatalytic. Furthermore, acids catalyze hydrolysis and alcoholysis and therefore the remedy proposed is continuous removal of the phosphonate over a basic resin (Amberlyst A-21). The examples in the patents illustrate that very stable systems can be obtained when the acidic decomposition products are removed continuously.

The thermal decomposition of phosphites with aldehydes is illustrated in Figure 16 [40]. In this experiment the ligands are heated for 23 hours at 160 °C in the presence of pentanal. The figure shows the percentages of ligand that have decomposed as measured by NMR spectroscopy. This was done in the absence of a rhodium catalyst and so the numbers present a lower limit for the decomposition. One can see that minor effects in the ligand structure can have a tremendous influence on the rate of decomposition or rather the rate of reaction with aldehydes. While simple hydrolysis or alcoholysis might have been expected as the major source for ligand decomposition we see that a study to what is actually taking place is worthwhile as an attempt to stabilize the catalyst system.

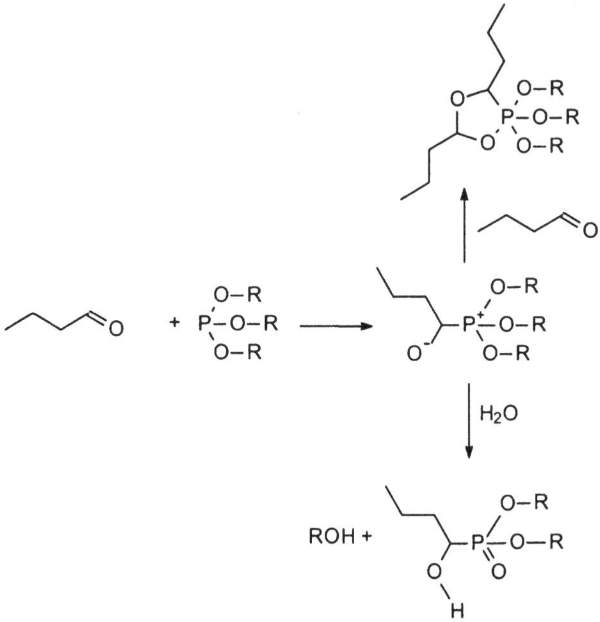

Figure 15. Reaction of phosphites with the aldehyde product

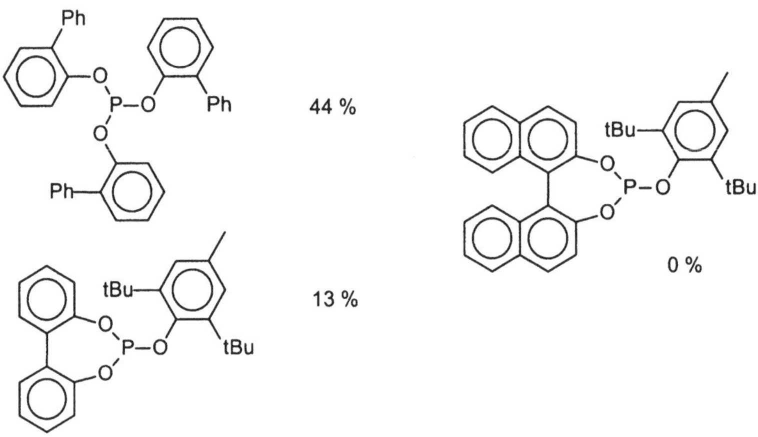

Figure 16. Stability of phosphites at 160 °C after 23 hours (see text for explanation; numbers give percentage of decomposition)

9.3.5 Formation of dormant sites

Dienes and alkynes are poisons for many alkene processes. In polyolefin manufacture they must be carefully removed as they deactivate the catalyst completely. Insertion of conjugated dienes is even much more rapid than insertion of ethene and propene. The resulting π-allyl species are inactive as catalysts.

In rhodium catalyzed hydroformylation the effect is less drastic and often remains unobserved, but surely diene impurities obscure the kinetics of alkene hydroformylation [42]. Because the effect is often only temporary we summarize it here under "dormant sites". Hydroformylation of conjugated alkadienes is much slower than that of alkenes, but also here alkadienes are more reactive than alkenes toward rhodium hydrides [43, 44]. Stable π-allyl complexes are formed that undergo very slowly insertion of carbon monoxide (Figure 17). The resting state of the catalyst will be a π-allyl species and less rhodium hydride is available for alkene hydroformylation. Thus, alkadienes must be thoroughly removed as described by Garland [45], especially in kinetic studies. It seems likely that 1,3- and 1,2-diene impurities in 1-alkenes will slow down, if not inhibit, the hydroformylation of alkenes.

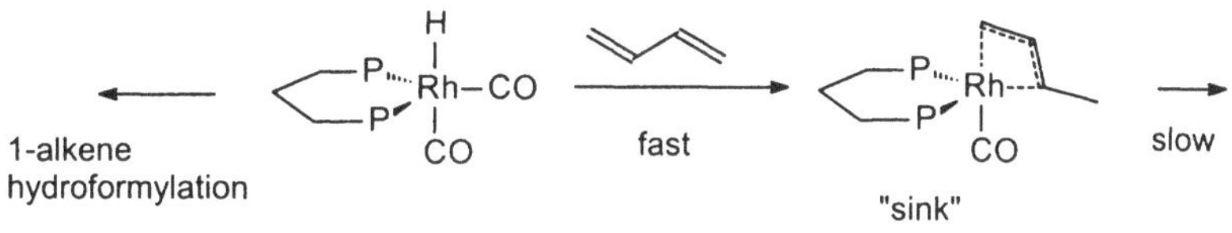

Figure 17. Reactions of diene to give π-allyl complexes

Dimer formation. Active, monomeric catalyst species may be involved the formation of inactive dimers. When this equilibrium is reversible it only leads to a reduction of the amount of catalyst available and it does not bring the catalysis to a full stop.

A well-known example is the formation of the so-called orange dimer from $HRh(PPh_3)_3CO$, already reported by Wilkinson [46]. Since then it has been reported by several authors [47 and references therein].

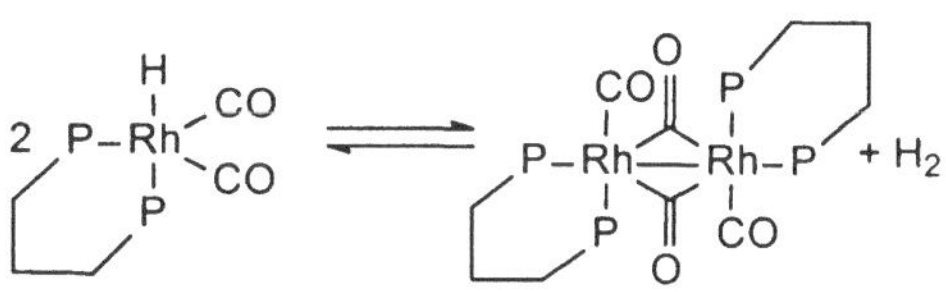

Figure 18. Dimer formation for hydroformylation catalysts

Since the hydroformylation reaction for most substrates shows a first order dependence in the concentration of rhodium hydride, the reaction becomes slower when considerable amounts of rhodium are tied up in dimers. This will occur at low pressures of hydrogen and high rhodium concentrations [47] (see Figure 18).

Work-up of hydroformylation solutions often leads to formation of dimers. It seems likely that in a liquid recycle of a continuous reactor rhodium will occur as such a dimeric species. Since the reaction with hydrogen is reversible it presents a means to recycle rhodium. Dimer formation is not restricted to phosphines as phosphites behave similarly [48].

When dimer formation becomes important one might attempt to destabilize the dimer relative to the monomer. For instance making the ligand very bulky might prevent dimer formation. Models show that the ligand size must be substantially increased to arrive at the desired effect. Another approach is so-called "site isolation" as was described by Grubbs [49]. His well-known example concerns a titanocene catalyst that is used as a hydrogenation catalyst. The intermediate titanium hydride is converted almost completely to a dimer rendering the catalyst with a low activity. Immobilization of the catalyst on a resin support prevents dimerization and an active catalyst is obtained.

Ligand metallation. In early transition metal polymerization catalysis often metalation of the ligand occurs leading to inactive catalysts. In late transition metal chemistry the same reactions occur, but now the complexes formed represent a dormant site and catalyst activity can often be restored. Work-up of rhodium-phosphite catalyst solutions after hydroformylation often shows partial formation of metallated species, especially when bulky phosphites are used [50]. Dihydrogen elimination or alkane elimination may lead to the metallated complex. The reaction is reversible for rhodium and thus the metallated species could function as a stabilized form of rhodium during a catalyst recycle. Many metallated phosphite complexes have been reported, but we mention only two, one for triphenyl phosphite and rhodium [51, 52] (see Figure 19) and one for a bulky phosphite and iridium [53].

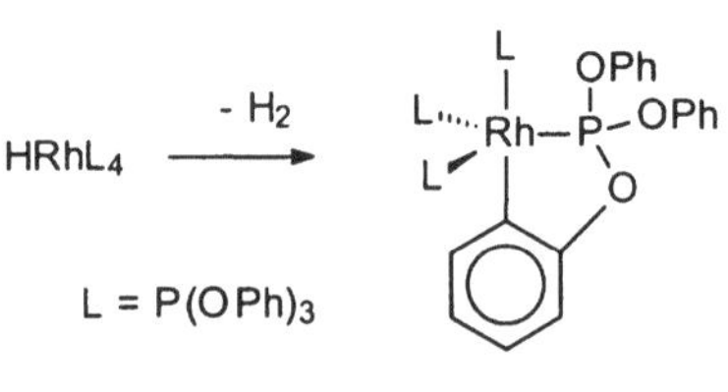

Figure 19. Metallation in rhodium phosphite complexes

9.4 Concluding remarks

Decomposition of organometallic catalysts under reaction conditions of catalytic processes lead to interesting organometallic chemistry. The study of this chemistry is often worthwhile either to find ways of circumventing the decomposition reactions or to find ways to work up the solutions such that the metal can be regenerated for further use and separated from the decomposed ligand fragments. From a stability viewpoint ligands having phosphorus as the donor atom are not very attractive, but for none of the metals active in hydroformylation (cobalt, platinum, palladium, rhodium) there seems to be an alternative other than carbon monoxide. Arsines lead to active catalysts, but it is not expected that they will form more stable complexes than phosphines. For several metal catalyzed reactions recently nitrogen ligands (Fe, Co, Pd) have been introduced which are often much more robust than phosphorus ligands. Rhodium hydride carbonyls do not seem to combine well with nitrogen ligands.

References

1 (a) Mieczynska, E.; Trzeciak, A. M.; Ziólkowski, J. J. *J. Mol. Catal.* **1993**, *80*, 189. (b) Buhling, A.; Kamer, P. C. J.; van Leeuwen, P. W. N. M. *J. Mol. Catal. A: Chemical*, **1995**, *98*, 69.
2 Coolen, H. K. A. C.; van Leeuwen, P. W. N. M.; Nolte, R. J. M. *J. Org. Chem.* **1996**, *61*, 4739.
3 Onada, T. *ChemTech*, September **1993**, 34.
4 Lazzaroni, R.; Pertici, P.; Bertozzi, S.; Fabrizi, G. *J. Mol. Cat.* **1990**, *58*, 75.
5 Vidal, J. L.; Walker, W. E. *Inorg. Chem.* **1981**, *20*, 249.
6 Bellet, E. M.; Casida, J. E. *Science*, December **1973**, *182*, 1135.
7 Garrou, P. E. *Chem. Rev.* **1985**, *85*, 171.
8 Garrou, P. E.; Dubois, R. A.; Jung, C. W. *ChemTech*, February **1985**, 123.
9 Cai, D.; Payack, J. F.; Bender, D. R.; Hughes, D. I.; Verhoeven, T. R.; Reider, P. J. *J. Org. Chem.* **1994**, *51*, 629.
10 Chini, P.; Martinengo, S.; Garlaschelli, G. *J. Chem. Soc. Chem. Commun.* **1972**, 709.
11 Dubois, R. A.; Garrou, P. E. *Organometallics*, **1986**, *5*, 466.
12 Harley, A. D.; Guskey, G. J.; Geoffroy, G. L. *Organometallics*, **1983**, *2*, 53.
13 Billig, E.; Jamerson, J. D.; Pruett, R. L. *J. Organomet. Chem.* **1980** *192*, C49.
14 Miller, D. J.; Bryant, D. R.; Billig, E.; Shaw, B. L. *U.S. Pat* 4,929,767 (to Union Carbide Chemicals and Plastics Co.) **1990**; *Chem. Abstr.* **1991**, *113*, 85496.
15 Herrmann W. A.; Kohlpaintner, C. W. *Angew. Chem. Int. Ed. Engl.* **1993**, *32*, 1524.
16 Goel, A. B. *Inorg. Chim. Acta*, **1984**, *84*, L25.
17 Sakakura, T. *J. Organometal. Chem.* **1984**, *267*, 171.
18 Abatjoglou A. G.; Bryant, D. R. *Organometallics* **1984**, *3*, 932.
19 Kong, K-C.; Cheng, C-H. *J. Amer. Chem. Soc.* **1991**, *113*, 6313.
20 Sisak, A.; Ungváry, F.; Kiss, G. *J. Mol. Catal.* **1983**, *18*, 223.

21 van Leeuwen, P. W. N. M.; Roobeek, C. F.; Frijns, J. H. G.; Orpen, A. G. *Organometallics,* **1990**, *9*, 1211.
22 Kikukawa, K.; Takagi, M.; Matsuda, T. *Bull. Chem. Soc. Japan,* **1979**, *52*, 1493.
23 Bouaoud, S-E.; Braunstein, P.; Grandjean, D. ; Matt, D. *Inorg. Chem.* **1986**, *25*, 3765.
24 Alcock, N. W.; Bergamini, P.; Kemp, T. J.; Pringle, P. G. *J. Chem. Soc. Chem. Commun.* **1987**, 235.
25 van Leeuwen, P. W. N. M.; Roobeek, C. F.; Orpen, A. G. *Organometallics,* **1990**, *9*, 2179.
26 Kaneda, K.; Sano, K.; Teranishi, S.; *Chem. Lett.* **1979**, 82.
27 Chan, A. S. C.; Caroll, W. E.; Willis, D. E. *J. Mol. Catal.* **1983**, *19,* 377.
28 Morita, D. K.; Stille, J. K.; Norton, J. R. *J. Am. Chem. Soc.* **1995,** *117*, 8576.
29 Segelstein, B. E.; Butler, T. W.; Chenard, B. L. *J. Org. Chem.* **1995**, *60*, 12.
30 Goodson, F. E.; Wallow, T. I.; Novak, B. M. *J. Am. Chem. Soc.* **1997**, *119*, 12441.
31 Pruett, R. L.; Smith, J. A. *J. Org. Chem.* **1969**, *34*, 327.
32 van Leeuwen P. W. N. M.; Roobeek, C. F. *J. Organometal. Chem.* **1983**, *258*, 343. *Brit. Pat.* 2,068,377, *US Pat.* 4,467,116 (to Shell Oil); *Chem. Abstr.* **1984**, *101*, 191142.
33 Billig, E.; Abatjoglou, A. G.; Bryant, D. R.; Murray, R. E.; Maher, J. M. *U.S. Pat.* 4,599,206 (to Union Carbide Corp.) **1986**; *Chem. Abstr.* **1989**, *109*, 233177.
34 Billig, E.; Abatjoglou, A. G.; Bryant, D. R. (to Union Carbide Corporation) U.S. Pat. 4,769,498, U.S. Pat. 4,668,651; U.S. Pat. 4,748,261, **1987**; *Chem. Abstr.* **1987**, *107*, 7392.
35 Burke, P. M.; Garner, J. M.; Tam, W.; Kreutzer, K. A.; Teunissen, A. J. J. MWO 97/33854, **1997**, (to DSM/Du Pont); *Chem. Abstr.* **1997**, *127*, 294939.
36 Sato, K.; Kawaragi, Y.; Takai, M.; Ookoshi, T. *US Pat.* 5,235,113, EP 518241 (to Mitsubishi); *Chem. Abstr.* **1993**, *118*, 191183.
37 Röper, M.; Lorz, P. M.; Koeffer, D. *Ger. Offen. DE* 4,204,808 (to BASF); *Chem. Abstr.* **1994**, *120*, 133862.
38 Brill, T. B.; Landon, S. J. *Chem. Rev.* **1984**, *84*, 577.
39 Werener, H.; Feser, R. *Z. Anorg. Allg. Chem.* **1979**, *458*, 301.
40 Billig, E.; Abatjoglou, A. G.; Bryant, D. R.; Murray, R. E.; Maher, J. M. (to Union Carbide Corporation) U.S. Pat. 4,717,775, **1988**; *Chem. Abstr.* **1989**, *109,* 233177.
41 Ramirez, F.; Bhatia, S. B.; Smith, C. P. *Tetrahedron,* **1967**, *23*, 2067.
42 van den Beuken, E.; de Lange, W. G. J.; van Leeuwen, P. W. N. M.; Veldman, N.; Spek, A. L.; Feringa, B. L. *J. Chem. Soc. Dalton Trans.* **1996**, 3561.
43 van Leeuwen, P. W. N. M.; Roobeek, C. F. *J. Mol. Catal.* **1985**, *31*, 345.
44 van Rooy, A.; de Bruijn, J. N. H.; Roobeek, C. F.; Kamer, P. C. J.; van Leeuwen, P. W. N. M. *J. Organomet. Chem.*, **1996**, *507*, 69.
45 Fyhr, C.; Garland, M. *Organometallics,* **1993**, *12*, 1753.
46 Evans, D.; Yagupsky, G.; Wilkinson, G. *J. Chem. Soc. (A)* **1968**, 2660.
47 Castellanos-Páez, A.; Castillón, S.; Claver, C.; van Leeuwen, P. W. N. M.; de Lange, W. G. J. *Organometallics,* **1998**, *17*, 2543.
48 Buisman, Thesis, Amsterdam, 1995.

49 Bonds, W. D.; Brubaker, C. H.; Chandrasekaran, E. S.; C. Gibsons, Grubbs, R. H.; Kroll, L. C. *J. Am. Chem. Soc.* **1975**, *97,* 2128.

50 van Rooy, A.; Buisman, G. J. H.; Roobeek, C. F.; Sablong, R.; van Leeuwen, P. W. N. M. unpublished results.

51 Parshall, G. W.; Knoth, W. H.; Schunn, R. A. *J. Am. Chem. Soc.* **1969**, *91*, 4990.

52 Coolen, H. K. A. C.; van Leeuwen, P. W. N. M.; Nolte, R. J. M. *J. Organomet. Chem.* **1995**, *496*, 159.

53 Fernandez, E.; Ruiz, A.; Claver, C.; Castillon, S.; Chaloner, P. A.; Hitchcock, P. B. *Inorg. Chem. Comm.* **1999**, *2*, 21.

Chapter 10

Novel developments in hydroformylation

Joost N. H. Reek, Paul C. J. Kamer, and Piet W. N. M. van Leeuwen

Institute of Molecular Chemistry, University of Amsterdam, Nieuwe Achtergracht 166, 1018 WV Amsterdam, The Netherlands.

10.1 Introduction

In this chapter we will describe several novel developments in catalytic hydroformylation that might be of importance for future applications. This involves novel catalysts that are highly active and selective, but more importantly new systems that are sustainable. These processes should be selective and efficient with a very small waste stream. A very good example of an environmentally benign process is the aqueous biphasic system developed by Ruhrchemie/Rhône-Poulenc [1]. The water-soluble ligand used (TPPTS: triphenylphosphine sulfonate) affords, in combination with a rhodium precursor, an active catalyst for the hydroformylation of propene that can be separated very easily from the organic phase and recycled (Chapter 7). This aqueous biphasic system can be applied only to substrates that are slightly water-soluble. Therefore, several alternatives for higher alkenes have been investigated in the past decade. In this chapter we first will discuss an interesting novel bimetallic active and selective catalyst system that does not operate via the common mononuclear Rh(I) hydride precursor as described in chapter 1. We will proceed with the description of systems that show novel concepts in catalyst/product separation and end with recent developments in supramolecular catalysis.

10.2 New bimetallic catalysts

It has been shown that a number of active sites of enzymes contain two metal ions that give high activity via a cooperative bimetallic mechanism

P.W.N.M. van Leeuwen and C. Claver (eds.), Rhodium Catalyzed Hydroformylation, 253–279.

[2]. Synthetic binuclear complexes as catalysts, aiming at high activity, have been subjects of intensive research [3]. However, the number of novel bimetallic catalysts that are active and actually operating via a bimetallic mechanism is extremely scarce. Stanley and co-workers reported a bimetallic rhodium complex that is an active and selective hydroformylation catalyst that was proposed to operate via a cooperative mechanism.

It has been shown that the tetraphosphine ligands **1** can simultaneously complex two metal centers [4]. From an X-ray structure determination of $Rh_2(nbd)_2$**(rac-1)**, it is clear that the two rhodium atoms are complexed in a square planar geometry forming 5-membered rings with the ligand and a metal-metal distance of 5.5 Å. The catalytic performance was compared with mononuclear complexes of ligands **2-5**. The activities of the rhodium norbornadiene (nbd) complexes of ligands **1-7** have been tested in the hydroformylation of 1-hexene [4] (Table 1). The results clearly show that $Rh_2(nbd)_2$**(rac-1)** forms a catalyst superior to all the other systems. Both the activity and the selectivity for the linear products are much higher. The large difference in activity between the bimetallic $Rh_2(nbd)_2$**(rac-1)** and monometallic $Rh_2(nbd)_2$**(2-5)** strongly suggests that the former complex acts via a different mechanism.

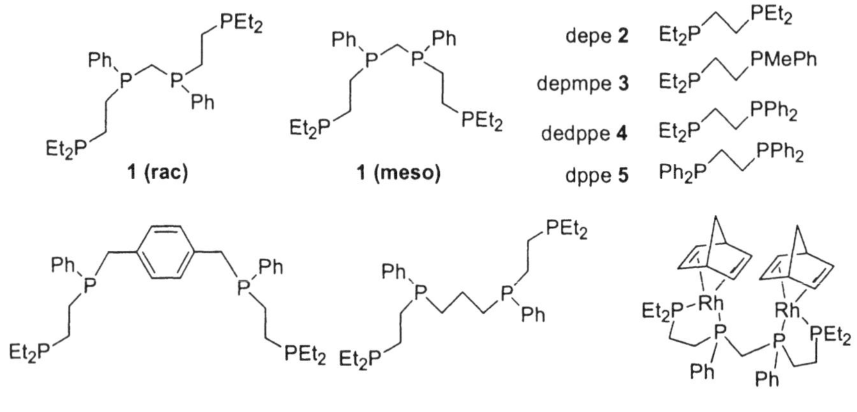

Stanley proposed a mechanism for the bimetallic hydroformylation that slightly deviates from the mechanism suggested by Heck and Breslow for the cobalt catalyzed reaction [5]. Spectroscopic evidence was found for the existence of unusual dicationic bimetallic rhodium(II) complexes, supporting the proposed mechanism [6] (see figure 1). The bimetallic cooperativity was thought to originate from the intramolecular hydride transfer from the rhodium hydride to the rhodium acyl species (in intermediate f), enhancing the elimination of the aldehyde from the acyl intermediate.

Table 1. Hydroformylation of 1-hexene using ligands **1-7**[a]

Ligand	Initial TOF ($mol.mol^{-1}.h^{-1}$)	Aldehydes (%)	l:b ratio	Isomerization (%)	Hydrogenation (%)
rac-1	636	85	27.5	8	3
meso-1	54	16	14	4	2
2	1	0.9	2	16	5
3	1.5	1.7	3	50	15
4	1.2	1.1	3	42	14
5	2.4	1.2	2.6	51	14
6	5.8	1.7	2.4	59	16
7	0	0	-	60	17

[a] Catalyst: $Rh_2(nbd)_2(BF_4)_2$ and ligand. Reaction conditions: 6.2 bar CO/H_2 (1:1); 1mM catalyst in acetone, 1M hexene, reaction performed at 90 °C for three hours.

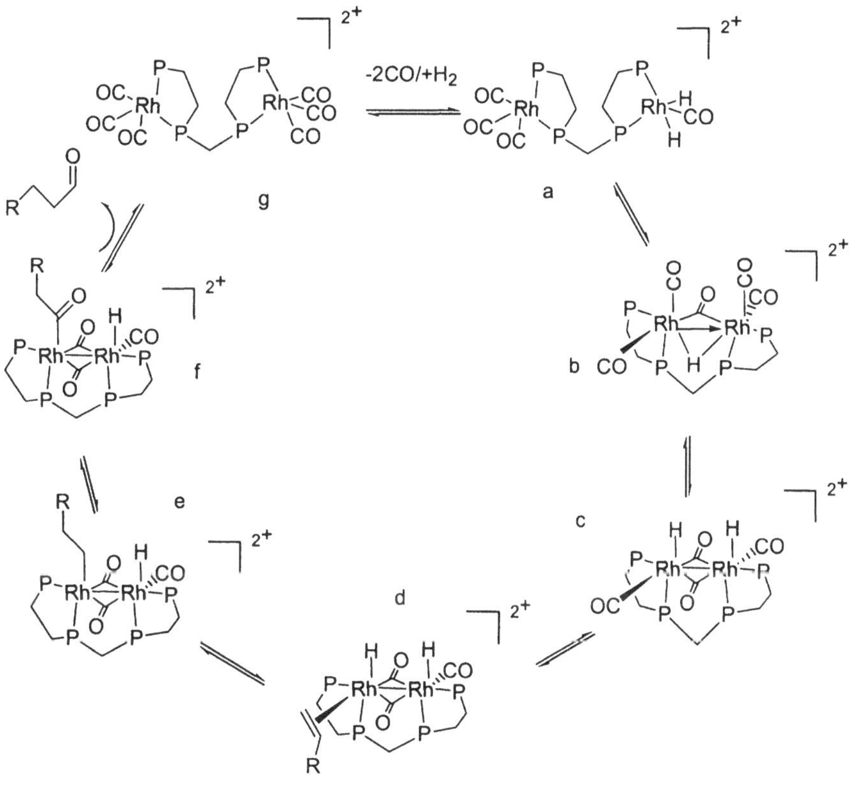

Figure 1. The proposed mechanism for the bimetallic hydroformylation using ligand **1**

The bridge between the two metal centers is too large in ligands **6** and **7**, which prevents the cooperation between the metal centers. Indeed activities and selectivities were found that were similar to the monometallic species. Another important indication supporting this bimetallic cooperativity was an observed alkene inhibition of the catalytic reaction. At higher concentration of the alkene a bis-acyl species can be formed, preventing the fast intramolecular hydride transfer.

10.3 Novel developments in catalyst separation

The great potential of homogeneous catalysis in the selective preparation of chemicals has been widely recognized and the number of industrial applications is growing enormously. If a general solution would be available for the often cumbersome separation of homogeneous catalysts from the product, the number of commercial applications would even further increase. A general solution allows the development of "general purpose operating units", which is of special interest since it can lower the costs of chemical processes. With this in mind a lot of investigators focussed on novel methods of separating and re-using homogeneous catalysts. As yet, there is not one general solution for this problem, but several interesting concepts have been reported. In this section of the most representative concepts will be discussed. As we will see, various attempts are based on extending the two-phase Ruhrchemie/Rhône Poulenc process in such a way that it may become applicable to larger organic substrates.

10.3.1 Micellar catalysis

One simple way to increase the solubility of organic substrates in aqueous solution is the addition of co-solvents to an aqueous two-phase system. This increases the reaction rate of the Rh/TPPTS catalyst, but at the same time the selectivity drops [7]. Furthermore, it can result in troublesome phase separation and leaching of the co-solvent into the product phase [7]. Instead of the addition of co-solvents triphenylphosphine can be added to the two-phase Rh/TPPTS system. The local concentration of the active rhodium complex at the interface will be increased, resulting in a rate enhancement in the hydroformylation of 1-octene by a factor of 10-50 [8]. The active species at the interface contains both TPPTS and triphenylphosphine ligands. Test reactions performed in methanol confirmed that this increase in reaction rate was due to promotion of interfacial catalysis. However, after recycling the catalyst Rh/TPPTS was found in the aqueous phase whereas the PPh_3 was in the organic phase and in the subsequent reaction with freshly added organic phase the activity of the catalyst is similar to that of the biphasic reaction in the absence of PPh_3.

Another way to improve the solubility of organic substrates in the aqueous phase is the addition of amphiphiles (also called surfactants or tensides) that form aggregates [9]. Above the critical aggregation concentration (CAC) amphiphiles form organized structures in aqueous solution in such way that the polar groups are pointing to the water phase and the organic parts of the molecules stick together (Figure 2). Dependent on the relative sizes of the head group and the organic part of the molecule one observes the formation of micelles, vesicles or bilayers. Organic substrates will dissolve in the organic part of the aggregates. Thus the addition of amphiphiles will lead to an increased solubility of the substrate, but a water-soluble catalyst (e.g. Rh/TPPTS) will be in a different phase. The combination of organic substrates with a non water-soluble catalyst induces high local concentrations of both components in the aggregate, which can result in impressive rate enhancements [10]. This type of concentration or medium effects introduced by micelles has been extensively studied [9]. The problem of catalyst-product separation still subsists since both the catalyst and the substrate will be extracted into the organic phase. Only recently amphiphilic ligands have been used in catalysis.

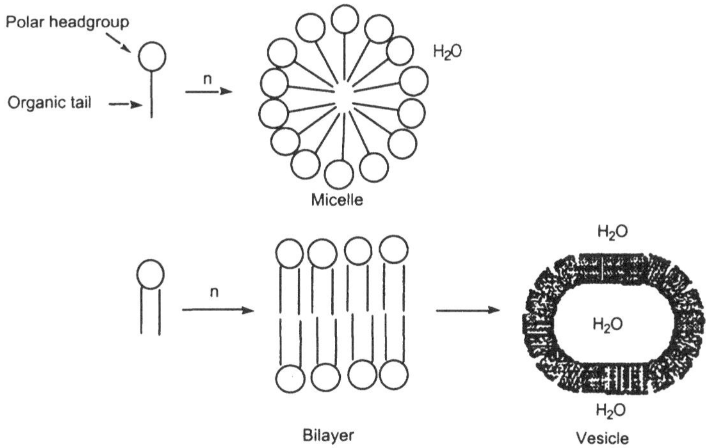

Figure 2. Schematic presentation of the aggregation of amphiphiles into micelles, bilayers and vesicles

Ligands that have an amphiphilic character and form the solubilizing aggregate are themselves very interesting, since they can combine the advantage of two-phase catalysis, i.e. easy catalyst separation, with a reasonable solubilizing effect on organic substrates. Fell [11] and Hanson [12, 13] reported the first compounds that combined these properties. Fell prepared a series of zwitterionic phosphines with different organic chain lengths (**8**) (see Figure 3) and he found an optimal catalytic performance using the octane derivative (n=7) at 75 bar pressure (CO/H_2=1:1). Hanson prepared the sulfonated ligands **9**, and showed with dynamic light

Table 2. *Two-phase hydroformylation of 1-octene comparing **9c**, **10b** and TPPTS*

Ligand	L/Rh	TOF ($mol.mol^{-1}.h^{-1}$)	Yield nonanals (%)	l:b
9c	2	190 [a]	38	2.4
9c	5	400 [a]	80	4.5
9c	9	435 [a]	87	8.1
TPPTS	2	70 [a]	14	2.2
TPPTS	5	145 [a]	29	3.1
TPPTS	9	220 [a]	44	3.5
10b	2	63 [b]	63	1.8
10b	5	76 [b]	76	3.7
10b	9	83 [b]	83	10.1
TPPTS	2	36 [b]	36	2.1
TPPTS	5	57 [b]	57	3.0
TPPTS	9	76 [b]	76	3.1

Reaction conditions: T=120°C, initial pressure (at 25°C) 14.3 bar (CO/H_2=1:1), [Rh]=0.0049 M, Rh/1-octene=500, $MeOH/H_2O$=1:1. [a]TOF after 1 h. [b]TOF after 5 h.

scattering that at room temperature these amphiphiles indeed formed micellar aggregates in aqueous solution.

The catalytic performance of the catalysts based on these amphiphilic ligands was compared with that of TPPTS (see table 2). The reactions were performed in aqueous methanol. High activity was observed for **9c** compared to TPPTS. In all cases the ligand concentration has a large effect on the selectivity and activity of the catalyst. The highest selectivity towards the linear product was found for the amphiphilic BISBI analogue **10b** (l:b=10). An effect of the co-solvent methanol on the selectivity of the reaction was observed for TPPTS, which has been attributed to an increased intra-complex ionic repulsion between the sulfonate groups. Using ligand **9c** these groups are further apart and as a consequence this repulsion is smaller. This results in much better selectivities towards the linear product compared to TPPTS. Under these conditions the BISBI analogue **10b** only performed slightly better than TPPTS and selectivity toward the linear product is lower than that observed for the homogeneous reaction. Catalysis in aqueous solution using these systems was not reported.

Amphiphilic diphosphines based on the Xanthene backbone have been synthesized (**11**) [14]. The large natural bite angle of ligands based on the Xanthene backbone has a beneficial influence on the regioselectivity of the homogeneous hydroformylation catalyst. Furthermore, the chelate effect results in a strong metal-ligand bond, which is of importance in recycling experiments. Compounds **11b** and **11c** form vesicles in aqueous solution. According to electron microscopy and light scattering experiments the size of these vesicles was around 100 nm. Furthermore, the aggregates were stable at temperatures up to 90 °C and they were able to dissolve organic

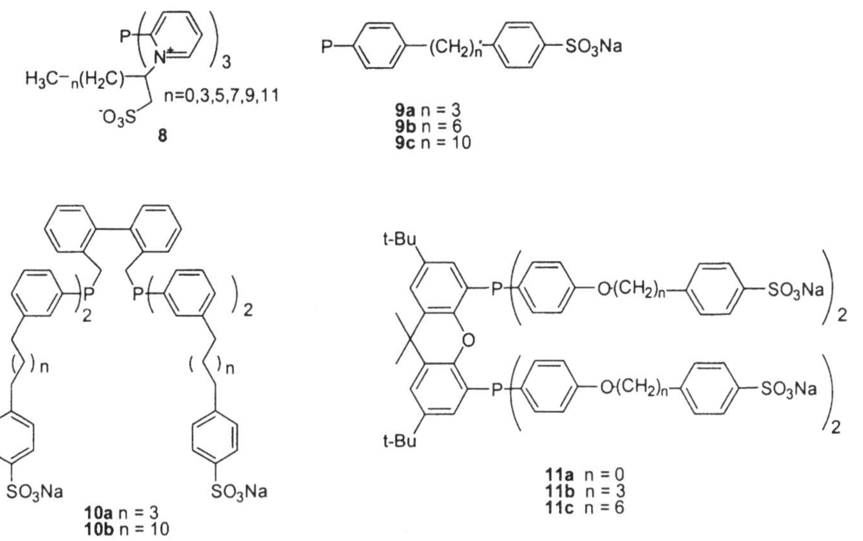

Figure 3. Some of the amphiphilic phosphine ligands that have been reported

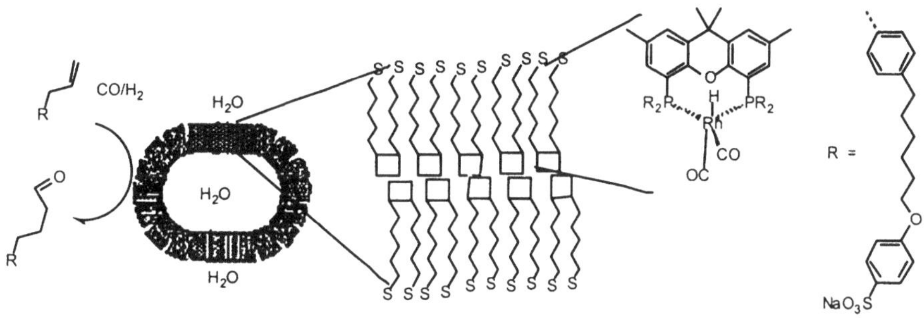

Figure 4. Schematic representation of catalysis using amphiphilic ligands that form vesicles

substrates in the organic phase of the aggregate.

These are obviously important properties for the application of these ligands in the hydroformylation (at elevated temperatures) of organic substrates such as 1-octene in aqueous solution (see Figure 4). Importantly, the addition of methanol resulted in the disruption of the vesicles. Hydroformylation reactions were therefore performed in aqueous solution, at 70 °C and 80 °C using 15 bar CO/H_2 (1:1). The rate enhancement observed for the aggregating systems was a factor of 14 and the high selectivity for the linear product (l:b = 50) usually observed for these type of catalysts was retained. Test reactions performed in aqueous methanol solution showed that the enhancement was only a factor of 3. Recycling experiments showed that the catalyst/product separation was good; the catalyst performance was retained completely in 4 consecutive runs and no rhodium leaching into the organic phase was detected. Surprisingly, no emulsions were formed on using **11b** and **11c**, allowing easy and fast separations.

10.3.2 Supported aqueous phase catalysis (SAPC)

Davis and Hanson developed a new concept of immobilizing homogeneous catalysts denoted as supported aqueous phase catalysts (SAPC) [15]. They reasoned that in aqueous biphasic catalysis the reaction mainly takes place at the interface. In order to increase this interface they used a high-surface-area hydrophilic support (figure 5). These materials have a thin film of water adhered to the surface, in which the water-soluble catalyst is dissolved. The reaction, performed in an organic solvent such as toluene, occurs at the water-organic interface. The supported catalyst has a

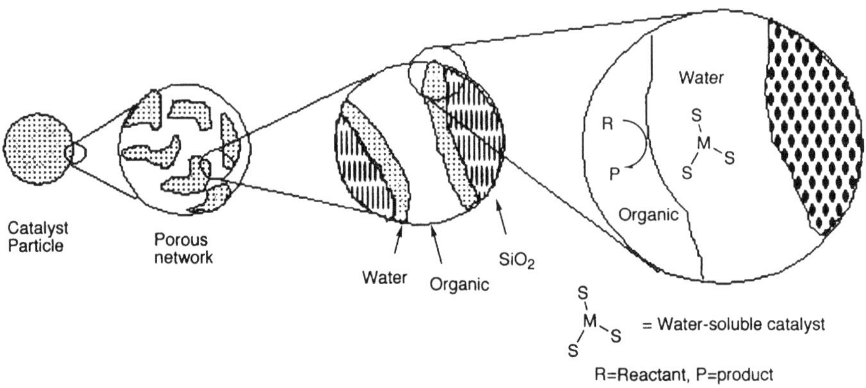

Figure 5. Schematic representation of the concept of supported aqueous phase catalysis

very large interfacial area, which results in very efficient catalysis for organic substrates. Furthermore, the catalyst stays completely on the support. The important issues are the generality of the concept, the robustness of the catalyst system and the influence of the thickness of the water layer [16]. This water layer has an enormous impact on the catalytic activity (figure 6). If this layer is too thin the activity of the catalyst is much lower due to a decrease of the catalyst mobility. The catalyst is bound to the silica resulting in a heterogeneous system.

This was verified by ^{31}P NMR spin relaxation measurements, which show that the spin relaxation declines with increasing water content [17]. It was observed in the hydroformylation of 1-heptene using a $HRh(CO)(TPPTS)_3$-SAPC that the TOF increases with a factor as high as 100 when going from 2.9 to 9 wt. % water (table 3).

Table 3. Hydroformylation performance of SAPC compared with homogeneous and biphasic systems.

Catalyst system	Substrate	T (°C)	TOF ($mol.mol^{-1}.h^{-1}$)
SAPC (2.9 wt.% H_2O)[a]	1-heptene	75	0.75
SAPC (9 wt.% H_2O)[a]	1-heptene	75	72
Homogeneous[b]	1-heptene	75	288
SAPC[c]	Mixture[h]	100	432[e]/432[f]/396[g]
Homogeneous[d]	Mixture[h]	100	1656[e]/1800[f]/1800[g]
Biphasic[c]	Mixture[h]	125	17[e]/5[f]/1[g]

[a] $HRhCO(TPPTS)_3$, P=7 bar (H_2/CO=1:1) [16d]. [b] $HRhCO(PPh_3)_3$ in toluene, P=7 bar (H_2/CO=1:1) [16b]. [c] $HRhCO(TPPTS)_3$, P=51 bar (H_2/CO=1:1) [18]. [d] $HRhCO(PPh_3)_3$ in hexane, P= 51 bar (H_2/CO=1:1) [18]. [h] The substrate was a 1:1:1 mixture of 1-hexene, 1-octene and 1-decene. [e] Heptanals. [f] Nonanals. [g] Undecanals.

At higher water contents of the support the water layer becomes too thick and the substrate has to diffuse into the water layer, or the catalyst has to diffuse to the interface. The result is a decrease in catalyst-product contact time leading to lower activities. This sensitivity towards water is a drawback of this otherwise attractive concept. Horváth performed experiments using substrates with different solubilities in water and showed that, under optimal conditions, this solubility did not influence the activity [18]. Furthermore, he performed a hydroformylation reaction in a continuous system and even under reaction conditions no leaching of rhodium complex was detected. The water obviously leaches if the SAPC is used in a continuous flow system, which in a practical application should be compensated for by using water-saturated organic solvents.

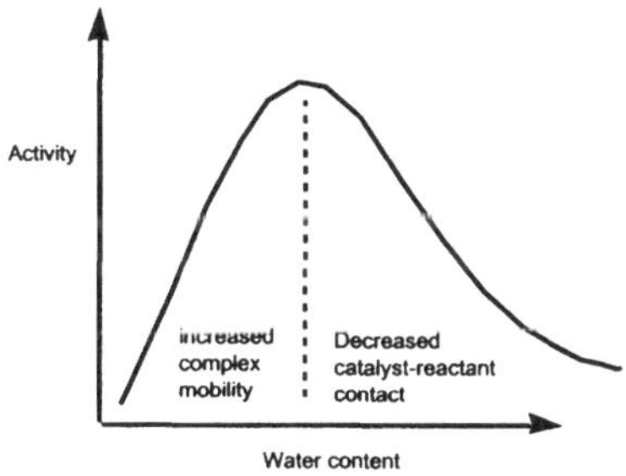

Figure 6. The influence of the water contents of the hydrophilic support on the relative catalytic activity

A water-soluble chelating diphosphine ligand (**12**) with a wide natural bite angle has been prepared [19a]. These ligands are known to form very selective catalysts for the hydroformylation of 1-alkenes, yielding mainly linear aldehyde as the product. These ligands were also studied as supported aqueous phase catalysts and it was shown that these ligands performed well as SAPC; they were much more selective than the SAPC known so far [19b]. Recycling experiments showed that these catalysts retained their activity and selectivity in at least ten consecutive runs, whereas under similar conditions the TPPTS based catalyst showed a reduced performance in the fourth run.

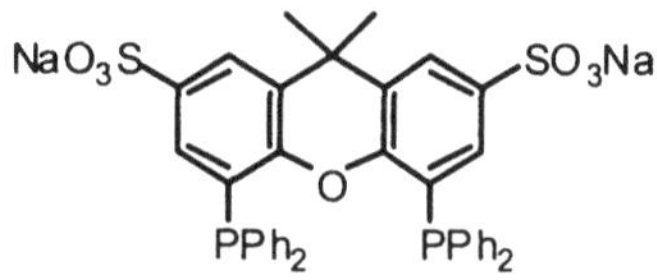

12

Mortreux compared the activity of the SAPC catalysts with that of the homogeneous analogue in the hydroformylation of methyl acrylate [20]. They observed an activity for the SAPC that was strongly dependent on the amount of water present in the system. More remarkably, the optimized activity of the SAPC was higher than that of the homogeneous systems. This effect was ascribed to the polar interactions between the substrate and the silica support. This effect was not observed for nonpolar substrates such as propene, which supported the hypothesis.

In order to show the versatility of the method Davis extended the concept to other hydrophilic liquids such as ethylene glycol and glycerol [21]. The reactions then take place at the hydrophilic-hydrophobic liquid interface. In this specific example the supported-phase concept was used for the asymmetric reduction using a ruthenium catalyst. The obtained enantioselectivity was higher than that of the SAPC, which was ascribed to the decrease in hydrolysis of chloro-ligand from the ruthenium complex. Naughton et al. used this supported homogeneous film catalysis concept for the hydroformylation reaction using TPPTS as the ligand [22]. The low molecular weight polyethylene glycol resulted in the formation of an effective film.

10.3.3 Hydroformylation in supercritical fluids

Supercritical fluids, especially supercritical carbon dioxide (scCO_2; T_c = 31°C, p_c = 73.75 bar, d_c = 0.468 g/mL), are receiving considerable interest as alternative reaction media [23]. scCO_2 is a non-toxic, cheap solvent, but more importantly it has "gas like" properties that may provide advantages compared to convential solvents. It is highly miscible with reactant gases, there is no liquid/gas phase boundary, it has a high compressibility, very low viscosity, and thus high diffusivity. Several homogeneously catalyzed reactions have been performed in supercritical fluids. The low solubility of the catalyst can complicate this approach. Furthermore, the high investment and operating costs caused by the relatively high pressures required for supercritical fluids is a serious disadvantage when it comes to commercial application for low-cost aldehydes.

Rathke was the first to perform a hydroformylation reaction in $scCO_2$ [24]. They studied the hydroformylation of propene using $Co_2(CO)_8$ as the catalyst. They found that the selectivity toward the linear aldehyde is slightly higher (88%) compared to that found in benzene (83%). Leitner prepared perfluoralkyl-substituted aryl-phosphines (**13-15**) in order to obtain $scCO_2$ soluble ligands (figure 7) [25]. Rhodium complexes of these ligands prepared with [(cod)Rh(hfacac)] were shown to have a much higher solubility in $scCO_2$ (a factor 150 compared to non-fluorinated complexes). At higher densities (ρ = 0.75 mL^{-1}) a complex concentration of 4.4 mM could be reached. This high solubility allowed the hydroformylation of 1-octene using rhodium phosphine complexes.

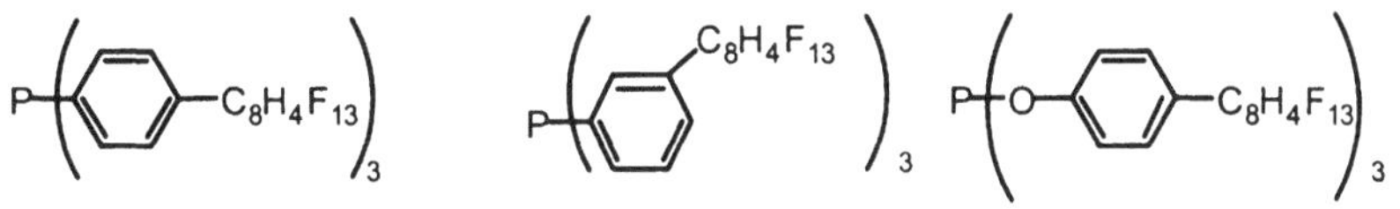

13 **14** **15**

Figure 7. The perfluoroalkyl-substituted aryl phosphines/phosphite prepared for the hydroformylation in $scCO_2$

Table 4. Hydroformylation of 1-octene in $scCO_2$.

Ligand	[Rh] (mM)	L/Rh	*Psyn* (bar)	*Ptot* (bar)	TOF ($mol.mol^{-1}.h^{-1}$)	Yield[c] (%)	l:b
-	0.11	-	20	200	122	92	1.6
13	0.13	10	20	200	97	99	4.5
14	0.13	10	20	200	108	93	5.6
15	0.13	10	20	200	23[b]	95	8.5
14[d]	0.66	4	45	210	14	97	3.2
14[e]	1.6	6	60	235	11	90	4.6
14[d]	0.64	13	45	210	2	13	5.5

[a] Conditions: [1-octene]=280 mmol, $V_{reactor}$=225 mL, *d*=0.62 g mL^{-1}, precursor [cod]Rh(hfaca)], T=65°C. [b] The reaction time was 88 h compared to approximately 20 for all other cases. [c] The total conversion to the aldehyde. [d] [1-octene]=192 mM, $V_{reactor}$ =25 mL. [e] [1-octene]=384 mM, $V_{reactor}$ =25 mL.

Initially they studied the activity of solely the catalyst precursor, [(cod)Rh(hfacac)], towards the hydroformylation of 1-octene, and found that it was remarkably active in $scCO_2$ [28]. The activity increased 5-fold compared to the reaction conducted under similar conditions in toluene. Moreover, the presence of branched aldehydes and virtually no internal alkenes indicate that in $scCO_2$ the activity towards internal bonds is high.

Below the critical temperature of CO_2, thus in the liquid state, almost no activity was observed, presumably due to the low solubility of the precursor in liquid.

On using phosphine ligands the selectivity of the reaction increased, up to 99% aldehyde was formed. The selectivity for the linear product increased upon changing the P/Rh ratio from 4 to 10; the linear to branched (l:b) ratio increased from 3.2 to 5.6. A further increase in P/Rh ratio did not improve the selectivity further, but a dramatic drop in activity was observed. Similar effects are observed for reactions performed in organic solvents, indicating that the reactive species is indeed the phosphine-rhodium complex. Interestingly, the phosphite ligand **15** also gives a high chemo- and regio-selectivity, whereas in organic solvents under similar conditions this type of catalysts gives a lot of isomerized products. β-H elimination of the rhodium alkyl complex results in the internal alkenes, whereas the CO insertion leads to the aldehyde (see figure 6, chapter 4). The high CO concentration in $scCO_2$ favors the CO insertion, which results in higher selectivities compared to the analogous reactions performed in organic media.

The catalyst-product separation was also studied. By changing the pressure and temperature after the reaction they created a phase separation from which the aldehydes could be collected as a colorless liquid (containing less than 1 ppm rhodium). The catalysts of ligands **13-15** were used in 5 consecutive runs without significant changes in the catalytic performance.

The concept can be extended to asymmetric hydroformylation of styrene in $scCO_2$ using BINAPHOS as the chiral ligand. The initial experiments using this ligand clearly show that the active catalyst in $scCO_2$ does not involve the chiral ligand [27]. The low solubility of the ligand was the major problem, since the catalyst precursor in absence of ligand also gave a high activity, but of course no enantioselectivity. Therefore a BINAPHOS

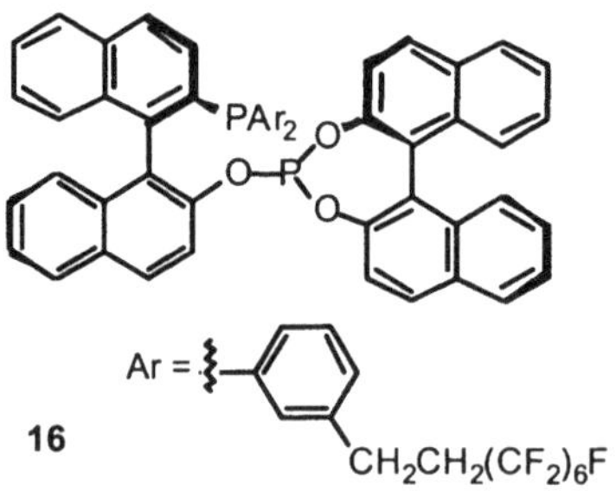

16

analogue was synthesized that contained perfluoroalkyl groups (**16**) [28]. Using this ligand very high enantioselectivities and regioselectivities were obtained in the hydroformylation of styrene. These results were independent of the phase behavior of the reaction mixture; the good performance was similar using temperatures above and below the T_c of CO_2. A slightly higher regioselectivity was also observed on using **16** compared to BINAPHOS in the homogeneous phase, which was ascribed partly to the ligand and partly to the solvent.

Erkey studied the hydroformylation of 1-octene in $scCO_2$ using several different fluoralkyl and fluoralkoxy functionalized aryl phosphines [29]. They found an increasing activity of the rhodium catalyst with decreasing basicity of the phosphine ($[3,5\text{-}(CF_3)_2C_6H_3]_3P$ > $[4\text{-}CF_3C_6H_4]_3P$ and $[3\text{-}CF_3C_6H_4]_3P$ > $[4\text{-}CF_3OC_6H_4]_3P$). The basicity was measured by the carbonyl stretching frequencies of the complexes.

The low solubility of the catalyst in $scCO_2$ is not necessarily a problem. Cole-Hamilton used rhodium trialkylphosphine complexes for the hydroformylation of 1-hexene in $scCO_2$ [30]. The rhodium concentration was about 6.5 mM with a P:Rh ratio of 6. The catalyst performed similarly in $scCO_2$ compared to experiments performed in toluene. The reaction rate decreased with increasing PEt_3, whereas higher P_{CO} or P_{H_2} resulted in an enhanced reaction rate. The selectivity for the linear product slightly increased on using $scCO_2$ compared to toluene (l:b = 2.4 and 2.1 respectively).

Although the $scCO_2$ approach is very elegant, one should keep in mind that it is a relatively expensive method, especially for the bulk chemical industry. For each recycling step the system has be depressurized and pressurized again with the newly added substrate. This process requires more powerful pumps and more capital expensive reactors than for example biphasic aqueous phase process. Furthermore, the low solubility of potential interesting substrates might hamper the commercialization of $scCO_2$ in the fine chemical industry.

10.3.4 Fluorous Biphase catalysis

The fluorous biphasic catalysis (FBC) is another elegant concept of immobilizing catalysts in a different phase, which shows similarities with the $scCO_2$ approach. In the early nineties some experiments were already described in the thesis M. Vogt [31a]. The concept was independently investigated and reported by Horváth in 1994 [31b]. The system is based on the immiscibility of fluorinated compounds with organic solvents, which is due to the polar C-F bond and low polarizability of fluor making this a "non-interacting" solvent. Ligands functionalized with long fluorinated alkyl groups, the so-called "pony tails", are soluble in the fluorous phase, whereas the substrates generally dissolve in the organic phase. Perfluoroaryl groups are less suitable for this purpose, since they offer dipole-dipole interactions and thus dissolve much better in organic solvents. The beauty of the concept stems from the temperature dependency of the miscibility of the solvents (figure 8). At room temperature the two phases are well separated, but at higher temperatures the fluorous biphasic system can become a homogeneous one-phase system. This allows one to combine a homogeneous reaction at higher temperature with catalyst-product separations at lower temperatures. The exact solvent behavior obviously depends on the solvents used [32]. Below the critical temperature phase separation takes place, but still significant amounts of perfluorosolvent are dissolved in the organic phase. This requires additional extraction or distillation. Other major points of concern are catalyst leaching, solvent costs, and contamination of the product with fluoro compounds.

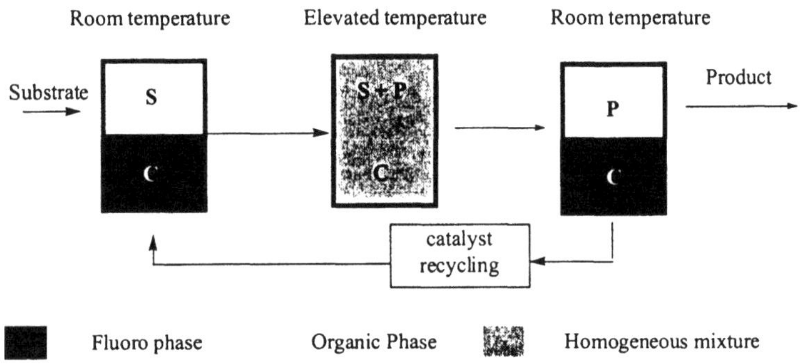

Figure 8. The general concept of fluorous biphasic catalysis

After the initial report on the concept Horváth performed a detailed kinetic study of rhodium catalyzed hydroformylation of 1-decene in which the fluorous-soluble ligand $P[CH_2CH_2(CF_2)_5CF_3]_3$ was compared with the non- fluorous-soluble ligands $P[(CH_2)_7CH_3]$ and PPh_3. Between the fluorous pony tails and the phosphine donor atom a bridge of two CH_2 groups insulates the two groups making the former two ligands electronically identical.

The reactions were performed in 50/50 vol % toluene/$C_6F_{11}CF_3$ at 100 °C and 11 bar CO/H_2 pressure, thus under these conditions in a homogeneous phase. For all three catalyst-systems a first order reaction rate in rhodium concentration was observed. A first order in 1-decene was observed for the systems based on the alkyl phosphines, but for the PPh_3 system the kinetics appeared to be more complicated. The effect of the P/Rh ratio and phosphine concentration on the selectivity and activity of the catalyst was studied (see table 5). As observed for homogeneous systems (see chapter 3 and 4), the P/Rh ratio does not affect the catalytic performance, whereas the absolute phosphine concentration largely influenced the selectivity and activity of the catalyst.

Table 5. Effect of the phosphine concentration on the selectivity and activity of the hydroformylation of 1-decene.[a]

[L] mM [b]	P/Rh	TOF ($mol.mol^{-1}.h^{-1}$)	l:b
22.1	27	2160	3.3
22.5	39	1764	3.3
41.3	76	1440	4.4
82.1	79	756	5.0
152.2	103	288	6.3
304.0	102	144	7.8
23	Not given	1908	3.2
23 [c]	Not given	22572	3.2
23 [d]	Not given	792	2.3

[a] Conditions: 50/50 vol. % toluene/$C_6F_{11}CF_3$, T=100 °C, pressure=11 bar CO/H_2 (1:1), initial [1-decene]= 1 M. [b] $P[CH_2CH_2(CF_2)_5CF_3]_3$ [c] PPh_3 [d] $P[(CH_2)_7CH_3]_3$.

At higher concentrations of the ligand a lower activity and higher l:b ratio was observed, which is in line with results previously obtained in organic solvents. The activity of the fluorous pony-tail phosphine ligand was similar to the alkyl phosphine, but remarkably the selectivity was identical to the PPh_3 system. No explanation was found for this effect. A minor effect of the fluorous phase on the catalytic performance of the Rh/PPh_3 was observed. The activity was approximately 30 % lower in 50/50 vol % toluene/$C_6F_{11}CF_3$ compared to toluene, and the l:b ratio slightly increased. This effect was attributed to differences in gas solubilities.

The fluorous biphase catalyst recovery concept was tested by performing nine consecutive reactions in a batch reactor. A total loss of 4.2% of rhodium was detected, and the decreasing l:b ratio suggested some ligand leaching. The total turnover number reached with the system was 35,000 mole of aldehyde/mole rhodium, with a rhodium loss of 1.18 ppm per mole of aldehyde. Further optimization of this system, i.e. the use of heavier fluorous solvents and longer pony tails, should decrease the amount of catalyst and fluorous solvent leaching.

10.3.5 Dendrimer supported catalysts

A recent development in catalysis is the functionalization of dendrimers with ligands that can give transition metal complexes suitable for catalysis [33]. It is anticipated that dendritic catalysts can have true advantages over their low molecular weight parent species [34]. Many times it has been stated that these types of catalysts are in between homogeneous and heterogeneous catalysts, combining the benefits of both systems. This should result in catalysts that are well-defined (precise distribution of catalytic sites), efficient (fast kinetics due to good accessibility), tunable by ligand design, and easily separated from the reaction mixture by nanofiltration (batch wise or continuously). The recent developments of novel dendritic catalysts suggest that this indeed is a very promising approach towards efficient recyclable catalysts, but more detailed information about these novel systems is required.

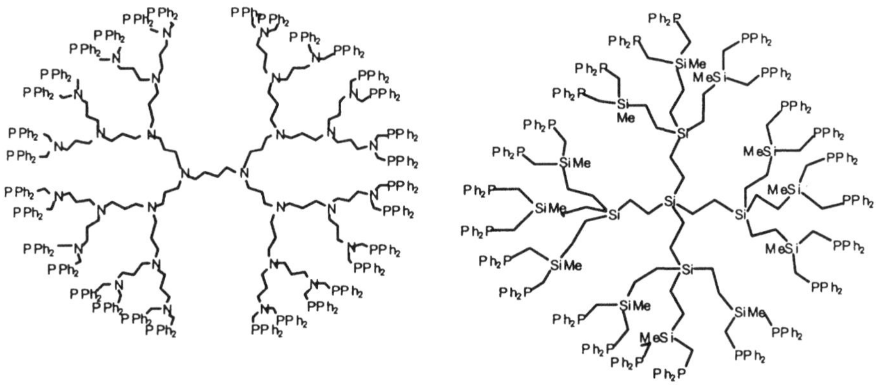

Figure 9. Phosphine functionalized dendrimers based on the polyamino dendrimers from DSM (left) and carbosilane dendrimers (right)

Reetz [35] and van Leeuwen [36] reported similar phosphine functionalized dendrimers, based on different types of dendrimers (figure 9). Initial experiments in continuous membrane systems using palladium complexes of these dendrimers as catalysts in the allylic substitution reaction [35, 36] and hydrovinylation [37] show that the systems based on the higher dendrimer generations are large enough for efficient separation by nanofiltration. Both phosphine dendrimers are active in the rhodium-catalyzed hydroformylation. The preliminary results suggest no significant differences in catalytic performance compared to the smaller parent compounds. Alper reported a similar diphosphine functionalized polyamidoamine (PAMAM) dendrimer that was anchored onto a silica support [38], a concept that was previously used for phosphite modified polymers grafted on silica spheres of 12 nM [39]. The dendrimer was actually synthesized on the support complicating detailed analysis of the system. The Rh-PPh_2-PAMAM-SiO_2 catalyst was shown to be an active catalyst for the hydroformylation reaction and, as expected, more selective towards the branched product. The difference in chain-length (n = 0, 2, 4) (see figure 10) had mainly an impact on the recyclability of the catalyst and it was suggested that this is due to steric congestion of the more compact dendrimers. Additional Soxhlet extraction experiments showed that rhodium leached under catalytic conditions; both CO and substrate promote metal leaching from the support. The phosphite-modified polymers grafted on silica did not show any rhodium leaching when the hydroformylation was performed under the proper conditions [39]. A constant conversion was observed in a continuous flow experiment that was performed for 10 days in benzene (residence time 3.3h).

Functionalized dendrimers offer new solutions to the catalyst/product separation problem of homogeneous catalysts and might give catalysts with unique novel properties. However, if these systems will ever be attractive alternatives for commercial processes remains to be seen. They have to compete with other systems consisting of cheaper polymer materials. Metal leaching and catalyst decomposition are serious problems that should be addressed. Especially the influence of the local very high concentration of metal sites in the case of periphery functionalized dendrimers can promote both these effects. The general problem of metal leaching can be solved by the proper choice of ligands, as will be clear from an example given in the next section.

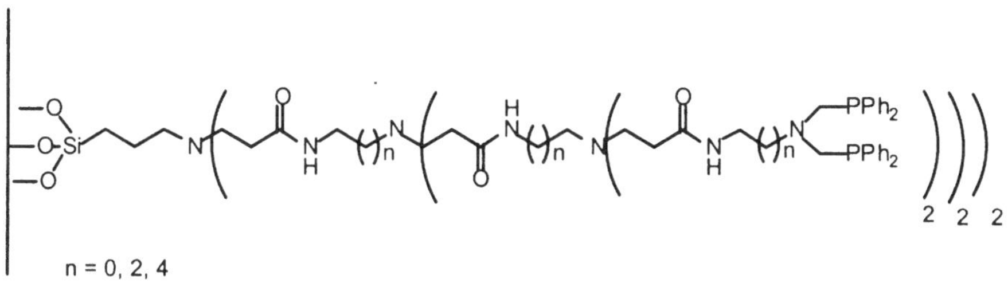

Figure 10. The phosphine functionalized PAMAM dendrimers anchored on a silica support

10.3.6 Novel developments in polymer supported catalyst.

Polymer supported catalysts have attracted a lot of attention already since the early sixties [40]. In principle, immobilized homogenous catalysts can combine high activity and selectivity with facile catalyst-product separation, but to date there are no commercial processes based on this concept. So far, most systems suffered from either metal leaching, low selectivity and/or low activity. However, there are some interesting developments in this field that are very promising, which will be discussed in this section. Also in the light of the current developments in combinatorial chemistry and rapid screening techniques these developments are of interest. An effective support for homogeneous catalysts can be very valuable for rapid synthesis and screening of novel catalysts. Moreover, the supported catalyst can directly be used as an immobilized homogeneous catalyst, which is easily separated from the reaction medium after the reaction by filtration.

The most widely studied polymers for catalyst-immobilization are organic polymers such as cross-linked polystyrenes [41]. The major advantages of these type of supports are the relative ease of functionalization and the wide range of physical properties fine-tuned by the degree of cross-linking. These polymers, however, suffer from major drawbacks as poor heat transfer abilities, poor mechanical properties (they are pulverized in the reactor) and solvent dependent swelling behavior. Inorganic supports such as silica, alumina, glasses, clays and zeolites do not suffer from these drawbacks.

An illustrative example of a hydroformylation catalyst immobilized on a highly cross-linked polystyrene support was reported by Nozaki and Takaya [42]. Several BINAPHOS ligands carrying vinyl groups have been prepared and cross-polymerized with ethyl-styrene and divinylbenzene (Figure 11). By using different ligands (**17b-17d**) the degree of freedom of the ligand in the polymer matrix was varied. The influence of the degree of cross-linking and the polymer/catalyst preparation method on the catalyst performance was investigated (table 6). Generally, the results obtained with the polymer-supported catalysts are similar to the homogeneous system (run 1).

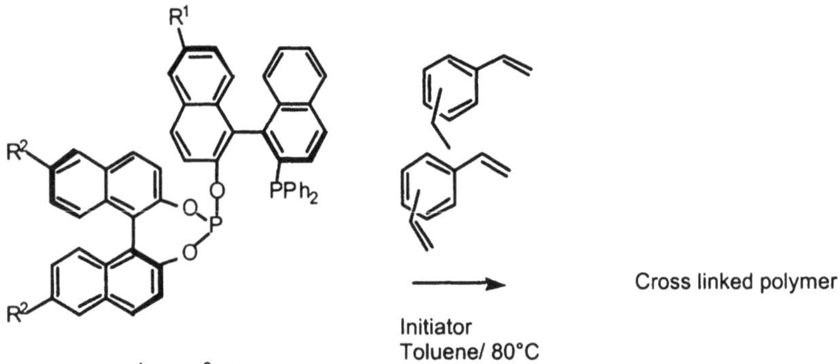

17a R^1=H, R^2=H
17b R^1=vinyl, R^2=H
17c R^1=H, R^2=vinyl
17d R^1=vinyl, R^2=vinyl

Figure 11. Chiral phosphine-phosphite ligands on a highly cross-linked polystyrene polymer

Two methods of catalyst-preparation were tested; the co-polymerization of the ligand and subsequent formation of the rhodium acetylacetonate complex (run 2-7) and co-polymerization of the rhodium acac complex of the ligand (run 8-11). This generally gave the same results, except for ligand

17d (cf. run 7 and 10). This ligand is frozen in the polymer matrix in a different conformation, since it is co-polymerized at three positions of the ligand. Some of these conformations are not able to form the rhodium complex that gives the enantioselective hydroformylation catalyst. In contrast, the other ligands have more freedom and can therefore form the proper catalyst after the polymerization. The degree of cross-linking has no significant influence on the selectivity of the reaction (runs 2, 4, 5, and 11). The reaction can be performed in hexane, but the ee was slightly lower. The recycling experiments clearly show the drawback of these types of polymers. The polymers were partly crushed by the stirring bar and approximately 50% of the initially charged rhodium was removed after the first run.

Table 6. Asymmetric hydroformylation of styrene using polystyrene supported rhodium catalysts based on (R,S)-BINAPHOS)[a]

Run	Catalyst	l:b	ee (%)
1	Rh(acac)(**17a**)	0.12	92
2	(PS-**17b**)-Rh(acac)	0.19	89
3[b]	(PS-**17b**)-Rh(acac)	0.25	81
4[c]	(PS-**17b**)-Rh(acac)	0.20	89
5[d]	(PS-**17b**)-Rh(acac)	0.23	84
6	(PS-**17c**)-Rh(acac)	0.12	89
7	(PS-**17d**)-Rh(acac)	0.14	68
8	(PS)-[Rh(acac)(**17b**)]	0.18	90
9	(PS)-[Rh(acac)(**17c**)]	0.11	97
10	(PS)-[Rh(acac)(**17d**)]	0.15	85
11[d]	(PS)-[Rh(acac)(**17b**)]	0.18	90

[a] Conditions: 6.20 mmol styrene in benzene, 0.0031 mmol catalyst, 20 bar CO/H_2 (1:1), 60 °C, 12 hours, conversions to aldehydes >99%. [b] Hexane as solvent. [c] Lower cross-linking degree. [d] Higher cross-linking degree.

The covalent anchoring of hydroformylation catalysts onto inorganic silica surfaces was already studied in the late seventies [43]. Chloromethyl silicone polymers were converted to the iodo-analog and subsequently reacted to the diphenylphosphinemethyl polysiloxanes. After treatment with $RhCl(CO)(PPh_3)_2$ these materials were used as catalyst in the hydroformylation of 1-hexene (68 bar CO/H_2 (1:1), 100 °C, benzene). Under these conditions, however, the metal dissociated from the support. The resulting rhodium carbonyl species was responsible for the activity leading to low selectivities.

Pantser pioneered the field of organo-functionalized silanes and polysiloxanes. 3-Chloropropyltrialkoxysilanes were converted into functionalized propyltrialkoxysilanes such as diphenylphosphine propyltrialkoxysilane [44]. These compounds can easily be used to prepare surface modified inorganic materials. Upon polycondensation or co-

condensation of these compounds novel functional polymeric materials were obtained that allowed transition metal complexation. This co-condensation, later also referred to as sol-gel process [45], appeared to be a very mild technique to immobilize catalysts. Two different routes towards these functional polymers can be envisioned (Figure 12). One can first prepare the metal complex and then do the co-condensation reaction (route I), or one can prepare the metal complex after the polymerization step (route II). The former route results in far better catalysts due to the higher structural homogeneity. Also more stable catalysts are obtained showing less metal leaching. Lindner performed a detailed study on the sol-gel processed ether-phosphines and their ruthenium complexes using solid state NMR spectroscopy [46]. The obtained different materials strongly influenced the catalytic performance in the ruthenium catalyzed hydrogenation.

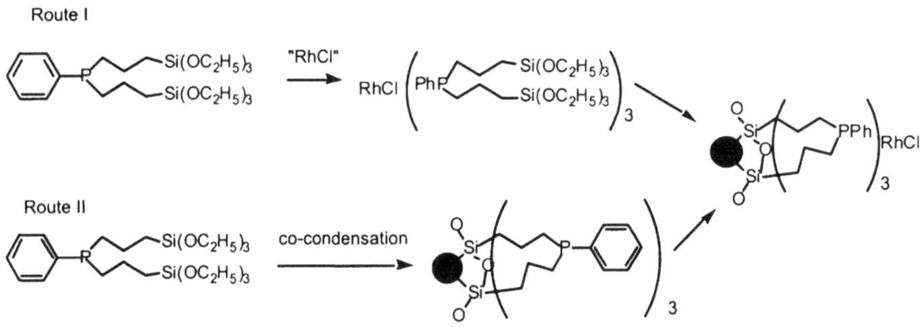

Figure 12. Two different routes to prepare sol-gel immobilized transition metal complexes

Pantser studied the hydroformylation of 1-octene using a polysiloxane-based rhodium phosphine-amine based catalyst in a trickle-bed reactor [47]. The selectivity for the linear product was low (l:b =1.5), but the catalytic performance was retained over a period of 1000 h, and almost no rhodium leaching was detected. Contrary to the homogenous phase reaction, the usage of an excess of phosphine ligands did not result in larger amounts of the linear product, which was attributed to the rigid structure of the polymeric material.

Diphosphine ligands with wide natural bite angles based on the Xanthene backbone have been functionalized with a propyltrialkoxysilane group (**18**) [48]. Silica immobilized rhodium catalysts for the hydroformylation of 1-octene were prepared via the sol-gel method. The catalyst system obtained is very selectively producing the linear product and on recycling no deterioration of the catalyst performance is observed for at least 8 cycles. Furthermore, the reaction could be performed in pure substrate giving reaction rates that are comparable with the homogenous system (80 °C, 50 bar CO/H_2 (1:1), rhodium:ligand=1:10).

These silica immobilized systems prepared via the sol-gel method are very promising organic-inorganic hybrid materials. The chemistry takes place at the interface of the materials suppressing the detrimental influence of the support [46c]. A series of network modifiers for the sol-gel process are available, which can be used to optimize the support. This tool to further improve the catalyst performance already proved to be valuable for several reactions [46c]. For the hydroformylation reaction this still needs to be explored.

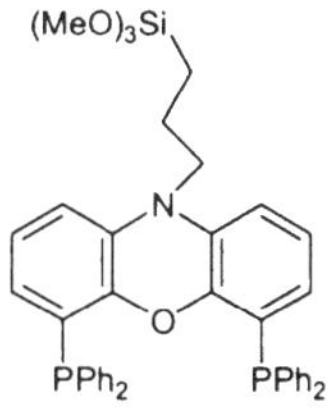

18

Bergbreiter introduced an attractive novel concept by immobilizing catalysts on "smart polymers"; these polymers show temperature dependent solubility properties [49]. Initially these novel systems were based on polyethylene polymers. At room temperature these polymers are completely insoluble in all solvents but they are soluble in several solvents at elevated temperature. Functionalized oligomers (M_n = 2000) also exhibited this type of behavior. Systems have been obtained that were catalytically active at higher temperatures with similar activity as the low molecular weight parent compounds and completely inactive and recyclable at room temperature. The concept was extended using block copolymers (PEO-PPO-PEO) and poly(*N*-isopropyl acrylamide) (PNIPAM) polymers. These materials phase separated at elevated temperatures and formed a homogeneous solution at lower temperatures. The concept remains the same, but now the catalysis takes place at room temperature and the recycling can be achieved at higher temperatures. Moreover, these materials are potentially useful in controlling exothermic reactions.

Similar to the fluorous biphasic concept, a system was developed that formed a homogenous phase at elevated temperature, but phase-separated at room temperature [49c]. It was found that a mixture of functionalized PNIPAM polymer, ethanol, heptane and water exhibited these properties. The hydrogenation of 1-octadecene and 1-dodecene using a phosphine functionalized PNIPAM with a rhodium precursor were taken as test reactions and the high activity was found was similar to that of $RhCl(PPh_3)_3$. At room temperature the mixture phase separated and the catalysis stopped since the catalyst is completely insoluble in heptane. The substrate is dissolved in the heptane allowing a facile catalyst/product separation without the loss of activity. The concept is obviously limited to substrates that show a strong preference for the heptane layer. So far, none of the smart polymer concepts has been tested for the hydroformylation reaction. However, the current rapid developments in polymer technology and the already described interesting systems makes the usage of smart functionalized polymers a promising development.

10.4 Supramolecular catalysis

Supramolecular chemistry has been a very popular research topic for three decades now. Most applications are foreseen in sensors and opto-electronical devices. Supramolecular catalysis often refers to the combination of a catalyst with a synthetic receptor molecule that pre-organizes the substrate-catalyst complex and has also been proposed as an important possible application. The concept, which has proven to be powerful in enzymes, has mainly been demonstrated by chemists that investigated hydrolysis reactions. Zinc and copper in combination with cyclodextrins as the receptor dramatically enhance the rate of hydrolysis. So far, the ample research devoted to transition metal catalysis has not been extended to supramolecular transition metal catalysis. A rare example of such a supramolecular transition metal catalyst was the results of the joined efforts of the groups of Nolte and Van Leeuwen [50]. They reported a basket-shaped molecule functionalized with a catalytically active rhodium complex that catalyzed hydrogenation reactions according to the principles of enzymes. The system showed substrate selectivity, Michaelis Menten kinetics and rate enhancement by cooperative binding of substrate molecules. The hydroformylation of allyl catachol substrates resulted in a complex mixture of products.

Reetz et al. developed water-soluble supramolecular transition metal catalysts that are based on functionalized β-cyclodextrins [51]. The β-cyclodextrin is a frequently used building block that is soluble in the aqueous phase and contains a hydrophobic cavity. Binding of organic substrates takes place in this cavity and is driven by hydrophobic interactions. Initial studies were focussed on selective hydrogenation and optimization of the spacer length between the catalyst and the cavity. The optimal supramolecular system (Figure 13) has been used to study hydroformylation reactions. Several striking differences in catalytic performance were observed for this supramolecular catalyst compared to the parent compound. The two-phase aqueous hydroformylation of 1-octene was very efficient; at 80 °C and 100 bar syn-gas pressure complete conversion was achieved in less than 18 h (TON=3172). Moreover, a complete chemoselectivity toward the aldehyde (>99%) was observed (no isomerization). The activity of this system was higher than the aqueous catalyst system based on TPPTS, even in the presence of phase transfer catalysts. Also internal alkenes such as 3-octene, cyclic alkenes and conjugated systems as styrene and 4-methyl-1,3-pentadiene could be hydroformylated.

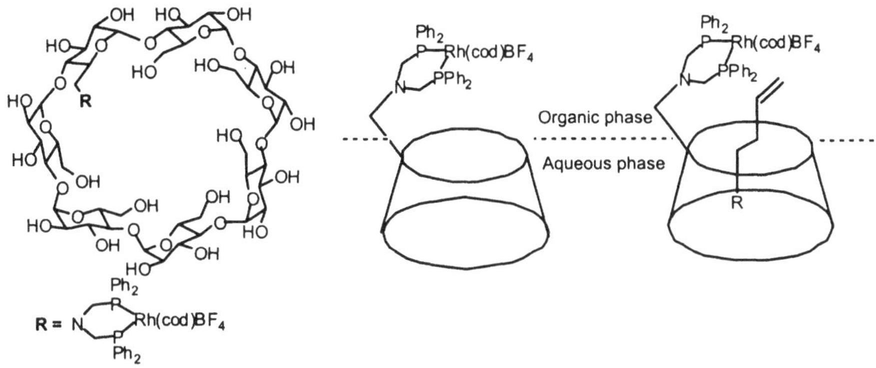

Figure 13. The functionalized cyclodextrin building block as supramolecular catalyst for two-phase catalysis. The actual catalysis is proposed to take place at the phase boundary

Interestingly, the selectivity for the linear product increased on using the supramolecular system; the l:b ratio was 3.2 compared to 1.5 for the parent catalyst. A better test for the selectivity of this supramolecular catalyst is allylbenzene, a substrate that easily isomerizes to methyl styrene. Again the hydroformylation reaction was very selective and a higher l:b ration was observed on using the supramolecular system. The presence of an excess of toluene, a guest molecule that competes for the cavity with the substrate, resulted in a lower regioselectivity. This indicates that the formation of the supramolecular complex is responsible for the higher selectivity.

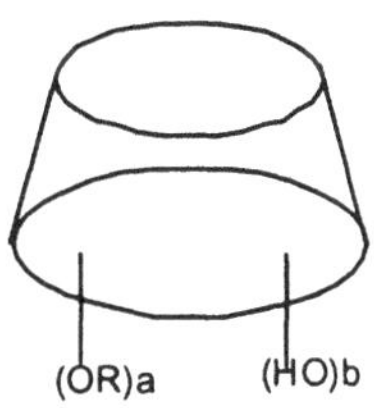

19

Initial experiments aiming at recycling of the catalysts showed that the supramolecular system was mainly in the water layer after separation from the organic phase. Reuse of the aqueous phase revealed that 50% of the catalytic activity was retained. Prior the work of Reetz several groups studied the effect of cyclodextrins as inverse phase transfer materials on the hydroformylation reaction [52]. The general idea is similar to that described above, but now the catalyst and the host molecule are not covalently linked to each other. The cyclodextrin transfers the 1-decene from the organic to aqueous phase where the hydroformylation takes place. Several cyclodextrins (**19**) have been used (see Table 7).

Table 7. Hydroformylation of dec-1ene in the presence of chemically modified cyclodextrins [52b].[a]

Type (R)	a	b	TOF	l:b	Aldehyde (%)
–	-		6	2.7	60
α (-)	0	18	6	3.2	85
γ (-)	0	24	6	2.5	66
β (-)	0	21	12	2.1	78
β (Me)	13	8	47	1.8	91
β (Me)[b]	13	8	83	1.9	95
β (Me)	21	0	19	2.5	57
β (COMe)	21	0	4	2.6	66
β (COMe)	14	7	29	2.6	57
β ($CH_2CHOHCH_3$)	6	15	20	2.0	84
β (SO_3Na)	9	12	4	2.8	69

[a] Reaction conditions: $Rh(aca)(CO)_2$ 0.16 mmol, TPPTS 0.8 mmol, cyclodextrin 1.12 mmol, H_2O 45 mL, dec-1-ene 80 mmol, undecane 4mmol, T=80°C, pressure=50 bar (CO/H_2=1:1), t =8h. a is the number of R groups and b is number of free OH groups. [b] 2.24 mmol cyclodextrin was added, full conversion after 6h.

The effect of the addition of unmodified cyclodextrins to a reaction mixture on the aqueous hydroformylation of 1-decene is rather small. The conversion was enhanced by a factor 2 upon the addition of β-cyclodextrin. The modified β-cyclodextrins improved the activity to a greater extent; in the optimized situation the activity increased with a factor of 14. These modified cyclodextrins are more efficient since they are soluble in both the organic phase and the aqueous phase thereby improving the efficiency of the substrate transfer process. Similar to the systems described above, the isomerization was suppressed under these optimized conditions. However, internal alkenes could not be hydroformylated using this system. Another striking difference is the selectivity of the reaction. The addition of the modified cyclodextrins resulted in a decrease of the l:b ratio (2.7 to 1.9), whereas Reetz reported an increase for his supramolecular system! So far no solid explanation has been found for this difference.

These types of supramolecular systems are expected to be too expensive for commercial applications as yet. The prices of cyclodextrins, however, are in the range of organic solvents and therefore this is not a limitation for commercialization. Furthermore, cyclodextrins are non-toxic, biodegradable and available in large quantities, which makes them suitable for commercial applications.

10.5 Conclusions

The above examples have shown that there is considerable activity in searching for other ways to achieve separation of product and homogeneous catalysts. The improvements needed are the same for all reaction types, also reactions other than hydroformylation. Most of the new techniques have in common that not only the product, but also the starting material and byproducts are separated from the catalyst. This mixture of organic products needs further separation, but as we have seen in Chapter 8 separation of catalyst and byproducts such as heavy ends is an important issue in hydroformylation of simple alkenes. There may be a future for immobilized "drop in" catalysts as described above in the hydroformylation of fine chemicals.

References

1 Kuntz, E. D. *CHEMTECH*, **1987**, *17(9)*, 570.

2 (a) Reedijk, J., Ed.; *Bioinorganic Catalysis*, Dekker, New York, **1993**. (b) Bertini, I.; Gray, H. B.; Lippard, S. J.; Selverstone Valentine, J., Eds; *Bioinorganic Chemistry*, University science books: Mill Valley, **1994**.

3 van der Beuken, E.; Feringa, B. F. *Tetrahedron*, **1998**, *54*, 12985.

4 Brousard, M. E.; Juma, B.; Train, S. G.; Peng, W, -J.; Laneman, S. A.; Stanley, G. G. *Science,* **1993**, *260*, 1784.

5 (a) Breslow, D. S.; Heck, R. F. *Chem. Ind.* **1960**, 467. (b) Breslow, D. S.; Heck, R. F. *J. Am. Chem. Soc.* **1961**, *83*, 4024.

6 (a) Matthews, R. C.; Howell, D. K.; Peng, W.-P.; Train, S.G.; Dale Treleaven, W.; Stanley, G. G. *Angew. Chem. Int. Ed. Engl.* **1996**, *35*, 2253. (b) Stanley, G. G. in *Catalysis by di- and polynuclear metal cluster complexes, Chapter 10*, Adams, R. D.; Cotton, F. A. Eds. Wiley-VCH, **1998**, New York.

7 (a) Hablot, I.; Jenck, J.; Casamatta, G.; Delmas, H. *Chem. Eng. Sci.* **1992**, *47*, 2689. (b) Monteil, F.; Queau, R.; Kalck, P. *J. Organomet. Chem.* **1994**, *480*, 177.

8 Chaudhari, R. V.; Bhanage, B. M.; Desphpande, R. M.; Delmas, H. *Nature*, **1995**, *373*, 501.

9 (a) Tascioglu, S. *Tetrahedron*, **1996**, *52*, 11113. (b) Christian, S. D.; Scamehorn, J. F., Eds. *Solubilization in Surfactant aggregates*, Marcel Dekker, New York, **1995**. (c) Fendler, J. H.; Fendler, E. J. *Catalysis in Micellar and Macromolecular Systems*, Academic press, New York, **1975**.

10 For example: Otto, S.; Engberts, J. B. F. N.; Kwak, J. C. T. *J. Am. Chem. Soc.* **1998**, *120*, 9517.

11 (a) Fell, B.; Papadogianakis, G. *J. Mol. Catal.* **1991**, *66*, 143. (b) Fell, B.; Papadogianakis, G. *J. Mol. Catal. A: Chem.* **1995**, *101*, 179.

12 (a) Ding, H.; Hanson, B. E.; Bartik, T.; Bartik, B. *Organometallics*, **1994**, *13*, 3761. (b) Ding, H.; Hanson, B. E. *J. Chem. Soc. Chem. Commun.* **1994**, *13*, 2747. (c) Bartik, T.; Ding, H.; Bartik, B.; Hanson, B. E. *J. Mol. Catal. A: Chem.* **1995**, *98*, 117.

13 (a) Ding, H.; Hanson, B. E.; Bakos, J. *Angew. Chem.* **1995**, *107*, 1728. (b) Hanson, B. E.; Ding, H.; Kohlpainter, C. W. *Catal. Today* **1998**, *42*, 421.

14 Schreuder Goedheijt M.; Hanson, B. E., Reek, J. N. H.; Kamer, P. C. J.; van Leeuwen, P. W. N. M. *J. Am. Chem. Soc.* **2000**, *122*, 1650.

15 Arhancet, J. P.; Davis, D. E., Merola, J. S.; Hanson, B. E. *Nature*, **1989**, *339*, 454.

16 (a) Davis, M. E. *CHEMTECH*, **1992**, August, 498. (b) Archancet, J. P.; Davis, M. E.; hanson, B. E. *J.Catal.* **1991**, *129*, 110. (c) Guo, I.; Hanson, B. E.; Tóth, I.; Davis, M. E. *J. Organomet. Chem.* **1991**, *403*, 221. (d) Arhancet, J. P.; Davis, M. E.; Hanson, B. E. *J.Catal.* **1991**, *129*, 94. (e) Yuan, Y.; Xu, J.; Zhang, H.; Tsai, K. *Catal. Lett.* **1994**, *29*, 387. (f) Fremy, G.; Monfier, E.; Carpentier, J.-F.; Castanet, Y.; Mortreux, A. *J. Catal.* **1996**, *162*, 339.

17 Bunn, B. B.; Bartik, T.; Bartik, B.; Bebout, W. R.; Glass, T. E.; Hanson, B. E. *J. Mol. Catal.* **1994**, *94*, 157.

18 Horváth, I. T. *Catal. Lett.* **1990**, *6*, 43.

19 (a) Schreuder Goedheijt, M.; Kamer, P. C. J.; van Leeuwen, P. W. N. M. *J. Mol. Cat.* **1998**, *134*, 243. (b) Sandee, A. J.; Slagt, V. F.; Reek, J. N. H.; Kamer, P. C. J.; van Leeuwen, P. W. N. M. *Chem. Commun.* **1999**, 1633.

20 Fremy, G.; Monflier, E.; Carpentier, J.-F.; Castanet, Y.; Mortreux, A.; *Angew. Chem. Int. Ed. Engl.* **1995**, *34*, 1474.

21 (a) Wan, K. T.; Davis, D. E. *Nature*, **1994**, *370*, 449. (b) Wan, K. T.; Davis, D. E. *J. Catal.* **1994**, *148*, 1.

22 Naughton, M. J.; Drago, R. S. *J. Catal.* **1995**, *155*, 383.

23 Jessop, P. G.; Ikariya, T.; Noyori, R. *Chem. Rev.* **1999**, *99*, 475.

24 Rathke, J. W.; Klingler, R. J.; Krause, T. R. *Organometallics,* **1991**, *10*, 1350.

25 Kainz, S.; Koch, D.; Baumann, W.; Leitner, W. *Angew. Chem. Int. Ed. Engl.* **1997**, *36*, 1628.

26 Koch, D.; Leitner, W. *J. Am. Chem. Soc.* **1998**, *120*, 13398.

27 Kainz, S.; Leitner, W. *Catal. Let.* **1998**, *55*, 223.

28 Francio, G.; Leitner, W. *Chem. Commun.*, **1999**, 1663.

29 (a) Palo, D. R.; Erkey, C. *Ind. Eng. Chem. Res.* **1999**, *38*, 3786. (b) Palo, D. R.; Erkey, C. *Organometallics*, **2000**, *19*, 81.

30 Bach, I.; Cole-Hamilton, D. J. *Chem. Commun.* **1998**, 1463.

31 (a) Vogt, M. PhD. Thesis, **1991**, Rheinisch-Westfälischen Technischen Hochschule, Aachen, Germany. (b) Horváth, I. T.; Rábai, J. *Science*, **1994**, *266*, 72.

32 (a) Grosse, A. V.; Cady, G. H. *Ind. Eng. Chem.* **1947**, *39*, 367. (b) Scott, R. L.; *J. Am. Chem. Soc.* **1948**, *70*, 4090. (c) Hildebrand, J. H.; Cochran, D. R. F. *J. Am. Chem. Soc.* **1949**, *71*, 22. (d) Dorset, D. L. *Macromolecules* **1990**, *23*, 894. (d) Scott, R. J. *J. Phys. Chem.* **1958**, *62*, 136.

33 For reviews on dendrimers see: (a) Tomalia, D. A.; Durst, H. D. *Top. Curr. Chem.*, **1993**, *165*, 193; (b) Issberner J.; Moors, R.; Vögtle, F. *Angew. Chem.* **1994**, *106*, 2507; (c) Fréchet, J. M. J. *Science*, **1994**, *263*, 1710; (d) Newkome, G. R.; Moorefield, C. N.; Vögtle, F., Dendritic Molecules-Concepts, Synthesis, Perspectives, WCH, Weinheim **1996**. (e) Majoral, J. P.; Caminade, A. *Chem. Rev.*, **1999**, *99*, 845. (f) Bosman, A. W.; Janssen, H. M.; Meijer, E. W. *Chem. Rev.* **1999**, *99*, 1665.

34 (a) Knapen, J. W. J.; van der Made, A. W.; de Wilde, J. C.; van Leeuwen, P. W. N. M.; Wijkens, P.; Grove, D. M.; van Koten, G. *Nature,* **1994**, *372*, 659. (b) Tomalia, D. A.; Dvornic, P. R. *Nature,* **1994**, *372*, 617.

35 (a) Reetz, M. T.; Lohmer, G.; Schwickardi, R. *Angew. Chem. Int. Ed. Engl.* **1997**, *36*, 1526. (b) Brinkman, N.; Giebel, D.; Lohmer, G.; Reetz, M. T.; Kragl, U. *J. Catal.* **1999**, *183*, 163.

36 de Groot, D.; Eggeling, E. B.; de Wilde, J. C.; Kooijman, H.; van Haaren, R. J.; van der Made, A. W.; Spek, A. L.; Vogt, D.; Reek, J. N. H.; Kamer, P. C. J.; van Leeuwen, P. W. N. M. *Chem. Commun.* **1999**, 1623.

37 Hovestad, N. J.; Eggeling, E. B.; Heidbüchel, H. J.; Jastrzebski, J. T. B. H.; Kragl, U.; Keim, W.; Vogt, D.; van Koten, G. *Angew. Chem. Int. Ed. Engl.* **1999**, *38*, 1655.

38 (a) Bourque, S. C.; Maltais, F.; Xiao, W. -J.; Tardif, O.; Alper, H.; Arya, P.; Manzer, L. E. *J .Am. Chem. Soc.* **1999**, *121*, 3035. (b) Bourque, S. C.; Alper, H.; *J. Am. Chem. Soc.* **2000**, *122*, 956. (c) Arya, P.; Rao, N. V.; Singkhonrat, J.; Alper, H.; Bourque, S. C.; Manzer, L. E. *J. Org. Chem.* **2000**, *65*, 1881.

39 Jongsma, T.; van Aert, H.; Fossen, M.; Challa, G.; van Leeuwen, P. W. N. M. *J. Mol. Catal.* **1993**, *83*, 37.

40 Hartley, F. R.. *Supported metal complexes, a new generation catalysts,* Reidel, Dordrecht, **1985**.

41 (a) Pomogailo, A. D.; Wöhrle, D. in *Macromolecule-Metal Complexes*, Ciardelli, F.; Tsuchida, E.; Wöhrle, D. Eds., Springer, Berlin, **1996**. (b) Keim, W.; Driessen-Hölscher, B. in *Handbook of Heterogenous Catalysis, vol 1,* Ertl, G.; Knözinger, H.; Weitkamp, J. Eds. WILEY-VCH, Weinheim, **1997**.

42 (a) Nozaki, K.; Itoi, Y.; Shibahara, F.; Shirakawa, E.; Ohta, T.; Takaya, H.; Hiyama, T. *J. Am. Chem. Soc.* **1998**, *120*, 4051. (b) Nozaki, K.; Shibahara, F.; Itoi, Y.; Shirakawa, E.; Ohta, T.; Takaya, H.; Hiyama, T. *Bull. Chem. Soc. Jpn.* **1999**, *72*, 1911.

43 Farrell, M. O.; van Dyke, C. H.; Boucher, L. J.; Metlin, S. J. *J. Organomet. Chem.* **1979**, *172*, 367.

44 Deschler, U.; Kleinschmit, P.; Panster, P.; *Angew. Chem. Int. Ed. Engl.* **1986**, *25*, 236.

45 (a) Monaco, S. J.; Ko, E. I. *CHEMTECH,* **1998**, 23. (b) Ingersoll, C. M.; Bright, F. V. *CHEMTECH,* **1997**, 26.

46 (a) Lindner, E.; Kemmler, M.; Mayer, H. A.; Wegner, P. *J. Am. Chem. Soc.* **1994**, *116*, 348. (b) Lindner, E.; Schneller, T.; Auer, F.; Wegner, P.; Mayer, H. A. *Chem. Eur. J.* **1997**, *3*, 1833. (c) Lindner, E.; Schneller, Auer, F.; Mayer, H. A. *Angew. Chem. Int. Ed. Engl.* **1999**, *38*, 2155.

47 Wieland, S.; Panster, P. *Catal. Org. React.* **1994**, *62*, 383.

48 Sandee, A. J.; van der Veen, L. A.; Reek. J. N. H.; Kamer, P. C. J.; Lutz, M.; Spek, A. L.; van Leeuwen, P. W. N. M. *Angew. Chem. Int. Ed. Engl.* **1999**, *38*, 3231.

49 (a) Bergbreiter, D. E.; *Catal. Today*, **1998**, *42*, 389. (b) Bergbreiter, D. E.; Zhang, L.; Mariagnanam, V. M. *J. Am. Chem. Soc.* **1993**, *115*, 9295. (c) Bergbreiter, D. E.; Liu, Y.-S.; Osburn, P. L. *J. Am. Chem. Soc.* **1998**, *120*, 4250.

50 (a) Coolen, H. K. A. C.; van Leeuwen, P. W. N. M.; Nolte, R. J. M. *Angew. Chem. Int. Ed. Engl.* **1992**, *31*, 906. (b) Coolen, H. K. A. C.; Meeuwis, J. A. M.; van Leeuwen, P. W. N. M.; Nolte, R. J. M. *J. Am. Chem. Soc.* **1995**, *117*, 11906.

51 Reetz, M. T.; Waldvogel, S. R. *Angew. Chem. Int. Ed. Engl.* **1997**, *36*, 865. (b) Reetz, M. T. *J. Heterocyclic Chem.***1998**, *35*, 1056. (c) Reetz, M. T. *Catal. Today,* **1998**, *42*, 399.

52 For hydroformylation reactions with cyclodextrins as additives: (a) Anderson, J. R.; Campi, E. A.; Jackson, W. R. *Catal. Lett.* **1991**, *9*, 55. (b) Monflier, E.; Fremy, G.; Castanet, Y.; Mortreux, A. *Angew. Chem. Int. Ed. Engl.* **1995**, *34*, 2269. (c) Monflier, E.; Tilloy, S.; Fremy, G.; Castanet, Y.; Mortreux, A. *Tetrahedron Lett.***1995**, *36*, 9481. (d) Tilloy, S.; Bertoux, F.; Mortreux, A.; Monflier, E. *Catal. Today*, **1999**, *48*, 245.

INDEX

Acetalization, 165, 166
acrolein acetals, 149
acrylonitrile, 153
activity, 207
aldol condensation, 207, 212, 223
aldol reaction, 157,177
alkylphosphines, 67
alkynes, 178
allyl alcohol, 217
amination, 168
amphiphilic ligands, 256, 259
aqueous two-phase catalysis, 256
Arbuzov rearrangement, 44, 245
asymmetric hydroformylation, 107-144
atropisomers, 109

BDPP, 134
Berry mechanism, 114
bimetallic catalysts, 253
BINAP, 3, 4, 134
BINAPHOS, 4, 7, 108, 124, 264, 270
binaphthyl, 125
BINAS, 193
biphasic hydroformylation, 135136
BIPHEPHOS, 147,170,178
BIPHOLOPHOS, 136
BISBI ligands, 7, 82-87, 136, 147, 258
BISBIS, 192
bisphenol bridges, 45
bite angle, 5, 8, 10, 82-99, 135
BPPM, 132
bridge length, 45
bulky diphosphite catalysts, 45, 108
bulky phosphite, 6, 7, 37-44, 151, 158
1,4-butanediol, 217, 225
2-butene, 46

Calix[4]arene based phosphites, 59
catalyst cost, 204
catalyst decomposition, 235-249
catalyst preparation, 17, 195, 233-234
catalyst recovery, 210-216
catalyst separation, 256
chelation control, 134
chiral cooperativity, 116
chiral diphosphines, 131
chiral diphosphites, 109
CHIRAPHOS, 134
CO dissociation, 74, 93, 100-102
cone angle, 8, 9, 40
continuous flow system, 261, 269
cooperative effect, 131

cosolvents, 200
cyclodextrins, 275

DEGUPHOS, 134
dendrimer supported catalysts, 268
detergent alcohols, 223
detergents, 201
deuterioformylation, 16, 24-27, 30-31, 33, 65, 84
dihydrofurans, 150
dihydropyrans, 151
dimer formation rhodium complexes, 54, 64, 72, 247
dimethoxyacrolein, 153
dimethylbut-1-ene, 18, 25
dinuclear species, 137
DIOCOL, 133
DIOP, 2, 78, 82, 84, 133
dioxaphosphepine, 120
DIPAMP, 2, 4
diphosphines as ligands, 76-102, 131-140
diphosphites as ligands, 44-59, 108-120
dirhodium species, 54, 65, 69,137
dissociation of CO, 74, 93, 100-102
distillative separation, 214
1,1-disubstituted alkenes, 149,164
dormant sites, 247
double-bond isomerization, 207
DPBS, 219
DPPB, dppb, 217
DuPhos, 3, 4

Economics, 198, 204
electron withdrawing substituents, 38, 85
electronic effects, 67, 79, 85-87, 89
electronic parameter, 8
electron-withdrawing phosphines, 79
enantiofacial selection, 128
energetics, 100-102
ethyl acrylate, 139,154
evaporation, 211-214
extraction, 211, 215-219, 220, 225
extraction *after* one-phase reaction, 216

Ferrocene diphosphine ligands, 79
fluorous biphase catalysis, 190, 265
fluxional behavior, 114

Gas recycle process, 213
glucal, 158,167
glucopyranose, 123

Heavy-ends, 207
Heck mechanism, 64
heteroatom-directed hydroformylation, 160
β-hydride elimination, 41, 80, 101-102
hydroaminomethylation, 172,177
hydrocyanation, 1,121
hydrolysis phosphites, 44

Immobilization, 129, 201
immobilized aqueous phase, 260
in situ IR reflectance spectroscopy, 68
in situ IR transmission spectroscopy, 28, 68-69, 90, 138
in situ NMR spectroscopy, 49-52, 68-69, 80, 91
indicator ligand, 39
internal alkenes, 55-59, 95, 263, 275
IR spectroscopic studies, 43

Josiphos, 4

Kinetic studies, 43, 47, 69-72, 97-102, 112
kinetics two-phase catalysis, 193-195
kinetics, fluorous phase, 266

Ligand effects, 1-14, 38, 66-68, 76-99
ligand loss, 209, 235-246
liquid recycle process, 213
long chain alkenes, 200

Manufacturing cost, 204
mechanistic studies, 22, 40-44, 48-55, 68-72, 96-99, 113-120, 124, 138
metal plating, 235
metalation, 240, 248
methyl methacrylate, 5
methyl oleate, 55-57
3-methylpentane-1,5-diol, 39
micellar catalysis, 256
monophosphines as ligands, 63-75
monophosphites as ligands, 37-44, 121
muscone, 149

Nanofiltration, 268
NAPHOS, 137, 193
natural bite angle, 82, 88, 96-99
NMP, 220
NMR spectroscopy, 49-52, 68-69, 80, 91
nylon monomers, 226

Organic synthesis, hydroformylation in, 145
orthometallation, 54-55, 210, 240
oxidation of phosphorus ligands, 209, 235-237

P-C bond cleavage, 209, 237-241
3-pentenoate esters, 58, 227
pH aqueous phase catalysis, 195
phase transfer, 275
phase-transfer catalyst, 225
phosphine ligands, 63-100
phosphine-phosphite ligands, 129-131
phosphites, 35-61, 109-124
phospholes, 67, 82
phosphonium intermediates, 242
phosphonium salt, 210
phosphorus ligand effects, 1
phyllantocin, 162
platinum systems, 107,132
P-O bond splitting, 210
polyketone, 4
polymer bound bulky phosphite, 40
polymer supported catalysts, 269-274
process aspects, 196-198, 203-227
propanal, 200
pyrrole, 174

Quaternary phosphonium salts, 210

Raffinate-II, 197, 220
regioselectivity, 72-75, 96-99
Reppe chemistry, 1
resting state, 40-43, 69-72, 74, 81, 96, 113, 138
rhodium containment, 208
rhodium leaching, 208
rhodium plating, 208
rhodium price, 204
ribose, 122
Ruhrchemie Rhône-Poulenc-Process, 191, 196

Silica immobilized systems, 273
silylenol ethers, 176
silylformylation,178, 181

smart polymers, 273
sol-gel process, 271
solution structures, 49-52, 68-69, 80, 91, 111, 113, 137
"standard" conditions, 70
steric effects, 38, 40, 46, 68, 113-120
stripping reactor, 212
styrene, 18, 20, 21, 26, 30, 107-142
substituted alkenes, 55-59
substrate-directed stereoselectivity, 155
sugar derivatives, 121
sulfonated ligands, 256-262
supercritical fluids, 262-263
supported aqueous phase catalysis, 260
supramolecular catalysis, 274

Terpenes, 167
Tischenko reaction, 207
tpp, 6, 63-75, 212, 213, 221
tppms, 5, 7, 221
tppts, 5-7, 191, 215, 221
TPPTS, see tppts
transition state geometry, 98
trifluoropropene, 18, 153
triphenyl phosphite, 38
triphenylphosphine, see tpp
tris(ortho *tert*-butylphenyl)phosphite, 37-44, 221
tris(2,2,2-trifluoroethyl) phosphite, 38
turn-stile mechanism, 53, 114, 138
two-phase catalysis, 190
two-phase reaction, 215

Unmodified rhodium catalysts, 15-33

Vinyl acetate, 18, 20
vinyl arenes, 152
vinyl aromatics, 107
vinyl esters, 150
vinylpyridine, 18, 153
vinylpyrrole, 153

Wilkinson, 64, 72
Wittig reaction, 177

Xantphos ligands, 87-95, 147, 201, 259, 262, 273
xylofuranose, 122

Catalysis by Metal Complexes

Series Editors:
R. Ugo, *University of Milan, Milan, Italy*
B. R. James, *University of British Columbia, Vancouver, Canada*

1.* F. J. McQuillin: *Homogeneous Hydrogenation in Organic Chemistry.* 1976 ISBN 90-277-0646-8
2. P. M. Henry: *Palladium Catalyzed Oxidation of Hydrocarbons.* 1980 ISBN 90-277-0986-6
3. R. A. Sheldon: *Chemicals from Synthesis Gas.* Catalytic Reactions of CO and H_2. 1983 ISBN 90-277-1489-4
4. W. Keim (ed.): *Catalysis in C_1 Chemistry.* 1983 ISBN 90-277-1527-0
5. A. E. Shilov: *Activation of Saturated Hydrocarbons by Transition Metal Complexes.* 1984 ISBN 90-277-1628-5
6. F. R. Hartley: *Supported Metal Complexes.* A New Generation of Catalysts. 1985 ISBN 90-277-1855-5
7. Y. Iwasawa (ed.): *Tailored Metal Catalysts.* 1986 ISBN 90-277-1866-0
8. R. S. Dickson: *Homogeneous Catalysis with Compounds of Rhodium and Iridium.* 1985 ISBN 90-277-1880-6
9. G. Strukul (ed.): *Catalytic Oxidations with Hydrogen Peroxide as Oxidant.* 1993 ISBN 0-7923-1771-8
10. A. Mortreux and F. Petit (eds.): *Industrial Applications of Homogeneous Catalysis.* 1988 ISBN 90-277-2520-9
11. N. Farrell: *Transition Metal Complexes as Drugs and Chemotherapeutic Agents.* 1989 ISBN 90-277-2828-3
12. A. F. Noels, M. Graziani and A. J. Hubert (eds.): *Metal Promoted Selectivity in Organic Synthesis.* 1991 ISBN 0-7923-1184-1
13. L. I. Simándi: *Catalytic Activation of Dioxygen by Metal Complexes.* 1992 ISBN 0-7923-1896-X
14. K. Kalyanasundaram and M. Grätzel (eds.), *Photosensitization and Photocatalysis Using Inorganic and Organometallic Compounds.* 1993 ISBN 0-7923-2261-4
15. P. A. Chaloner, M. A. Esteruelas, F. Joó and L. A. Oro: *Homogeneous Hydrogenation.* 1994 ISBN 0-7923-2474-9

Catalysis by Metal Complexes

16. G. Braca (ed.): *Oxygenates by Homologation or CO Hydrogenation with Metal Complexes*. 1994 ISBN 0-7923-2628-8
17. F. Montanari and L. Casella (eds.): *Metalloporphyrins Catalyzed Oxidations*. 1994 ISBN 0-7923-2657-1
18. P.W.N.M. van Leeuwen, K. Morokuma and J.H. van Lenthe (eds.): *Theoretical Aspects of Homogeneous Catalysis*. Applications of *Ab Initio* Molecular Orbital Theory. 1995 ISBN 0-7923-3107-9
19. T. Funabiki (ed.): *Oxygenases and Model Systems*. 1997 ISBN 0-7923-4240-2
20. S. Cenini and F. Ragaini: *Catalytic Reductive Carbonylation of Organic Nitro Compounds*. 1997 ISBN 0-7923-4307-7
21. A.E. Shilov and G.P. Shul'pin: *Activation and Catalytic Reactions of Saturated Hydrocarbons in the Presence of Metal Complexes*. 2000 ISBN 0-7923-6101-6
22. P.W.N.M. van Leeuwen and C. Claver (eds.): *Rhodium Catalyzed Hydroformylation*. 2000 ISBN 0-7923-6551-8

KLUWER ACADEMIC PUBLISHERS – DORDRECHT / BOSTON / LONDON

**Volume 1 is previously published under the Series Title: Homogeneous Catalysis in Organic and Inorganic Chemistry.*

9 781402 004216

FSC
www.fsc.org
MIX
Papier aus verantwortungsvollen Quellen
Paper from responsible sources
FSC® C105338

If you have any concerns about our products,
you can contact us on
ProductSafety@springernature.com

In case Publisher is established outside the EU,
the EU authorized representative is:
Springer Nature Customer Service Center GmbH
Europaplatz 3, 69115 Heidelberg, Germany

Printed by Libri Plureos GmbH
in Hamburg, Germany